Tropical Bioproductivity

This book investigates the fundamental role that tropical bioproductivity – or more specifically net primary productivity – has played in shaping the global geographies of food, finance, governance and people.

The book examines the basic astronomical and thermal properties of our planet to illustrate the dynamic nature of the tropics and how the region resides at the very heart of global energetics, driving the environmental flows that shape planetary climate and bioproductivity. The author explores how the region's relatively small, but hyper-productive, land area provided the groundswell for the economic, social, political and demographic changes that fuelled empires, European colonialism and nation-building. Also covered are discussions on how the critical intake of capital needed to fuel the industrial and technological revolutions driving modern globalization was first expropriated from the tropics by harnessing the region's natural productivity and biological crop diversity and then transforming it into tradeable commodities using the inhabitants' labour and knowledge. With modern tropical nations accounting for the bulk of people living in poverty and registering some of the highest income disparities, the author presents cross-cutting evidence showing that their histories and the persistence of expropriating institutions have fostered anocratic tendencies, poor governance, unorthodox financial flows and mass migration.

Tropical Bioproductivity cuts across vast geographies, topics and histories to deliver a readable narrative that links people, places and events with the environmental mechanics of our planet. It will be of interest to students and researchers in the areas of environmental studies, economics, history, agriculture, anthropology and geography.

David S. Hammond is Principal, NWFS Consultancy, based in Ashford, Kent, UK and Portland, Oregon, USA. He has undertaken environmental projects in a dozen countries in Latin America, Asia and Africa for clients such as WWF-UK, UNDP, UK DFID, IUCN and the Inter-American Development Bank. He was previously at the Iwokrama International Centre in Georgetown, Guyana, CABI Bioscience and Imperial College, UK.

Earthscan Studies in Natural Resource Management

For more information on books in the Earthscan Studies in Natural Resource Management series, please visit the series page on the Routledge website: www.routledge.com/books/series/ECNRM/

Tropical Bioproductivity
Origins and Distribution in a Globalized World

David S. Hammond

Routledge
Taylor & Francis Group
LONDON AND NEW YORK

from Routledge

First published 2019
by Routledge

2 Park Square, Milton Park, Abingdon, Oxfordshire OX14 4RN
52 Vanderbilt Avenue, New York, NY 10017

Routledge is an imprint of the Taylor & Francis Group, an informa business

First issued in paperback 2020

British Library Cataloguing-in-Publication Data
A catalogue record for this book is available from the British Library

Library of Congress Cataloging-in-Publication Data
Names: Hammond, D. S. (David Scott), 1965– author.
Title: Tropical bioproductivity : origins and distribution in a
globalized world / David S. Hammond.
Description: Abingdon, Oxon ; New York, NY : Routledge, 2019. |
Series: Earthscan studies in natural resource management |
Includes bibliographical references and index. |
Identifiers: LCCN 2018048145 (print) | LCCN 2018059590 (ebook) |
ISBN 9780429488733 (eBook) | ISBN 9781138594609 (hardback :
alk. paper)
Subjects: LCSH: Tropical plants–Ecology. | Tropical plants–
Economic aspects. | Ecology–Tropics. | Tropics–
Economic conditions.
Classification: LCC QK936 (ebook) | LCC QK936 .H36 2019 (print) |
DDC 581.70913–dc23
LC record available at https://lccn.loc.gov/2018048145

ISBN: 978-1-138-59460-9 (hbk)
ISBN: 978-0-367-66273-8 (pbk)

Typeset in Baskerville
by Wearset Ltd, Boldon, Tyne and Wear

Contents

Preface

This book attempts to understand the role of the tropics and their bioproductivity in shaping human society and in particular their role in globalization. It asks three questions: why are the tropics bioproductive, how is this bioproductivity distributed across the tropics and in relation to the extra-tropics and how has tropical bioproductivity been important to the history of globalization? This is not a book about environmental determinism – a much derided term suggesting that people fail to develop due to unfavourable environments. On the contrary, it attempts to show how the superior bioproductivity of the tropics did not engender a need by tropical societies to go global, but drove Europe to catapult global society into a fast-lane of globalization because the tropics offered biological prizes that it did not have, but needed in order to achieve greater demographic, economic and cultural growth. Whether we like it or not, the fate of many regions of the tropics by the sixteenth century was to be rapidly subsumed into European history. This collision drastically altered cultures and landscapes in the tropics, but also reflexively, fundamentally altered European cultures and landscapes as well. The relationship was reciprocal but inequitable. In the final chapters, the book examines how, if at all, this historic linkage has played out in modern tropical nations, whether bioproductivity has proven a benefit or burden in a globalized economy and what root principles need to be pursued to allow tropical nations to escape both the poverty and the middle-income traps that they currently face.

One might suggest that the brush is too broad in painting this picture. Environmental science, history, economics, botany, geography, political science, philosophy and anthropology are rare compatriots in part because some rely mainly on numbers and others rely mostly on words. Handling such diverse topics might have been better achieved by assembling a group of world experts in their fields, but then the project would be just that – an assembly of disjointed perspectives missing the key connections. Without doubt, I have missed some of these too but travelling across boundaries has also revealed, I hope, some interesting relationships. If globalization teaches us anything, then it is that our world is made up of reservoirs, concentrations and storehouses and that change occurs through fluxes, flows or exchanges between these pools – I have chosen to focus mainly on the latter in this book.

Facts are better than opinions, but neither is necessarily true. I have relied extensively on both numbers and words, many being documented centuries ago and cross-checked from alternate sources and metrics wherever possible. But antiquity has a way of funnelling information towards singular sources and I cannot attest to the veracity of each source: some may be treated as fact, but are actually opinion. This uncertainty – the fact–opinion duality in historical narrative – must be both the bread and bane of all historians, though I do not claim to be one. In every instance, I have relied on original sources whenever available and then considered more recent analyses and treatments in their absence or where further clarification or cross-checking was needed.

Numbers can reflect motive as much as words. Data can be used to impress, to report success, to advance careers, to cover failure, to seek changes in policy or allocation of resources, or to forebode spectacular profit or imminent danger. I cannot attest to the full accuracy of the many environmental, economic, agricultural and demographic datasets employed in this book. I have provided caveats to these data where they have been highlighted by others, but have purposefully avoided the sort of in-depth coverage of data and methods that is more typical of purely scientific research publications. Generally speaking, historical data are less accurate than modern data due to advances in measurement technologies and fewer sources available for verification. This could play on certain patterns or trends presented in some of the chapters. Comparisons between datasets collected contemporaneously may contain bias, but differences are less likely to be affected by changes in accuracy. For example, a concern over the difference between net primary productivity data calculated from ground-based methods compared to remotely sensed (satellite) approaches is more relevant to absolute amounts than to comparisons between the bioproductivity of different regions and ecosystems. Data, wherever constructed from proxy measurements, represent the pool of estimates made by multiple authors when these are available. A fine example is the very wide range of estimates constructed by various authors in determining the role of sugar in the daily caloric intake of British citizens. I have taken the average of this pool in estimating its changing role.

Nothing is built in a vacuum. I wish to express my professional gratitude to the many scientists and historians that have responded to information and publication requests. I would like to extend my thanks to various persons that sparked my interests in complex tropical landscapes and histories, including Hardy Eshbaugh for introducing me to the chilli pepper and tropical ethnobotany; Tim Whitmore (1935–2002) for his guidance in forest land use and ecology; Vincent Florens, Claudia Baider, Sharveen Persand and Sujit for introducing me to Mauritius' environment and the world of endangered island ecosystems; Ramon Perez-Gil and Dennis Breedlove for their guidance in understanding milpa agriculture and the biodiversity of Chiapas and José, Ignacio, Pedro, Broads, Lagadou, Daniel, George, Roxroy and others for their great assistance with work in the fields, forests and rivers of Mexico, Colombia and Guyana. I wish also to thank the Maritime Navel Museum at Greenwich, London, the British Library and the Archivo de las Indias in Seville for access to their facilities, as well as the Bank of England, National Oceanic and Atmospheric Agency (NOAA), European Space Agency (ESA), World Wide Fund for Nature (WWF), the World Bank, the Food and Agriculture Organization of the United Nations (FAO) and Oregon State University (OSU) for making available important datasets used in producing results presented in this book.

Part I

Structure, origins and distribution

1   Two tropics

Cartography has always had a distorting effect on the way we perceive our planet. It is precisely the distortions introduced by cartographers in their fifteenth-century maps of the world that led Christopher Columbus in his belief that sailing west would lead directly to the spice riches of the Orient, setting the stage for several hundred years of ensuing war as the competing powers fought over the trade in tropical commodities – a path that has proven to profoundly shape our modern, global society, as we shall explore more clearly later in this book. These cartographic misconceptions continue to this day. Many school children still see the island of Greenland as exceptionally large. It appears larger than the continent of Australia, when it is actually less than a third of its size. This has to do with the universal adoption of the Mercator map and the way in which an imperfect sphere is converted into a rectangular map using a special type of cylindrical projection that was suitable for use in navigating the majority of low-latitude shipping routes of the day, but hopelessly distorts proportions across the sub-polar regions of our planet. Using the Tropics of Cancer and Capricorn to delimit the tropics is equally distorting, but for different reasons. Their position, and the extent of the tropics, has nothing to do with maps.

In Greek, a turn or turning point is *tropos* and in the world of European languages, the tropics derive their modern name from this root. It is an interesting descendancy. If asked how best to describe the tropics, most of us would probably point out the constant, high temperatures or the plants, like palms, that rely entirely on this thermal constancy to prosper. For much of the planet's population, the word denoting the "tropics" builds on this perception. In Chinese Mandarin and Japanese the characters combined to symbolize the "tropics" translate literally as "hot belt". A similar outcome is reached in Hindi, where "tropic" is derived from the combination of symbols for "tepid" and "zone" or "girdle". The linguistic approach taken in the East and West to describe the tropics is obviously very different. In the West, it was guided by the early observations that at a certain latitude, the Sun would slowly rise in the sky across one half of the year until it reached a point overhead on the summer solstice, only to turn about on that day and roughly re-trace its earlier ascent over the latter half of the year. The name reflects the fascination attached to celestial cycles by early Middle Eastern and European astronomers. In the Far East, the etymological origin is, by contrast, distinctly terrestrial.

Neither approach is inherently more comprehensive, more accurate, than the other in describing the waist of our planet. These different approaches, celestial and terrestrial, are two sides of the same coin. They reflect on cause and effect, and the inextricable link between the astronomical interactions that govern the size and position of our planet's tropical zone and the relatively high temperature of its climate. But there are important, sometimes surprising, variations in the thermal behaviour across the tropics and these do not always fit with the cartographic depiction. It is this dynamism, both celestial and thermal, that places the entire island of Madagascar, not just that part north of the Tropic of Capricorn, firmly in the tropics. More importantly, it was also this connection that drove the early European explorers to the New World, creating a crucible of global economic competition, and sparked the race to control the abundant resources of tropical nations that continues to this day. The consequences have been profound, not only to the planetary environment, but to the social and economic prospects of a club of nations beset with the jewels in the crown of global environmental value – an immense propensity for bioproductivity and the very high levels of biodiversity that appear to march in step with this profligate primary production of biomolecules.

The Tropic of Gemini

If the American author Henry Miller had also been an amateur astronomer he might have chosen to title his famously controversial book *The Tropic of Gemini*. Although the title appears to have little bearing on its content[1] the namesake for the title he chose, *The Tropic of Cancer*, is more commonly recognized as the boreal, or northern, boundary of the tropics. This imaginary circle describes the northernmost latitude where the Sun, at its zenith, can be observed directly overhead at the summer equinox. Further north of this line, the Sun's highest point in the mid-day sky drops lower towards the horizon and never quite reaches a point directly overhead. The same can be said of its antipodes, the Tropic of Capricorn. Together, these parallel circles form the edges of a broad tropical belt that stretches around the mid-riff of our planet. The problem with Mr. Miller's title, and the continued use of Cancer and Capricorn, can be traced to their origins. Archaeological evidence suggests that the 12 common zodiacal constellations were organized into an astrological system of timekeeping by the ancient Babylonians around the fifth century BC. This system attached the position of the Sun at the four cardinal points in a year, the solstices and equinoxes, to the constellations that it appeared to intersect on those days. At the time, the Babylonians would have correctly observed that the Sun's track in the sky (the solar ecliptic) came in contact with the constellation of Cancer on the (northern) summer solstice sometime in the latter half of June. Equally, the solar ecliptic would have placed the Sun in front of Capricorn during the December winter solstice more than 1,500 years ago. The Babylonian scholars cleverly organized the celestial hemisphere into a timekeeping device that provided an accurate means of tracking time throughout the year. What they failed to understand was that this annual periodicity is subject to change at greater timescales and that the solar ecliptic would not remain permanently pinned to the constellations.

Four hundred years later it was the Greek astronomers, most notably Hipparchus and Ptolemy, who first documented that our planet does not simply spin on a fixed axis, but that the axis itself must be moving. They understood this by noting how the position of the stars had changed from the time of measurements recorded by their predecessors. It transpires that there are a number of cycles at work, but the Greeks recognized early one of the greatest of these manoeuvres: axial precession. Precession is best described as a wobble – the type we observe as a spinning top loses speed. The vertical axis of the fast-spinning top migrates from a perpendicular position relative to the surface towards a position parallel to the surface. As this happens, two imaginary cones are formed along the axis of spin. The base circles of these cones define the wobble and the time it takes for the axis to travel 360 degrees along the circle is the precessional cycle. We are currently moving 1 degree along this circle in an average person's lifetime, about 71 years. Considering this rate, the time it takes to complete a full wobble amounts to 25,560 years. Some 400 years after the Babylonian zodiac was established, it became clear that its position relative to the Sun was not fixed, but cycled through the 12 constellations in tandem with the Earth's wobble. At that rate, this places the current solar ecliptic in contact with Taurus (Gemini prior to 1990) and Sagittarius at the summer and winter solstices. In deference to long-established convention (but really to avoid all sorts of confusion), I reluctantly maintain this reference through-out the chapter. It may seem a point of historical trivia, but the underlying dynamical behaviour driving the need for a name change to two of the most widely known global geographic features highlights the difference in the way we can see the tropics as a fixed, geographic zone or as a dynamic consequence of celestial interactions and varying surface features that lead to changes in the distribution of global bioproductivity.

The celestial tropics are shrinking

The dynamic nature of planetary motion is continuously altering the geographic extent of the tropics. The Tropics of Cancer and Capricorn are currently positioned at 23.44 degrees north and south of the equator.

But their true position is changing based on a slow oscillation in the tilt of our planet relative to the Sun, referred to as obliquity. Obliquity is calculated by the difference between the angle of the line passing through the north and south poles – the same axis on which our planet spins – and another line running through our planet that is perpendicular to the ecliptic plane. The ecliptic plane describes the path of the Earth as it orbits the Sun, or, conversely, the position of the Sun in the sky as seen from the Earth. The wobble that causes precession of the equinoxes, as described earlier, is a consequence of this axial tilt. Without a tilt, the planet would spin but without its characteristic wobble, and there would be no long-term change in the direction the planet faces on the equinoxes or solstices. The same push-and-pull battle between gravitational forces that causes our planet to wobble over a 27,000+ year cycle simultaneously causes it to behave in a way that slowly changes its tilt relative to the Sun. This occurs more slowly than precession, taking approximately 41,000 years to complete a full cycle from the upper to lower limits of 24.5 and 22.1 degrees and back again.[2] Over hundreds of millennia, this pattern takes a signal form, as a repeating series of waves with ascending and descending phases. We are currently in the descent, which means that the axial tilt is lessening and our planet is moving towards the ecliptic plane. Figure 1.1 illustrates the oscillatory nature of obliquity and precession, the differences in their periodicities, and how these behave more like a trend than an oscillation at smaller timescales. Transcribed onto the surface of the planet, this pattern reveals the general effect of obliquity on the position of the Tropics of Cancer and Capricorn in relation to the surface of our planet. This movement equates to a little more than one-fifth of an arc-second, or about 6 metres, per year on average. But extended over the full breadth of the planet this amounts to around 5,500 square kilometres of surface area moving outside the tropics each year – an area twice the size of Luxembourg, slightly larger than the state of Delaware or the country of Trinidad and Tobago. As the current phase of the cycle continues to compress that part of our planet exposed to an overhead Sun, the surface area leaving the tropics each year will accelerate. This is due to our planet not being a proper sphere, but more ellipsoidal in shape. This has the effect of growing the circumference of the Tropics of Cancer and Capricorn at a faster rate as these boundaries slide towards the equator in tandem with a decrease in tilt. The anticipated motion over this cycle indicates that the next minimum obliquity should occur around 12,030 AD and then the tropics will abruptly turn course and begin marching back towards the poles. By then, the tropics will have contracted by slightly more than 5 per cent of their current area. This is a small fraction perhaps, but equivalent to a massive 10.2 million square kilometres, or an area larger than Canada.

Obliquity interacting with the ellipsoidal form of our planet determines the extent of the tropics by regulating how much of our planet's surface is

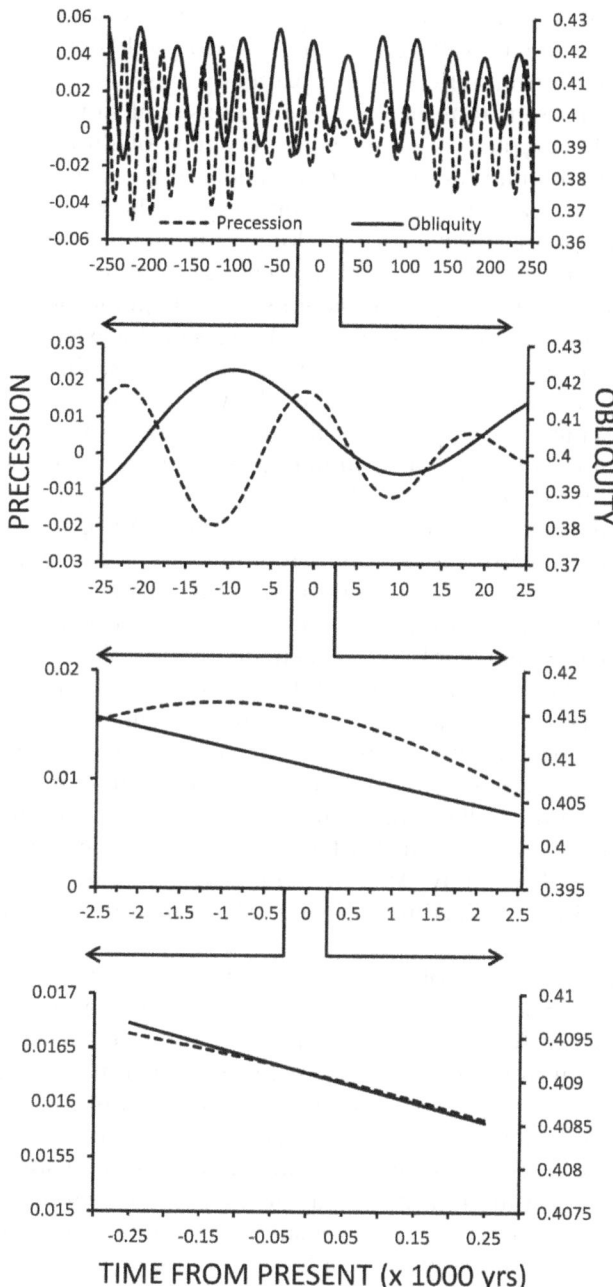

Figure 1.1 The behaviour of precession and obliquity decomposed into four
logarithmic timeframes between 250,000 and 250 years before (–)
and ahead (+) of present time.

exposed to more intense overhead sunlight. If we could position ourselves outside the solar system and accelerate the pace of time we would see a spectacle composed of many different dances, performed in harmony, each deriving its motion from differences in planetary mass and the forces altering these over different time periods. Some consist of slow, languid movements that only complete a cycle once every tens of thousands of years. Precession and obliquity, as we have described, are the most relevant motions, but others, such as the shape, or eccentricity, of the Earth's orbit around the Sun can play critical roles over longer timelines. Still others move more quickly, such as lunar nutation,[3] creating more frequent, less predictable, but substantively smaller inflections. The shape of the planet also changes over time and this contributes to changes in the extent of the tropics. Simply put, it is thought that this occurs as the distributions of water and ice are altered over the long term by the variation in the Earth's orientation and distance from the Sun and over the near term through oscillatory behaviour of very large environmental circulation systems, such as the El Niño Southern Oscillation (ENSO) or Pacific Decadal Oscillation,[4] or through changes in ice mass at the polar limits. All of these can "squeeze" or "release" the Earth to alter the amount of surface area falling within the tropics – they alter the geodetic state of our planet. Together, the continuous push-and-pull of gravitational forces exerted by the celestial bodies in our solar system combines with changes in the planet's shape through time to alter both the amount and distribution of solar radiation intercepted by our planet. This can clearly be seen by calculating the insolation at the current position of the Tropic of Cancer and then projecting the amount both in the past and future at different timescales, as seen in Figure 1.2. In the most immediate timeframe at the bottom of this graph, the amount of radiation received at this latitude is declining in line with the descending phase of obliquity. At larger timescales, changes to insolation become less periodic, reflecting the composite signal of overlapping cycles.

The tropics then, again, are not simply a cartographic construct. Nor is this area static. Unlike many other global map objects born from the great age of European exploration, the Tropics of Capricorn and Cancer were not imaginary lines put in place to better assist us in expanding transport and communications, to tell us where we are when faced with an infinite horizon, or, as was the case with the establishment of the Prime Meridian and fixed degrees of longitude, to parcel continuous time on a spinning globe to aid and abet growing European maritime power. In its most basic, physical form, the tropical zone is a composite signal formed from interacting celestial motion and mass. The result is a geographic area stalwart in its contribution to the global energy budget, but continuously changing in size; a puppet in fact – riding on a planet with mass and motion of its own, but suspended by gravitational strings, with our larger celestial neighbours in the collective role of puppet-master.

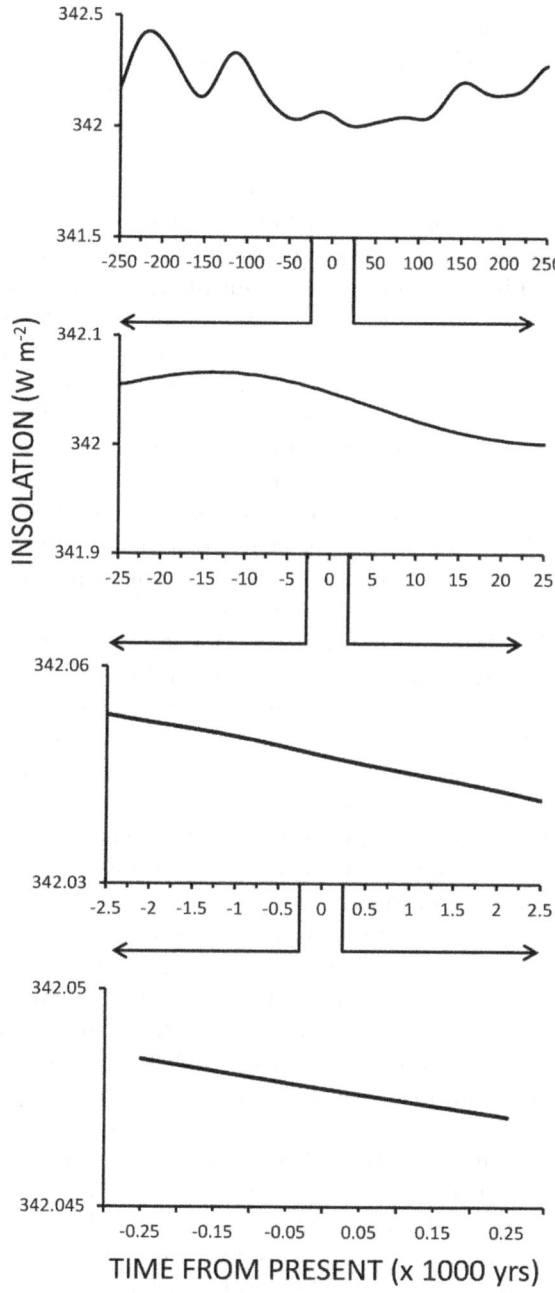

Figure 1.2 The change in total insolation intercepted by Earth as a function of Milankovitch cycling decomposed into four logarithmic timeframes between 250,000 and 250 years.

The thermal tropics are widening

The Greek astronomers and their predecessors had good reason to turn to the sky in describing the tropics. The motion of our planet manifestly shapes variation in the timing and duration of solar irradiance. It also drives the re-distribution of the energy imported through this irradiance and thus the potential bioproductivity of the Earth's surface. In turn, the amount of solar radiation received and re-distributed each day and throughout the year dictates temperatures, how much they vary and, in part through this, global patterns of bioproductivity. However, while the mechanics of gravitational interactions between our planet and its celestial neighbours are the primary cause of variation in the amount of energy received across the Earth's surface, other factors work vigorously to alter this blueprint. These factors then also create aberrations, distortions, often referred to as anomalies, in the global energy distribution. To understand the thermal effect of this variation, we can return to the Far East and their planet-bound, terrestrial definition of the tropics.

Faced with describing the tropics, virtually everyone I have ever asked employs "hot" or "warm" in their response, apart from a few unruly colleagues that invariably throw in a more expansive, but fine-tuned biological, chemical or geological description. On one hand, temperature is what we sense, what we feel. It is how our neural network communicates to our brain the differences between the ambient conditions and the exchange of energy that takes place between our surroundings and that of our own internal, self-regulated, thermal state. But temperature measured relative to how we sense differences, our skin temperature, is not the same as that measured using other approaches, commonly referred to as the surface temperature. Air temperature is strongly affected by the height above the planet's surface as clear air density declines at a constant rate up through the first ten kilometres of the atmosphere – the troposphere. This rate of decline, the adiabatic lapse rate, tells us what temperature we can expect at any given height, or altitude, above the surface at any given latitude. This of course is a critical factor in explaining the cooler temperatures we experience in ascending mountains on a clear day anywhere on the planet. Measuring surface temperature typically involves a single station positioned at a fixed point above the ground or, in the case of measurements taken remotely aboard aircraft or satellites, integrates the measurement over a pre-defined slice of air resting at the base of the troposphere closest to the ground. However, a critical issue in making comparisons between surface temperature datasets arises when values are derived from different altitudes, thereby altering, among other influences, the effect of lapse rate on the integrated temperature reading. Consequently, surface temperatures observed through different techniques may not yield the same results.

Calculating the lapse rate impact of elevation on temperature will typically get you very close to a measured reading when the atmosphere is clear and still. Yet we know that ideal conditions, when they occur, rarely remain across most regions of the planet. Wind and rain invariably take up a far larger slice of the year than many of us living in the rain belts would like, while bringing much anticipated, but all too brief, relief to those living in more arid environments. But these disturbances, long and short, create extraordinary deviations from surface temperatures as measured under ideal conditions. The difference between temperature measured on a clear, still day and that taken during a wet and windy day is apparent to anyone who has experienced the listless effect of tropical humidity or the bitter bite of high-latitude windstorms. Both conditions – high humidity and high wind – alter the surface temperature profile by changing the way in which energy near the Earth's surface is moved about. As a result, they alter our perception of temperature too. Those of us having experienced high tropical humidity know the stupefying effect it has in transforming the paradisiacal into purgatorial. Conversely, the freeze of a cold mid-winter's night can become practically cryogenic as the "wind-chill factor" drives the temperature we feel far below the actual air temperature near ground level. The difference between how we feel the heat or cold under these conditions and the actual temperature illustrates the difference between skin and surface temperatures. Most modern efforts to character-ize temperature regimes have evolved sophisticated procedures to remove the varying effects of wind and humidity to reveal more comparable surface temperatures. Here, surface temperature, and how it is shaped by conditions within the lower troposphere, is always used in describing the thermal characteristics of the tropics. The interplay between surface tem-perature and humidity plays a particularly important role in defining ter-restrial bioproductivity potential and efforts to classify the tropics into distinct vegetation growth zones and climate regimes have proven just how difficult it can be to separate temperature from humidity in particular while revealing that when it comes to the factors engendering elevated bioproductivity, more is not necessarily better.

The tropics are classified

An orderly approach to understanding how the distribution of tempera-ture defines the tropics – as it deviates from that delimited by the Tropics of Cancer and Capricorn – can be found through climate classification, and its important relationship with plant growth. Perhaps the best known, and most well-received, classification system was developed by the Russian-German climatologist Wladimir Köppen over a 50-year period from 1884 to 1936. Köppen was one of the earliest climatologists during a time when very little of the mechanics governing planetary climate was understood. His system classified global climate into five broad types, according to tem-perature and precipitation (Köppen 1884, 1918). This was a sensible approach since both variables are intimately related to the amount of incoming solar radiation and biological productivity, as we will see later. Forever improving on his initial system, Köppen continued to refine the classifications and their geographic distributions until his death, at which time his friend and colleague, Rudolf Geiger, continued the work. His system, along with those later developed by others, remains a lynchpin in the general understanding and communication of global climate distribution and continues to be updated today (Rubel and Kottek 2010).

Köppen defined tropical climates as those experiencing average temperatures exceeding 18°C (64°F) in every month of the year. Virtually everyone on the planet, except the relatively few living above the polar circles and at the higher elevations of the greatest mountain chains, experiences mean monthly temperatures exceeding this limit. But these are seasonally driven by our planet's wobble and not sustained throughout the year. Applying Köppen's tropical limit to global temperature measurements derived from modern, satellite-borne instrumentation, averaged over the period 2000 to 2010, indicates that residents of Los Angeles, Miami, Cairo and Karachi were living in the tropics, at least for this decade. While all of these cities are known for their high temperatures, the surprise in this fact is that these cities are located up to 10 degrees of latitude north of the celestially defined limit – the Tropic of Cancer. In Asia, Köppen's classification extends only marginally beyond the Tropic of Cancer and noticeably less than across the Middle East, North Africa and North America. This limit also extends beyond the Tropic of Capricorn in the southern hemisphere, but only in South Africa, Madagascar, Western Australia and southern Chile is it seen to penetrate poleward to a similar extent, reaching cities such as Durban, Carnarvon and Antofagasta. The poleward extension of the secular tropics beyond their celestial boundary seems more the rule than the exception for both north and south limits, but only at regional scales.

Others followed Köppen's footsteps in developing classificatory schemes for global climate. Perhaps best known among these were the water-balance approach developed in the late 1940s by Charles Thornthwaite, a Professor of Climatology at John Hopkins University, and the life zone characterization scheme assembled by Leslie Holdridge, a tropical botanist and climate scientist based in Costa Rica. Thornthwaite was convinced that a basic water-balance approach – considering the variation in water arriving in an area as rainfall and leaving through evapotranspiration, absent surface run-off and storage – was the most accurate means of classifying climatological gradients since these were intimately linked to plant growth and through this, vegetation types. Temperature was rightly considered an indirect driver of the water balance, but unlike Köppen's approach, it was not employed directly in delimiting tropical climate. In his seminal 1948 article describing this two-pronged approach to climate classification (Thornthwaite 1948),[5] Thornthwaite indicated that he considered an average temperature of 23°C (73.4°F) to best delimit a tropical, megathermal condition from the more subtropical and temperate, meso-thermal environments experiencing lower potentials for evapotranspira-tion due to increasing seasonality of day length and solar radiation.

A year prior to Thornthwaite's publication, Leslie Holdridge presented his life zone classification system (Holdridge 1947). This depicts the global climate as a series of graduated units ranging along a triangulated contin-uum of rainfall and evapotranspiration, but with the addition of tempera-ture as a direct factor sorting life zone types along latitude and elevation. His triangular diagram depicting life zone compartments along rainfall, temperature and evapotranspiration gradients became an indispensable tool in standardizing the description of local climate. Holdridge, like his contemporary Thornthwaite, chose a much higher minimum average temperature, 24°C (75.2°F), in defining the tropical limit than their pre-decessor Köppen. In many ways, it is unclear why Holdridge and Thorn-thwaite chose this higher thermal limit in characterizing the secular tropics. Holdridge clearly recognized Köppen's tropical boundary by inserting a unique "critical temperature line" at the 18°C limit running across the belt of subtropical life zones to separate these from warm tem-perate regions. This would have been sensible in distinguishing subtropi-cal zones that rarely experience lowland frosts, such as southern Florida or southern Yunnan Province in China, from the true lowland tropics that never experience such events. Holdridge also characterized thermal limits using an unusual temperature profile. He adopted a "biotemperature" that he believed would better shadow the limits to plant growth by averag-ing only those temperatures between 0°C and 30°C. At higher latitudes, this was meant to filter out the impact of long periods of sub-zero temper-atures when there was no effective plant growth, but dealing with the tropics proved more complicated.

If we apply Thornthwaite's and Holdridge's tropical limits to the distri-bution of global land temperature as measured from satellites, we see a fundamental problem arising in their classification of the thermal tropics – most of the equatorial regions fall below their thermal minimum (Figure 1.3). This effectively excludes these core areas from the tropical zone designated to differentiate these very same areas from those experiencing wider ranging, lower temperatures. Using these average annual temperat-ures in Holdridge's system, most equatorial forests are classified as sub-tropical. Köppen, despite having developed his system more than a half-century prior, adopted a lower thermal limit that is more consistent with the full range of average temperatures across the tropics. It could be that Thornthwaite's and Holdridge's limits reflect a difference in the way that temperature is measured by satellite sensors compared to the more traditional, station-based thermometers that prevailed at the time – this is clearly a potential source of variation. But reading Holdridge's *Life Zone Ecology*, it is clear he knew that temperatures were lower at the equator than at the Tropics of Cancer and Capricorn, but in the absence of good

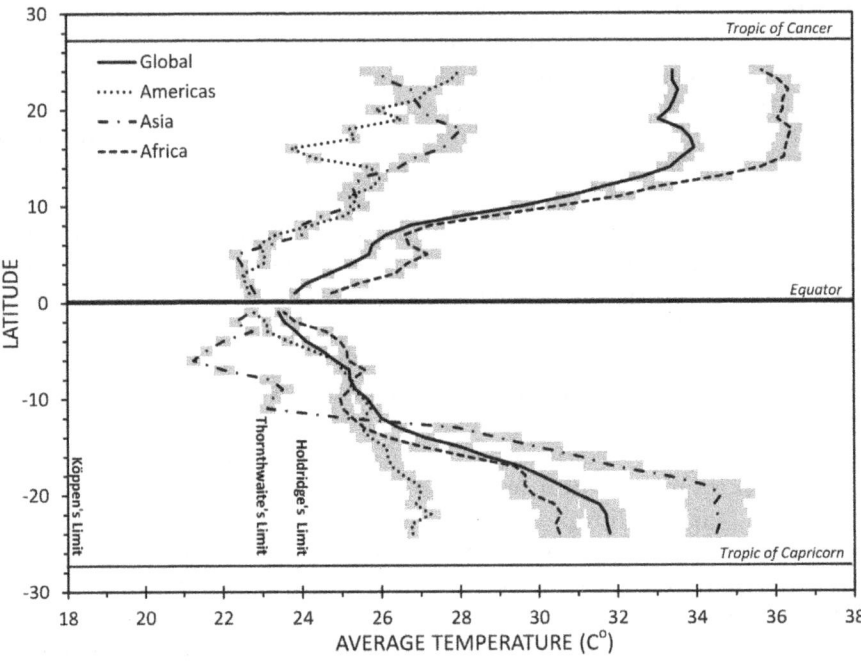

Figure 1.3 The pattern of average temperatures by latitude as recorded in 2000s across the global tropics and in each continent and referenced by the minimum "tropical" threshold temperature adopted in the three main climate zone classification schemes.

Note
Grey bars represent +/– 1 standard error of mean.

ground station coverage, it may be that he and Thornthwaite, without the advantage of remote-sensing, did not anticipate the degree of difference between these. Of course, what is said for them can also be said for Köppen. Could it be that they simply hinged their classification on average annual temperatures across the tropics, rather than discriminating, as Köppen did, regional differences by placing a minimum threshold on average monthly temperature? Examining the distribution of average annual temperatures over the same period, using the same data, unfortunately doesn't resolve this issue. Adopting an annual average temperature as the thermal limit simply extends the thermal distribution of the tropics poleward by 10 to 15 degrees latitude as high summer temperatures combine with more modest winter temperatures at the mid-latitudes to raise the average. The equatorial tropics, however, remain outside the thermal limit of Holdridge and Thornthwaite whether we adopt a monthly average or an annual average as the basis for discriminating the thermal minimum.

Distinguishing between areas consistently above this temperature and those that vary above and below tells us a great deal about thermal seasonality within the tropics and how this varies across the tropics, a fact that is often overlooked in describing the "hot belt". In retrospect, the minimum temperature limit placed on the thermal tropics by Köppen appears an amazingly good choice when we consider the relative dearth of solid climatological information available when he constructed his classification. His acumen led him to a system that accurately reflected transitional temperature gradients at a time when segmenting, separating and compartmentalizing the natural world into categories and classes really did drive the bulk of scientific thought. It turns out that for his successors, delimiting the outer bounds of the thermal tropics was not the main sticking point to tropical climate and life zone classification, but rather addressing an unanticipated dip in temperature from its centre to edge. As time has passed, Thornthwaite and Holdridge have kept much company in viewing the tropics as a thermally constant zone reaching from the equator to the Tropics of Cancer and Capricorn or declining along a gradient from equator to pole. Neither appears to be the case if we adopt a land temperature approach. We can see the magnitude of average temperature increase more clearly using the data collected on board NASA's CERES Terra satellite over the decadal period from 2000 to 2010 (Figure 1.3). These data allow us to see land surface temperature for the tropics as a whole and by regions. They are derived through a relationship between the amounts of radiation detected twice-daily at various band widths by the on-board MODIS spectrometer.[6] While very different from the traditional thermometric approach to measurements, they combine a much-needed departure from the disparate error margins of spatially unbalanced station-based monitoring with unparalleled pantropical coverage. To their detraction, the data have only been collected since the turn of the millennium, constraining their use in assessing longer-term patterns in temperature variation across the tropics.

The curves in Figure 1.3 illustrate the spatial decline in average global land temperature from the edge to centre of the tropics. We see in the composite curve (solid black line) a minimum pantropical temperature at the equator at 24°C. If we recall the dilemma presented by these modern satellite-borne data to Thornthwaite's and Holdridge's efforts to classify tropical climate, we can see that their chosen cut-off point fits relatively well with the global average equatorial temperature. It may be that they utilized this global average to render their classificatory limit for the region. But considering how land temperature changes within each region separately also reveals that there are considerable differences in the shape and depth of this decline across the tropics. The African tropics are considerably warmer than those in Asia or the Americas. This is particularly pronounced in the northern, or boreal, tropics where the equatorial region is squeezed between the vast Saharan and Kalahari deserts. Only in the outermost band of the southern, or austral, tropics is there a similarly rapid rise in temperature away from the equator as the deserts that make up the massive Australian Outback begin to take up an overwhelming share of the land area.

This then is the source of Thornthwaite's and Holdridge's dilemma. Their limits work nicely when considering temperature on a pantropical average basis (Figure 1.3). But this single average belies a large portion of the land area straddling the equator in Asia and the Americas that drops below their assigned limits and, consequently, falls into a subtropical classification. These anomalous regions are clearly visible in Figure 1.3. Köppen's limit varies from year to year in its poleward extension from the equatorial regions. In "warm" years it extends well beyond the Tropics of Cancer and Capricorn, absorbing many southern parts of the United States and China into the thermal tropics. During "cool" years, it contracts towards – and sometimes equatorward of – these celestially defined limits. But taking the average of all months over this same period, from 2000 to 2010, we can smooth out these important, but short-lived, variations. It yields a thermal tropical zone that is amazingly consistent with the invisible limits put in place by the gravitational interactions between our planet and its celestial neighbours, but it is important to remember that these too fluctuate.

The thermal tropics – the "hot belt" ascribed through East Asian languages – it seems is not necessarily more thermally constant than some extra-tropical regions. It expands and contracts from year to year, hovering around – not on – the celestial boundaries. Nor is it, as one might expect, the warmest at its centre – the equator. Land temperatures in areas as much as 10 degrees poleward of the tropical limits can average at or above those near the equator. Satellite-derived data also suggest a surprising amount of variation between tropical regions, most particularly between Africa and the other two main tropical regions in the Americas and Asia. The tropics then perhaps are most simply described as a frost-free zone that, on average, is warmer throughout the year than other similarly sized areas outside the tropics.

The wet and dry tropics

The tropics are differentiated by a minimum temperature threshold from the extra-tropics, but between the two outer boundaries the variation in temperature can be significant. This variation is relatively small compared to the extra-tropics for the reasons related to precession and eccentricity of our planet, but spatial and seasonal variation in precipitation is a much larger source of environmental change in the tropic zone. It is not my intent here to review the processes that govern precipitation patterns and there are many excellent volumes, such as Robinson and Henderson-Sellers (1999) and Hartmann (2015), focusing with great clarity on these as a fundamental component of our planetary climate. But understanding the distribution of bioproductivity, its origins and how it has surreptitiously stewarded globalization through to modern times relies in no small amount upon the factors impacting where and when rain falls across the tropics.

There are three major processes that dictate global patterns of tropical rainfall. Two operate at a global scale and one regionally. The most prominent feature at the regional scale is the action of a rapid rise in elevation. Mountain chains on all three continents act to "trap" moisture by forcing warm, moist air masses to rise, cool and precipitate as they move upslope, a process referred to as orogenic lifting. Cloud and elfin forests often form in the elevational band where this moisture condenses while the downslope run-off sustains high water tables in the adjoining lowland regions below. This rapid rise depletes the colliding air mass of its energy, leaving it relatively cool and dry as it descends the opposite side of the mountain range, often creating a "rain shadow" effect. The second process is the dynamic fluctuation in coupled atmospheric and oceanic conditions in the tropical Pacific related to ENSO. Changes in the state condition of ENSO create global changes in seasonal rainfall, particularly in northern South America, Central America and across the Pacific to South-East Asia (Ropelewski and Halpert 1987). The process is oscillatory in nature, characterized by recurrent migration of high sea surface temperatures and rainfall across the tropical Pacific in an accordion-like manner. Like the opposite slopes of a mountain chain, when one side of the Pacific under ENSO receives greater than average rainfall, the other is receiving less than normal. Other factors can shape rainfall variation in these regions of course, but ENSO, when in a non-neutral state can grow an imbalance between the east and west that dominates rainfall not only in the Pacific, but across other oceanic sectors as well. It also fundamentally shapes how energy is transferred poleward, but more about this important dynamic in the next chapter.

Orogenic lifting and ENSO create spatially anomalous changes in rainfall levels, but the largest segregating impact on rainfall within the tropics is governed broadly in the same way that our planet's angle, wobble and orbit interact with the Sun to determine the amount of solar insolation at a given latitude. The consequence is a tropical zone that is extremely hot and dry at its margins and warm and wet near the equator. Figure 1.3 illustrates the rise of temperatures towards the tropical margins and we can see the general inverse effect on rainfall in Figure 1.4. This graph represents the average annual rainfall as it varied with latitude between the years 2000 and 2010. The data are derived from NASA-JAEAs' Tropical Rainfall Measuring Mission (Kummerow *et al.* 1998). The minimum level of rainfall across the tropics occurs close to the margins at 20–25 degrees for both marine and terrestrial sectors, but you will notice that terrestrial rainfall near the Tropic of Capricorn is around 50 per cent greater than the

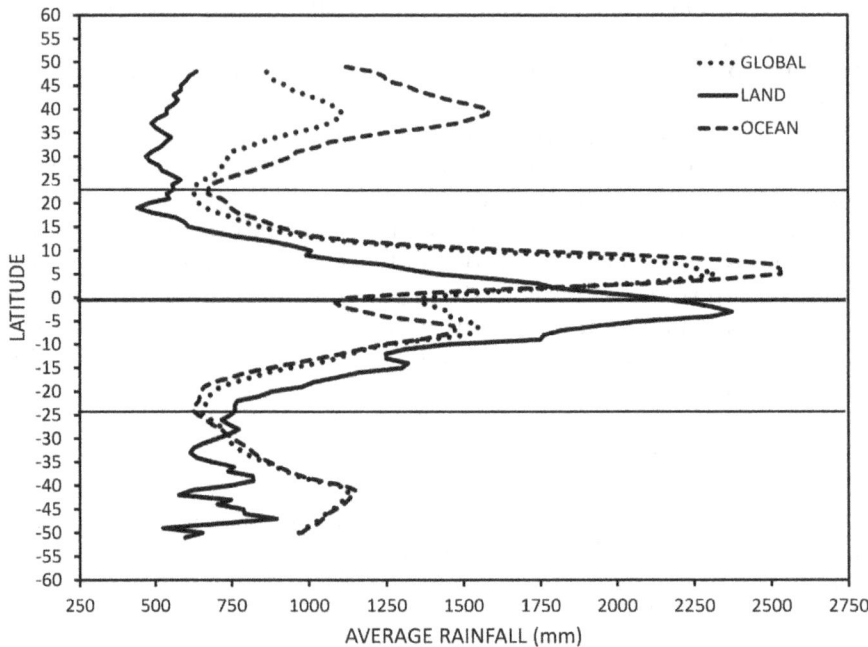

Figure 1.4 The average annual rainfall across tropical oceans and land by latitude
   derived from satellite observations between 2000 and 2010.

amount registering near the Tropic of Cancer. With much larger land
area along the latter, a continentality effect reduces the mitigating effect
of onshore flow of moisture along coastlines. This is most pronounced
over North Africa where evaporation is extreme and strong westerlies carry
off any available moisture to create the world's largest desert. This belt of
aridity extends across Western Asia.

   Over the oceans, the reduction in rainfall at the outer edges of the
tropics is more symmetric because evaporation is not subject to this geographic change in moisture availability. Peak average rainfall over land
and sea are also not spatially coincident. The oceanic peak is coincident
with the average position of the Inter-Tropical Convergence Zone (ITCZ).
Over the oceans, this position remains fairly stationary relative to its movement over the continents. Again, energy fluxes over land are more dramatic over time compared to those over oceans and this broader range
drives the ITCZ to migrate over a broader range of latitude compared to
oceans. You can also see a dip in rainfall over the equatorial oceans caused
by the equatorial counter-currents. These currents emerge through
surface wind displacement, allowing water to rise from depth, lowering sea
surface temperatures and evaporation. The global pattern of rainfall

described in Figure 1.4 is a precursor to understanding the distribution of tropical bioproductivity. Importantly, it also describes a broad, arid belt that separates the equatorial wet tropics from the moist, temperate latitudes. As I discuss in later chapters, the dry tropics are symptomatic of the global processes that geographically stratify the distribution of bioproductivity and it is this stratification that played a critical role in the emergence of globalization. Both terrestrial and marine environments are subject to this stratification, but making comparisons between similarly sized areas of ocean and land in the tropics is more difficult than one would think.

The blue tropics

This is due to the fact that tropical land is relatively scarce. The distorting effects of gravitational pull and spin on the shape of the Earth have left the tropical zone with more surface area than would be expected if our planet was a perfect sphere. Compression at the poles along the axis of spin, much like squeezing an orange between your hands, has left our planet with a middle-aged spread. This spread protrudes further outward at the waist of our planet, increasing the surface area that is in line with the plane of the Sun's irradiance (the solar ecliptic). As a consequence, the planimetric area[7] of the tropics, at around 203.6 million square kilometres, accounts for nearly 40 per cent of our planet's surface, but is bound within only 25 per cent of its latitudinal range. Yet despite this large surface area, the tropics contain only 34 per cent of the world's land area. The relative scarcity of tropical land is further appreciated when we consider that this accounts for a mere 10 per cent of the planet's surface. We can see this more clearly by calculating the amount of surface area allocated to land and ocean within each degree of latitude using a Mollweide, or other similar equal-area, cylindrical map projection. This does a very good job of translating the ellipsoid shape of our planet into a two-dimensional, map-like sheet without distorting the relative proportions of land and ocean, unlike the Mercator projection. If we examine the resulting distribution as a profile, seen in Figure 1.5, it is clear that tropical land area falls very short of the zone's relative contribution to total surface area.[8] This shortfall is not restricted to the tropics, but part of a general decline in land area southwards from the northern mid-latitudes, disappearing entirely between –55 and –65 degrees to form the great Southern Ocean. As in the tropics, land is remarkably scarce in the southern extra-tropics. Accounting for a mere 15 per cent of global land area, Antarctica alone accounts for more than half of the area south of the Tropic of Capricorn. Not surprising then that nearly 70 per cent of global land is situated at latitudes north of the Tropic of Cancer. While this has not always been the case through the long geological history of our planet, the effect of this current geodesic imbalance on the relationship between the tropics and the extra-tropics is omnipresent.

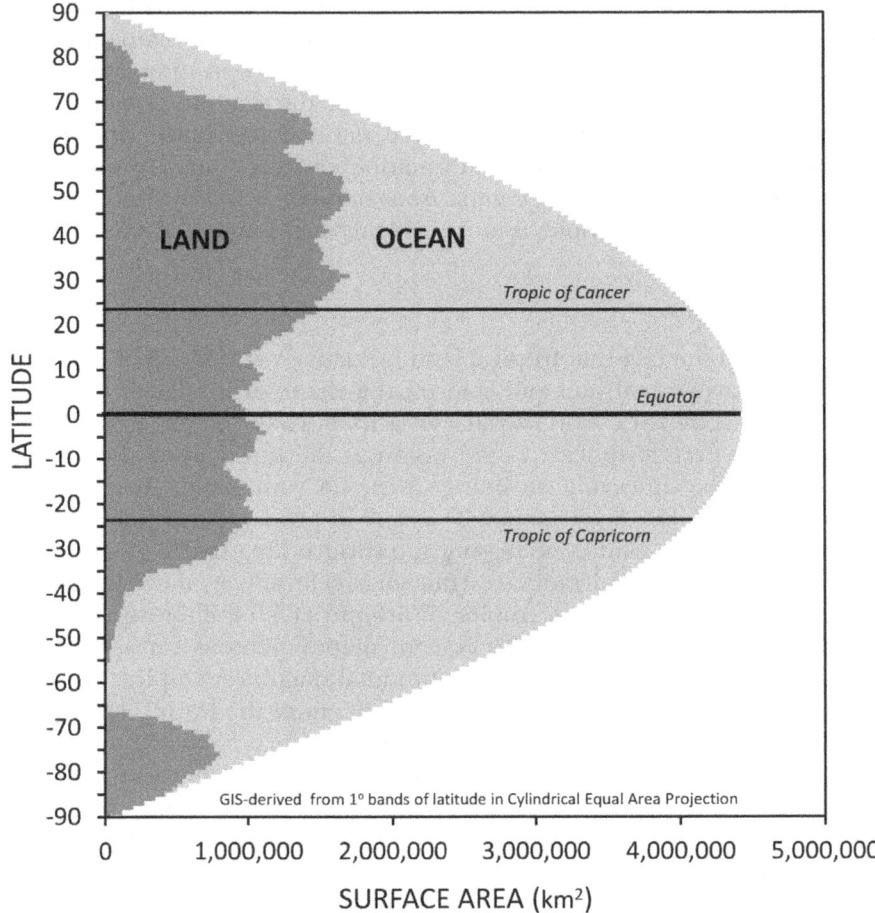

Figure 1.5 A depiction of Earth's land and ocean area as distributed by latitude.

Central to this effect are the tropical oceans. Estimated at around 153 million square kilometres, they currently account for about 42 per cent of global marine area. This equates to nearly one in every three square kilometres of our planet's surface. Combined with the relative dearth of land, the land-to-ocean ratio of the tropics at 1:3 is considerably higher than the extra-tropics at 1:2. This imbalance is caused by a striking difference in land distribution between north and south. In the north half of our planet, land and ocean are nearly equal in area across the extra-tropical regions. The southern extra-tropical oceans, by contrast, occupy 85 per cent of the surface area of this region. Again, this fairly obvious observation, gleaned from any map or model of Earth, is quantified in the profile diagram of Figure 1.5. The primary consequence of this asymmetry rests with the enormous exposure of the tropics to the Sun, the capacity of the region to absorb and convey energy and the impact of the relative occurrence of these two tropics – terrestrial and marine – on the distribution of bioproductivity across our planet.

Notes

1 Many of Miller's items of personal correspondence with friends suggest the title of his most famous work has more to do with the image of Cancer, and its zodiacal connection with the crab, than a celestial connection with the solar ecliptic.
2 These limits depend on the gravitational forces attributable to the Sun and planets within our solar system and their interaction. Over billions of years, the obliquity of Earth may exceed these limits as gravitational forces change with alterations in the state condition of the Sun, planets and their orbital geometries.
3 *Nutation* in this context describes a smaller variation in the precession-driven wobble of the Earth's polar axis. Current understanding attributes the main component of this motion to the interaction of the Sun and Moon on the distribution of surface water, called *tidal forces*, accounting for our planet's imperfect shape (and thus distribution of mass). A motion of this type has been calculated to cycle every 18.6 years, but imperfectly.
4 Longer-term changes in the distribution of moisture occur mainly as an indirect consequence of variation in our planet's obliquity, precession and eccentricity that alter the distribution of insolation. This has the effect of altering the extent and thickness of polar ice over glacial (Ice Age) and inter-glacial (present) stadia. Changes in the amount of polar ice then alter the distribution of planetary mass, impacting the extent of the equatorial "bulge". Near-term shifts in water due to oscillatory behaviour in the major oceanic and atmospheric circulations are also suggested to impact mass distribution, but at smaller timeframes and magnitudes of change (Chang and Tapley 2004).
5 This work was seminal in many ways, most notably in its introduction and use of potential evapotranspiration (PET). He would later further develop the rainfall-PET approach into a more detailed water balance, a widely used hydrological concept (Thornthwaite and Mather 1955).
6 The algorithm employed to translate the spectrometric readings into land surface temperatures is described in Wan (1999). The accuracy of the translation is placed at ± 0.5 °C. Here, data presented as averages for each month between January 2000 and 2010 were averaged and partitioned by each degree of latitude. The values for each raster cell within each degree were then averaged to produce the temperature curves with measures of dispersion denoting spatial variation.
7 *Planimetric* refers to the smooth surface area without consideration of elevational effects. This is a conservative measure of area.
8 The measure of global surface area varies depending on assumptions made regarding the "true" shape of our planet, how this is transformed in the process of converting from a three to a two-dimensional depiction, and the size of the unit used to estimate area. Estimates here are derived from 1-degree bands of latitude in Cylindrical Equal Area Projection and calculated in a geographic information system (ArcGIS Pro). Land area distribution is based on data from the Shuttle Radar Topography Mission (SRTM) processed into a digital elevation model (DEM) by the Jet Propulsion Laboratory. SRTM only provides

coverage between 60 degrees north and 60 degrees south. The data used have a precision of 3 arc-seconds (SRTM3 "finished"). ASTER GDEM data, provided through NASA's LP DAAC were utilized for regions poleward of SRTM limits. ASTER GDEM data are newer, with a higher precision of 1 arc-second, but with less effort to reconcile anomalous elevational patches. Thus a higher error rate is expected compared to SRTM3.

2   Planet's powerhouse

The Forum, cradle of Greek philosophical and scientific debate, was filled for the final showdown between two legends. Aristotle, the renowned Greek polymath, was squaring off against a contemporary known for his originating work on atomic theory. Leukippos of Miletus insisted that a vacuum, what we would call space, existed as an entity separate from that of material bodies and Aristotle begged to differ. "Horror vacui" Aristotle is supposed to have rejoined (in ancient Greek of course), delivering the founding broadside in a debate surrounding space and matter that continues to this day. We know from the root laws of thermodynamics that change takes place only where state conditions are transformed, releasing energy in the process. We also know that energy flows from high to low states: hot to cold temperatures, high to low pressures. It follows that where the flow, or flux, of energy is small or absent, there is little if any change in the state condition. But the second law of thermodynamics tells us that this state cannot remain indefinitely and all stable conditions (at a low-energy state) invariably descend into chaos (through an influx of energy). The famous late-nineteenth-century French chemist, Henri Le Châtelier formally summed up this view with his equilibrium law: a change in one of the variables that describe a system at equilibrium produces a shift in the position of the equilibrium that counteracts the effect of this change. This is the crux of Aristotle's response.

But where nature simply abhors a vacuum, evolution despises. Planetary change – physical, biological and social – evolves around the availability of energy. Many of the evolutionary currencies – the dynamism of the physical landscape, the abundance and diversity of life and the complexity of society – lose their value when deprived of the energy necessary to impart change. Life prospers most where these forces deliver the natural means for production – water, light and nutrients wrapped up in a warm envelope. When they are absent – or only available outside the envelope such as in deep caves or at the polar ice caps – the abundance and diversity of life diminishes. We can also think about our modern cities in this context – their frantic pace of activity and continuous consumption. One quickly recognizes that cities are simply energy consumption hotspots evolving as food, water and materials flood in, only to re-radiate outward this energy transformed through work as structures, devices, ideas, trends and data. What would the modern city be if the flux of energy, in all of its forms, slowed or stopped? As the respected historian Ferdinand Braudel noted: "A world economy always has an urban centre of gravity" (Braudel 1984). The global flow of energy manifests itself through environmental, economic and demographic fluxes and the tropical region holds a pivotal position in driving these through its shaping influence on global bioproductivity. But to understand this substantive role we need to consider where energy originates and the three exogenous sources of energy driving the evolution of our planet: geothermal, mechanical and, above all, solar. This chapter is about these sources and the role of the tropics as the planet's powerhouse.

Geothermal energy – our planetary dynamo

Geothermal power is born from the physical evolution of our planet. It is most visually apparent in the steaming hot springs, erupting geysers, streaming lava flows and eruptions that characterize the most active volcanic regions of our planet. But these are merely the end-products of a process that builds, and then releases, energy flowing from the mantle to the planet's surface through convection of viscous rock towards relative weak points in the overlying crust. These weak points, commonly referred to as fault lines, run across the surface of the planet to form boundaries between two adjacent pieces, or plates, of crust. At some plate boundaries, new crust is being created, while at others it is being destroyed. Combined, this conveyor-like process of creative destruction, known as plate tectonics, very slowly overturns large parts of the planet's surface. The release of the bulk of geothermal energy occurs as the state condition of the material changes at these boundaries: either as rock moves from liquid to solid (cooling) or as it moves from solid to liquid (melting). Driving the entire process is the massive decay of the radioisotopic materials – uranium (mainly U-238), thorium-232, and potassium-40 – that heat the mantle material (Rama Murthy *et al.* 2003).[1] Radioisotopic decay under the extreme pressure conditions of the planet's interior is the root source of geothermal energy. Since the 1950s, geophysicists have been wrestling with exactly how much thermal energy is shunted from the inner planet to the surface through tectonic activity. Henry Pollack and colleagues of the University of Michigan arrived at an estimate of 87+/−2 milliwatts per square metre for mean global heat flow based on an extensive analysis of data from more than 20,000 measurement sites (Pollack *et al.* 1993). This amount, the equivalent needed to power two to three LEDs, is comparatively small. But expressed on a global basis, by considering the instantaneous flux of geothermal heat from the entire surface area of the planet, it amounts to about 44 terawatts – enough power to run global civilization for 2.5 years at current annual consumption rates.

Expressing an amount of geothermal energy conveyed to the surface as a global average, however, overlooks important spatial variation in tectonic activity. All parts of the planet's surface do not participate equally in the cycle of crustal growth and loss, leaving intact some island-like areas of very old crust surrounded by much larger areas of more recently formed rock. These islands of crustal strength, or cratons, consist of dense rock formed during the early, more active part of our planet's crustal evolution in the Precambrian era, more than 550 million years ago. Some of these areas are well known by geologists for the tremendous age of their rock, such as the 3.5+ billion-year-old Acasta Gneiss of northern Canada, the Jack Hills of Western Australia and the Kaapvaal in South Africa. Cratonic areas of ancient rock are uniquely continental features, rendering the terrestrial part of our planet's skin typically much older, and thicker, than the much larger portion forming the sea floor. As a consequence, rock age and thickness increases as distance from the nearest plate boundary increases. This general, but not universal, feature of the tectonic cycle of creative destruction strongly shapes the variable flux of geothermal energy across the planet (Stein and Stein 1992). Continental land masses, particularly cratonic regions, are relatively poor sources in comparison to mid-oceanic seams where new crust is being formed. One result of this spatial disparity in geologic history is that estimating an average flux for the planet masks considerable variation in heat flow at regional scales. This variation can range from nearly zero in large cratonic areas up to 350 milliwatts per square metre emanating from the most active and youngest oceanic crustal zones. Pollack *et al.* estimate that the largest geothermal pockets arise from the fault zones of the southern (sub-)tropical oceans where the Earth's crust is thin and tectonic forces creating new crust are most active.

The massive tsunamis that brought uncontainable destruction and a desperate loss of life to the shores of Banda Aceh, Indonesia in 2004 and coast of Miyagi and Iwate provinces, Japan in 2011 characterized the release of tremendous mechanical energy attached to the planet's tectonic features. Earthquakes and tsunamis typify the destructive end of the tectonic conveyor where crustal plates battle it out with the inevitable result that one is consumed, or subducted, below the other. The energy released through these crustal movements is highly concentrated, but ultimately owes its destructive power to the same geothermal engine that delivers heat through volcanoes and crustal growth zones. Consequently, this mechanical action is not an independent, exogenous source of energy. It is merely a transformation of energy from the geothermal to mechanical. Mechanical energy is released when a force acts to alter the rate of movement. Similar to the release conveyed through earthquakes and tsunamis, most mechanical energy on our planet is ultimately sourced from a transformation of geothermal or solar sources.

Consider the mechanics of the car engine. It is fuelled by petroleum, a refined form of fossil carbon. This carbon was formed through compression of organic material, mainly zooplankton and algae, deposited millions of years ago. This base material was formed as a direct and indirect product of photosynthesis. In effect, then, all petroleum-fuelled engines are solar-powered, but through a very round-about process defined by an interval between energy source and sink measured in millions of years, rather than a matter of years for biofuels, and milliseconds for solar cells. Consequently, fossil fuel reserves are not an exogenous source of energy – merely a storage product of the immense work performed by plants in transforming carbon dioxide into chemical energy using sunlight. Running along similar deconstructions, we can source all carbon-based energy pools such as natural gas, oil shale, coal, peat, wood and any other form of biomass, such as algae, to sunlight. Direct forms of mechanical energy, such as wind and wave energy are also sourced invariably from incoming solar radiation, through differences in pressure created by gradients in the amount of energy reaching the surface of our planet, but more about these later in the chapter. To search for a true exogenous source of mechanical energy we must shift our gaze from the planet's interior towards the sky and back to the same force that drives the change in the orientation of the Earth and the size of its tropical belt – gravity.

Mechanical energy – the Moon, the Sun, the stars

We often think of tsunami and tidal wave as different words describing the same destructive swells that hit Indonesia and Japan. It may be that their connection has grown from observing the rapidly ascending water of a tsunami, and interpreting its behaviour as similar to a rising tide. But tidal waves are used in a much broader sense to describe a group of important wave-like changes in the depth of oceanic waters, pressure in the atmosphere and height of the Earth's crust. These wave-forms, or more accurately their dissipation, represent the only form of energy that cannot be traced back to solar or geothermal provenances. Most people understand the rise and fall of water attached to tidal movements from their personal observations of changes in the width of beaches or water height in relation to piers, docks and other coastal structures. But the process that creates the pulse of tidal movements along local coastlines is driven by interactions at vastly larger scales. The primary drivers of tidal behaviour, the Moon and Sun, interact with the centrifugal force generated through Earth's spin to distort the depth of oceanic waters. These competing gravitational forces pull on the oceans in a non-stop tug-of-war with the force of the tug being greatest on waters closest to the Moon. These are pulled away from the planet with greater force than those furthest away, creating a bulge on the side facing the Moon. As the Moon orbits, the gravitational force it exerts on the oceans drags the tidal bulge around the planet, creating a high-tide peak as it passes. High tides are visible as the march of this giant wave is interrupted by land, where it dissipates fully. But as the interaction of forces pulls on the oceans to create a bulge, it also pulls the Earth's mass slightly towards the Moon, leaving water on the other side of the planet stationary and thus further from the surface. This differential effect of gravity's attraction creates another tidal bulge on the diametrically opposite face of the planet. The visible outcome is a twelve-hour, or semi-diurnal, periodicity in the rise and fall of the tides as the two waves move in unison around the planet. The outward force exerted by the Earth's spin further shapes both bulges, as does the position of the Sun in relation to the Moon. The track of the Moon's path in relation to the Earth's tilt situates the peak of the tidal bulge north of the equator, but this bulge changes over time too as the Moon's orbit varies, adding further variation to the expected tides. An independent gravitational effect of the Sun on our oceans also creates tidal effects with different periodicities and as the Earth's orbit around the Sun varies, these too change. All of these resonant components of tidal wave movement are compounded as the lunar and solar-driven waves return to meet with the remnants from a prior cycle, further confounding the long-term stability of tides that would be predicted from simple lunar effects alone. The combination of these different waves, a form of harmonics, can create measurable variation in the height of tidal advances at scales ranging from hours to decades. The most notable among these are the unusually high spring and relatively low neap tides, when the Sun is aligned with and positioned at a right angle to the Moon, or further pronounced tides when the Moon is closest in its orbit to the Earth (at "perigee") and also aligned with the Sun to create the massive perigeal spring tide.[2] This last type, occurring a dozen or so times a year, is a major cause of coastal flooding.

It is the Moon's gravitational pull and the twin tidal bulges that it creates, however that acts as the overwhelming source of mechanical energy. This semi-diurnal, lunar component accounts for more than two-thirds of the change in sea-level elevation, or tidal range, at most locations. But some areas experience only a single high tide each day, or none at all, where effects of the twice-daily components are weak or nullified by interactions with abrupt changes in topography (particularly islands, that bend the tidal bulge's advance) and the bending force applied by the planet's spin – the Coriolis effect. The resulting formation is a tidal gyre, a series of waves rotating around a central point. This is similar to the effect of a large boulder in creating the sheltered waters of an eddy in a fast-flowing white water river. But the scale of tidal movements is massive, fundamentally driving the behaviour of the oceans globally, rather than just locally. The global flux of energy driven by tidal movements is fairly well established, estimates placing this at around 3.75 terawatts, of which approximately 2.5 terawatts is attributable to the twice-daily lunar component.[3] This is enough work to run global civilization for about 2.5 months. Lunar and solar gravity also act to distort the solid Earth shape, called the earth tide, in a similar, but less dramatic, way to that attenuating the more fluid oceans. Similarly, changes in sea height push on the atmosphere above, generating small wave-form changes in pressure. However, both of these sources of mechanical energy are trivial in comparison to the ocean-borne tidal flux. The earth tide has been estimated to generate an energy flux of 0.11 terawatts, less than 3 per cent of the amount attributed to ocean tides (Platzman 1985; Ray *et al.* 2001). The energy sourced from atmospheric tides is similarly insignificant, amounting to a mere 0.1 per cent of the global ocean tide flux (Platzman 1991). Thus, the overwhelming bulk of energy generated by gravitational attraction is dissipated through the ocean tides and the major portion of this is attributable to interactions with the Moon.

Oceanic tidal energy dissipates mainly through bottom friction in the run-up along coastlines, but also as the waves encounter deep-ocean topographic features. As waves move through the deep oceans and drag across solid features, or encounter shallow shoals along the coastlines, their amplitude is squeezed, releasing the energy carried mechanically through the tides as transformed heat. The global distribution of mechanical energy released through this dissipation depends primarily on the initial amplitude generated by the pull of the Moon and Sun. This range is most easily understood as the sum of the various wave-creating components accounting for the observed change in ocean height at any given point at any given time. To form a global picture of energy release, these components are typically modelled to identify the amplitude of ocean height change as a result of each harmonic component's contribution. Models are then calibrated with tidal gauge measurements or, more recently, satellite altimetry. The results from various efforts to model the effects of the most important semi-daily lunar component (called "M2") indicate that about one-third of energy dissipates as tidal waves encounter topographic features of the deep sea floor. The remaining two-thirds are released during the run-up along the coastlines (Egbert and Ray 2001). Expressed as a power density, dissipation amounts are small. Zones of intense dissipation due to friction can be found along some coastlines, but this amounts to less than 250 milliwatts per square metre. Deep sea drag typically generates less than a fifth of this amount – just enough to run a single LED. More importantly, the dissipation of tidal wave energy is not faithful to where it is first formed by the action of the Moon and Sun. Dissipation hot spots, characterized by immense tidal ranges up to 11 metres (36+ feet), are primarily located along sub-polar coastlines of eastern Canada, the British Isles, Alaska, Korea and Patagonia. The most obvious reason for a preponderance of tidal energy along sub-polar shores is attached to the simple fact that there is more coastline at these higher northern latitudes than elsewhere.

But tidal forces are exceptionally complicated by the harmonics of their interaction with the shape, orientation and distribution of coastlines and the size of the adjoining continental shelves where most dissipation takes place. Much of the mechanical energy driving large tidal ranges in the north cannot be accounted for by the expected response to the pull of lunar and solar gravity at these latitudes. Some oceanographers anticipate that this unexplained energy must be generated elsewhere, based on the results of their hydrodynamic models (e.g. Le Provost and Lyard 1997). This is then conveyed poleward where ocean height is lowered or unresponsive to the main, semi-daily tidal effects. Due to the differential manner in which the lunar force acts on seawater height, our planet's gravity suppresses tidal wave formation nearer to the poles while lunar and solar forces increase it in the tropics. In another Aristotelian example of movement from a high- to a low-energy state, the energy attached to the tropical tidal bulge propagates poleward through the same series of sweeping, circular gyres that create localized areas experiencing no tide.[4] The propagation of tidal waves across neighbouring gyres is propelled like a baton relayed from one pirouetting ballerina to the next, dynamically maintaining equilibrium in the distribution of mechanical energy across our planet. The vast southern oceans are thought to act as the main source of this energy. With relatively little shoreline in this region to capture the ocean tide, tidal waves are longer-lived than those generated at latitudes with high shoreline densities. Conveyed to the north via the tropics, a small portion of this energy dissipates on its journey as it passes over the ridges and seamounts formed at tectonic plate boundaries and island-arc chains. The remainder contributes to the formidable high tides seen along the coasts of the sub-arctic region. The tropics zone, for its part, acts as both producer and broker, conveying the mechanical energy generated from its own waters combined with that produced further south. The amounts may seem so small as to scarcely merit attention, but these transfers are thought to be very important in maintaining the global circulation of oceanic waters (Munk 1997), a process that allows the tropics to disgorge their mighty surplus of energy each year towards the energy-starved poles and, in particular, towards the European continent where their impact on the region's role in globalization and colonization has been profound, as we will see in later chapters.

Solar energy – one source to rule them all

Geothermal venting and the tides combine to supply 47 to 48 terawatts to our planet's energy budget. This is a sizeable amount of energy in comparison to our collective day-to-day requirements, but dwarfed by the amount delivered directly from the Sun as incoming short-wave radiation (ultraviolet, visible and infrared). The amount of this radiation that is initially intercepted by our planet can be calculated rather simply by employing the fact that when energy is radiated outward from a point source such as the Sun, the quantity, or density, of the flux is inversely proportional to the square of the distance from the source. This holds true regardless of the intensity of the source emission. For example, only a quarter of the energy reaching a distance of 10 metres from a photographer's 60-watt or 100-watt flash-bulb will make it to a point 20 metres away. Applying this inverse square law in determining how much solar radiation reaches the Earth's outer atmosphere yields a constant amount of 1,366 watts per square metre.

This value, often referred to as the solar constant, however, is not really constant. The amount of radiation emitted by our Sun can vary by a monthly average of 1–2 watts per metre squared, due to increasing and decreasing sunspot activity. Sunspots are temporary areas of relatively low radiation emerging onto the Sun's surface. They are formed by shifting magnetic fields within the Sun that impede the uniform flow of radiation from its core. Sunspots, however, are encased by a ring of very high radiation flux as energy flow is concentrated at the margins of these localized magnetic impedances, creating a surge in the amount of radiation emitted by the Sun rather than suppressing it. Consequently, the more sunspots, the higher the total radiation emitted by the Sun. As sunspot numbers vary, so too does the amount of solar radiation received by our planet. This occurs most prominently in relation to an 11-year cycle, although longer-term cycles in activity are thought to take place and can yield longer periods of reduced or heightened solar irradiance. One such period, the Maunder Minimum, occurred from 1645 to 1715 and coincided with anomalously low sunspot numbers and a record of severe winters and low temperatures in Europe and North America. Rivers, lakes and even the Baltic Sea froze over to unprecedented levels and accounts of widespread starvation were rife. The impact of this "Little Ice Age" was immortalized in the works of many prominent Dutch, British and Flemish painters of the time. Other factors, such as pronounced volcanic activity,[5] must have conspired with changes in solar activity to create the period of low temperatures during this time since the expected variation in solar output doesn't fully explain changes on our planet over this period, but the temperature–sunspot timeline is strongly coincident.

The amount of solar radiation intercepted by Earth is also sensitive to variation in the distance to the Sun. Recall in Chapter 1 how precession and obliquity alter the orientation of the planet in relation to the Sun, and thus the timing and latitudinal position of the Tropics of Cancer and Capricorn. A third, much longer-term effect on Earth–Sun positional geometry is linked to a cyclical change in the shape of our planet's orbit. Unlike precession or obliquity, the cycle of change in orbital shape, or eccentricity, alters the absolute amount of solar radiation intercepted by the planet rather than its surface distribution. It does this through an exaggeration of our planet's ellipsoidal orbit. As the orbit is stretched at two ends, becoming more elongated, the point during the year at which the planet is furthest from the Sun, called aphelion, becomes more distant.

Consequently, solar radiation reaching our planet declines during this phase. Combined with an increase in obliquity and a precessional orientation facing away from the Sun, the polar regions receive considerably less sunlight, the planet cools and an Ice Age ensues.[6]

More recent fluctuations in the amount of solar radiation intercepted by our planet are now being monitored continuously by satellite-borne radiometers. A first series of missions operated from the 1970s to the 1990s, when a new series of observers were launched. Combined, observations have been made over three 11-year sunspot cycles. However, observations made aboard the most recent mission vehicle, launched in 2003, have recorded significantly lower levels of solar radiation in comparison to earlier radiometric readings. Employing a more accurate radiometer design, the Total Irradiance Monitor (TIM) recorded a total solar radiation value of 1,360.8+/−0.5 watts per metre squared – a value lower by $4.6\,W\,m^{-2}$ than other satellite-based radiometric measurements made over the most recent period of minimum sunspot activity (Kopp and Lean 2011). This emphasizes how environmental observations are increasingly dependent on instrumentation, and that the employment of an instrument does not portend infallibility. The TIM record has not yet extended over multiple sunspot cycles. The difference in radiometric readings due to instrumentation changes suggests a corresponding decline in the longer-term average value of solar radiation intercepted by our planet over multiple sunspot cycles. But it does leave open the possibility that the swing in radiation flux attached to changing solar activity over the sunspot cycle is greater than previously thought. It depends on the instrument's response to changes in solar output. The latter effect would alter the results of current global energy models by increasing the amplitude of swings in the amount of incoming solar radiation over solar cycles. However, it is far more likely that increased instrument accuracy will lead to a general shift downward in readings across both high and low periods of sunspot activity. Either way, a decline in the expected amount of solar radiation imparts a direct negative effect on the amount of global temperature change that can be attributed to the source supply in global energy budget calculations. The varying solar constant gives us the maximum amount of solar energy available. But only a fraction of this is ultimately injected into the planetary life zone, or biosphere. To understand how this gross figure is whittled down to the actual amount available at the planet's surface, and how it is subsequently transported across the planet, we need to take a look at how the atmosphere interacts with the unending flux of radiation from the Sun.

The Earth's atmosphere is not particularly dense. Compared to the Jovian atmosphere, or that of the other gas giants in our solar system, it exerts less than one-thousandth of the surface pressure created by their thick, hydrogen-laden layers. If, however, our planet, like Mars, did not retain much of an atmosphere, virtually all of the solar radiation

intercepted by Earth would reach the surface. Mars' climate is comparatively much colder with average temperatures over most of the surface well
below freezing, but rising to about 20°C (70°F) during a few summer days
in some low-lying regions within its own tropical belt before dropping
below freezing during the night. In part, Mars' lower temperature is
explained by its more distant orbit. The amount of incoming solar radiation reaching Mars is less than half of that intercepted by our planet. An
absence of any significant atmosphere ensures that temperatures are
closer to those expected from its distance alone. Bombarded by harmful
ultraviolet radiation, while freezing in permanent winter-like temperatures, a thinning atmosphere on Earth would present a formidable challenge to all life except those organisms adapted to the most extreme living
environments.

A mirror, prism and sponge

This outcome, where an increase in solar energy due to atmospheric thinning results in much lower surface temperatures, might seem counter-
intuitive. But with planets, the amount leaving the system is contingent on
the composition and density of the atmosphere. In the case of our planet,
the atmosphere is sufficiently dense to disrupt the free-flow exchange of
energy between surface and space. In effect, the atmosphere acts as a great
mirror, a prism and a giant sponge all at once. Like a mirror, it reflects a
minor portion of the incoming short-wave radiation back into space. Like
a prism, it scatters a similar amount through the atmosphere. It does this
while soaking up both incoming radiation and large amounts of out-going
heat emitted from the surface like a sponge, storing this energy, and then
re-radiating it in all directions as long-wave radiation (mid-infrared heat).
As a result, the sky is brighter and our planet warmer than would occur
were the atmosphere thinner, like that on Mars.

The atmosphere modulates the planet's temperature through a
complex series of absorption and emission reactions of its various constituents. Six of these, molecular oxygen, ozone, water vapour, carbon dioxide,
nitrous oxide and methane, are naturally occurring in trace quantities, but
also increasingly sourced as by-products of human agricultural and industrial activity. They combine with a distinctly human-made class of molecules, chlorofluorocarbons (CFCs), to variously absorb radiation across
the full spectrum of ultraviolet, visible and infrared wavelengths that make
up our planet's energy flux. The different spectral components, or bands,
of solar radiation simultaneously engage in a wide-ranging, never-ending
game of hide-and-seek with the various atmospheric gases due to their
varying absorption responses. Ozone readily absorbs radiation in the ultraviolet spectrum, water vapour mops up far-infrared radiation, while carbon
dioxide, methane, nitrous oxide and water vapour absorb short-wave
infrared (heat) at intermittent wavelengths.[7] The atmosphere is virtually

transparent to energy in the visible light spectrum, with very little absorption taking place. The spectrum that we see as light almost always wins the game by avoiding absorption, although suspended particulates, such as soot, smoke, dust and sulfates, can intercede and substantively reduce the amount of visible light reaching the Earth's surface.[8] As a result of our atmosphere's selective properties, a little more than half of the total amount of solar energy intercepted by our planet is currently received at the surface as short-wave radiation.[9] The remaining portion is absorbed or reflected back into space by the atmosphere, particularly clouds. An additional 4 per cent of the short-wave radiation making its way to the surface is immediately returned to space. This is mainly in the form of sunshine reflected directly from relatively bright land features, such as sand and snow, or aquatic surfaces, such as rivers, lakes and oceans, or at dawn or dusk when the Sun's rays are arriving at more acute angles and bouncing off the Earth's surface.

Virtually everyone has experienced the glare emitted from these surfaces on a sunny day. The reflectivity of surfaces, or albedo, is much more than a strain on our eyes, however. It plays a critical role in shaping the global energy budget and through this, climate. Surfaces with high albedo, such as snow, ice and sand, return much if not most of the incoming radiation they intercept at the same wavelength as it arrived, in part to the atmosphere where it is scattered and absorbed by clouds, and in part further out to space. This minor portion of solar energy reaching our planet's surface is consequently not absorbed, transformed nor later emitted at longer infrared wavelengths. Neither is the quarter of solar energy that fails to reach the surface, having been directly reflected back to space by high albedo clouds, or through scattering in the atmosphere. The remaining half of incoming solar radiation is absorbed by surfaces with much lower albedo, where it is transformed. This transformation is critical to the environment we live in. A vast amount of this energy is converted into chemical energy by green plants, macro-algae and phytoplankton[10] through their abundant production of the extremely important biomolecule, chlorophyll. As a result, areas characterized by high densities of photosynthesizing life tend to have considerably lower albedos, since they absorb most of the radiation at certain wavelengths.[11] The final product of the work performed by autotrophic organisms is the starting point of the natural production that underpins virtually all of our food supply and the diversity of life. Changes in this primary production fundamentally influence our collective resource base and its ability to sustain more complex life. Tropical forests are the most productive ecosystems on the planet and register some of the lowest terrestrial albedos.

The portion of absorbed radiation that is not converted chemically is also transformed, but primarily into longer-wave radiation as heat. Heat sits within the infrared portion of the electromagnetic spectrum, at longer wavelengths than the UV and visible light that accounts for the bulk of incoming solar radiation. This is subsequently released either directly through conduction and convection, or indirectly through the evaporation and convection of water. Many man-made features, such as tilled fields, buildings, roads and parking lots have extremely low albedos, since they absorb most of the incoming visible and infrared radiation and emit high levels of long-wave radiation very quickly. If you have parked your car in a large asphalt parking lot in the tropics or on a mid-summer day in the mid-latitudes, you will have experienced first-hand the transformation from short-wave to long-wave radiation in the form of heat. Water is also an excellent energy absorber, but it does not emit energy as quickly as solid substances. The amount of this long-wave radiation emitted from the surface is the fundamental input to the global climate system. If our planet was completely covered in ice, its albedo would be nearer to one and virtually no long-wave radiation would be emitted. The short-wave incoming radiation would simply be bounced back into space and global temperatures would be considerably lower. On the other extreme, if the surface was close to being a perfect absorber – for example, paved with asphalt – nearly all of the solar radiation would be trapped, and then re-radiated at longer wavelengths. Both extremes, ice-box and hot-house, would coincide with extremely low and high global temperatures, but only due to our atmosphere. Without it, ice or asphalt would not make a great deal of difference. The chemical composition of our atmosphere acts as a gate-keeper to the global powerhouse, since the relative mix of various molecules, with different energy absorption characteristics, determines how much of this out-going radiation is let through to space, and how much is retained and recycled between the surface and atmosphere. Our atmosphere acts as a selectively permeable lid, putting a cap on the amount of energy leaving the planet directly. When its composition changes, so too does its permeability, and the balance struck between the energy received and that subsequently emitted back into space changes. The same can be said of changes to surface albedo that occur when terrestrial ecosystems, such as forests or deserts are altered, snow and ice fields expand or contract, and cloud cover increases or decreases. If we review the processes that impinge collectively on the planet's solar energy budget, three main forcing factors are apparent: (1) fluctuations in solar output that determine how much energy is available for interception by Earth, (2) our planet's position in the cycles of obliquity, precession and eccentricity that affect how much of the Sun's output reaches the planet and how this is distributed across the surface and (3) the spatial distribution of inter-related planetary features, such as surface albedo, bioproductivity and atmospheric composition, that regulate the distribution and balance of energy absorption and re-emission.

Tropical dominance

The role of the tropics in the accountancy of our planet's energy budget is of such importance that to overlook its contribution would be akin to forgetting all of recorded human history since the ancient Sumerians, except the first 500 years. Nearly 85 per cent of the total net radiation[12] reaching the planet each year is absorbed across the 40 per cent of surface area situated between the Tropics of Cancer and Capricorn. In absolute terms, this averages between 12.5 and 13 petawatts – enough energy to run global society for 786 years at current consumption rates. Contributions from geothermal and mechanical sources measure less than 0.5 per cent of that arriving through solar radiation. Taking a glimpse (in Figure 2.1) at the current distribution of net solar radiation absorbed at the planet's surface, as measured in 2010 through the CERES sensors,[13] it is immediately clear that the contribution of the tropics to the global energy budget is immense. Figure 2.1A illustrates the strong, orderly decline in net solar

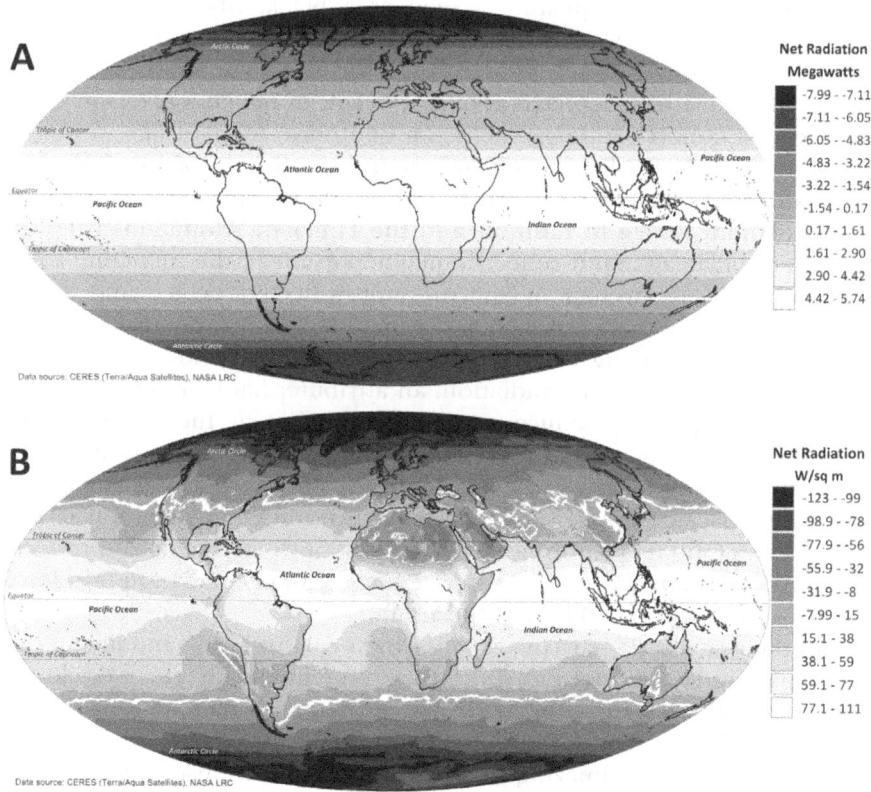

Figure 2.1 (A) Planetary pattern of incoming solar radiation as depicted by latitudinal averages. (B) Total incoming solar radiation for the year 2010.

radiation from the equator towards the poles. It shows the average of values at each degree of latitude. Averaging out local variation at latitude allows us to focus on the distribution of net radiation in relation to the Earth's main axis of spin. The impact that the obliquity of this axis has on the distribution of solar radiation over the course of the year is apparent to people living at higher latitudes as a widely varying day length over the year. This seasonal effect translates into an annual net deficit in the energy budget at our planet's surface at these higher latitudes as the amount lost during the boreal and austral winters surpasses the surplus accrued during much longer summer days. The threshold between surplus and deficit, seen as a pair of horizontal lines in Figure 2.1A, is currently struck at approximately 30 degrees north and south of the equator.

Latitudinal averages reveal the over-arching effect of celestial dynamics on net solar radiation, but also belie important variation due to more localized differences in albedo. The impact of varying albedo on the dividing line between a positive and negative annual net radiation budget is apparent by the way that the threshold between surplus and deficit takes on a more wave-like appearance, seen as a black, ribbon-like zone in Figure 2.1B. Features with very high albedo create smaller-scale distortions in the larger north-to-south gradient of absorption. The most prominent among these features, depicted as black pockets of net energy loss in 2.1B, are the vast desert sands spread from southern Afghanistan westward across the Rub-Al Khali of Saudi Arabia and the Sahara of North Africa. Other smaller pockets are created by the perennial snow and ice fields of the Karakoram Range in Kashmir and the Himalaya Mountains fringing the Tibetan Plateau. Sand reflects a major portion of the incoming solar rays, but only about half of the amount reflected by ice and snow. However, unlike ice and snow fields, sandy deserts are relatively permanent features, reflecting radiation constantly. Ice and snow reflect four-fifths of the incoming solar radiation, an attribute that combines with the effects of obliquity to reduce net solar radiation at the poles during periods of clear summer weather when absorption should be at its greatest. In the high mountains of Asia, this has a similar reducing effect. On clear days, glacial ice and snow combine to reduce net radiation levels, while heavy cloud cover during intervening periods intercepts and reflects incoming radiation. Combined, they reduce net solar radiation to a level far below that anticipated by latitude alone.

Cloud cover plays a particularly important role in locally modifying the latitudinal gradient of net radiation. In some regions, the effect of clouds is similar to that of ice, snow and sand in that they dampen absorption over a specific region throughout the year. The cloudiness over a select few of these areas is amazingly persistent – in some instances, to such extent and predictability that it appears to have shaped long-standing social and cultural traditions. The cloud and fog cover of Sichuan province in China provides an interesting example. This condition is of such unmovable duration that it led the seventh-century writer Liu Zongyuan to tell the tale of how cloud cover led to a rather unusual behaviour in its resident canines. During the very few periods when the Sun peeked through, dogs would become confused and agitated, being unfamiliar with the direct rays of sunlight reaching their cloud-enshrouded homes. They would then begin to bark loudly and in unison, as if a stranger was in their collective midst. The tale set to prose by Master Zongyuan gave birth over time to the familiar Chinese idiom "Sichuan dogs will bark at the Sun" – an expression still in use today to indicate how ignorant people respond when discovering something of common knowledge for the first time. Deep and persistent clouds across Sichuan, home of the giant panda, seem to have been a dominant feature of this area for millennia – to such extent that the tropical province immediately south, Yunnan, literally means "south of the clouds", lending further testament to the long-standing nature of Sichuan's legendary cloudiness. This persistent feature is clearly visible as a very large area of lower net radiation over south-east China (see Figure 2.1B). The immense cloud cover in this region is unique for this latitude – it occurs nowhere else on the planet. The overwhelming majority of land immediately north of the Tropic of Cancer experiences some of the lowest levels of cloud cover globally. Sichuan falls within a belt of very low average cloud cover more commonly characterized by vast stretches of barren sand and rock that make up some of the best known desert regions (see Figure 2.1A). This is made even clearer from a depiction of the average monthly cloud cover for the planet, in this case for the year 2010, in Figure 2.2. We can see the anomalous nature of the very high Sichuan cloud cover in comparison to other regions at the same latitude immediately north of the Tropic of Cancer. The same holds true for the comparable land area south of the Tropic of Capricorn.

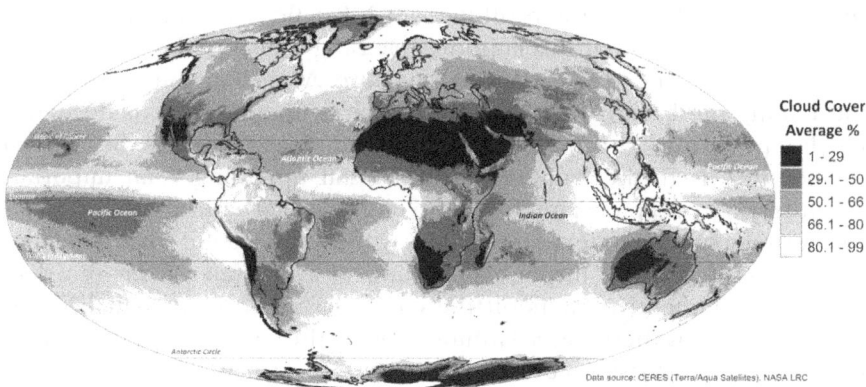

Figure 2.2 Planetary pattern of cloud cover as depicted by latitudinal averages.

Transmission of energy from the tropical powerhouse

Why does Sichuan then have such an unusual preponderance of cloud cover? While there are many factors influencing atmospheric behaviour in the region, much of Sichuan's weather is a simple accident of geography and a thermal transfer from the tropics. Warm, moist air flowing north-westward from the South China Sea is halted over the area by the sprawl-ing eastern rim of the Tibetan Plateau, creating a stationary body of dense cloud over the province as further moisture piles into the region from the east.[14] The origins of Sichuan's cloudiness, and Zongyuan's proverb, rest in the flux of energy from the tropical waters of the South China Sea and its capture through occlusion of its westward movement by the Tibetan highlands. As the warm, moist air moves on shore, it is met with a cooler air mass, resulting in persistent low cloud cover and fog. The south-eastern United States fall within the same belt of latitude as Sichuan province and are similarly located along the eastern edge of a large continental land-mass yet experience only half the average cloud cover. Much of this is due to topography and, in the case of the south-eastern US, the absence of any significant elevational gain west of the region that would block westward air movement at these latitudes, as the Tibetan Plateau does in relation to Sichuan.

The onshore flow of moisture from tropical waters into Sichuan is a small fraction of the energy conveyed through South-East Asia on its way northward aboard the strong west Pacific current, called Kuroshio. This south-to-north current shunts vast amounts of warm water northward as the westward-moving North Equatorial Current is deflected by the eastern margin of the Asian continental shelf. This breaks up and surrounds Japan before the bulk of the water is again deflected, this time eastward towards Alaska and British Columbia, where it cools while releasing energy. This then in part makes the waters off the North Pacific coast of North America warm relative to land at this latitude, releasing moisture that quickly con-denses to cloud and rain as it moves on shore. As the current moves south-ward along the Pacific coast, it does not warm up as quickly as needed to reach equilibrium with the much warmer land temperatures. Sea surface temperature is further suppressed by coastal upwelling – an important overturning of deep oceanic water driven by coastal wind flow that "peels-off" warmer surface waters westward away from the Pacific coast – creating a "void" that is filled through upwelling of cold water from the deep ocean. Consequently, waters off the coast of California are much cooler than expected at these lower latitudes due to the southward movement of its eponymous coastal current and the cooling effect of upwelling. Warming as it moves westward a few degrees north of the equator, water is once again shunted across the tropical Pacific to complete an enormous circuit of surface water, the Great North Pacific Gyre, and the export of energy from the tropics towards the pole.

This export of latent heat stored in tropical surface water as it travels on the Kuroshio Current is replicated across all of the oceans, north and south. Most famously perhaps, warm water is conveyed northward in the Atlantic aboard the Gulf Stream, the powerful current travelling along the eastern seaboard of the United States. Benjamin Franklin, the masterful empiricist, author, scientist and intellect of eighteenth-century America, was one of the first to formalize our knowledge of the tremendous flux of water heading north from the Caribbean. In 1770 he wrote:

> Vessels are sometimes retarded, and sometimes forwarded in their voyages, by currents at sea, which are often not perceived. About the year 1769 or 70, there was an application made by the board of customs at Boston, to the lords of the treasury in London, complaining that the packets [regular mail and freight ships] between Falmouth and New York, were generally a fortnight longer in their passages, than merchant ships from London to Rhode Island, and proposing that for the future they should be ordered to Rhode-Island instead of New York. Being then concerned in the management of the American post-office, I happened to be consulted on the occasion; and it appearing strange to me that there should be such a difference between two places, scarce a day's run asunder, especially when the merchant ships are generally deeper laden, and more weakly manned than the packets, and had from London the whole length of the river and channel to run before they left the land of England, while the packets had only to go from Falmouth, I could not but think the fact misunderstood or misrepresented. There happened then to be in London, a Nantucket sea-captain of my acquaintance, to whom I communicated the affair. He told me he believed the fact might be true; but the difference was owing to this, that the Rhode-Island captains were acquainted with the Gulf Stream, which those of the English packets were not.... We have informed them that they were stemming a current, that was against them to the value of three miles an hour and advised them to cross it and get out of it; but they were too wise to be counseled by simple American fishermen ... I then observed that it was a pity no notice was taken of this current upon the charts, and requested him to mark it out for me, which he readily complied with, adding directions for avoiding it in sailing from Europe to North-America.
>
> (Franklin 1786)

Franklin goes on to describe the anomalous warm temperatures of the current, and how thermometric readings could be used to navigate it. But the subtle significance of his early entry into maritime studies was that he recognized that tropical oceans were disgorging themselves of excess energy through swift currents such as the Gulf Stream. Franklin's work also represents one of the earliest formal reconnaissances of oceanic gyres and their causes. Although the navigational advantages of the Gulf Stream current to trans-Atlantic transport and trade were well known to some, it is clear from correspondence of the day that most were unaware. Yet it was, after all, the Gulf Stream's southerly counter-current, the Canary Current, that set Columbus unintentionally on his way to the New World.

The gyrations of waters in the great oceans of our planet, powered by differences in temperature and salinity and aided by the consistent wind direction of the easterlies and westerlies, were the primary source of energy driving the expansion of the early European empires and the globalized trade in commodities that ensued. The Brazil Current aided early European discovery of the South Atlantic and the Straits of Magellan – the only route to reach the Pacific directly from the Atlantic before the opening of the Panama Canal. It was the South Indian and West Australia currents that aided the Portuguese, Dutch and English traders in their out-bound voyages to the Far East. They would then ride the Agulhas and Mozambique currents, conveying warm waters southward towards the Antarctic from the equatorial reach of the Indian Ocean, back to the South Cape of Africa. On ships laden with coffee, tea and spices, they would return to Europe aboard the cold Benguela Current moving northward along the west coast of Africa.

The steady, swirling exchange of warm and cold waters between the tropics and poles as energy is exported from the perpetually oversupplied equatorial belt and the seasonally undersupplied poles has fundamentally shaped the oceans' circulation and, as an offshoot, the long-standing connection between the immense bioproductivity of the tropics and the intense economic activity of the extra-tropics. Of course the dynamics of this system and the transfer of energy from the tropics poleward aboard warm, tropical waters described here is an oversimplification. The process is considerably more complicated. Forces resulting from the interaction of vertical (hydrostatic) pressure gradients and horizontal (geostrophic) deflection caused by the spinning motion of the planet interact with one another, the geometry of land mass and sea bed, the friction these impart and numerous other factors to drive complex motion in the seas. These same forces influence the transmission of energy in the atmosphere as well, only much quicker; two systems – one oceanic, one atmospheric – are coupled through a continuous exchange of energy driven by complex processes overwhelmingly governed by the seasonal, poleward pull of Aristotle's "vacuum". The coupled effect of ocean and atmosphere creates a powerful engine driving the transmission of energy across the planet.

A child of the tropics

The climatological phenomenon that is popularly known as "El Niño" is an example of these complex dynamics that, until only relatively recently, was virtually unknown despite now being recognized as a major driver of exchange processes in the tropical seas and of weather variability across the planet. The "Christ Child" event is now well-covered by news and media outlets, portending drought, fires and floods for many living in the most sensitive regions. We now know through improved environmental monitoring that severe El Niño events occur every five years or so. Early stories about a periodic collapse of the anchovy fisheries off the coast of Peru around Christmas time emerged as early as the mid-nineteenth century but were certainly not recognized as a consequence of changes taking place in tropical waters across a 16,000-kilometre expanse of the equatorial Pacific (Glantz 2000). In fact, most of the scientific community was oblivious to the extraordinary changes that periodically affected our largest ocean up to the late 1960s. One meteorologist, however, had taken notice of odd fluctuations in the atmospheric pressure at the eastern and western ends of the tropical Pacific around the turn of the nineteenth century. His name was Gilbert Walker, a physicist and statistician by training that had taken up an appointment to the Indian Meteorological Service in 1903 to apply his skills with time-series analyses in an attempt to better understand variability in the Indian monsoon. A failure of the tropical monsoon, an intense seasonal wet-season that accounts easily for the bulk of annual rainfall across much of the country, had caused a horrible famine in 1899 and the British colonial office was determined to discover the cause. As part of the investigations, Walker, now Director General of the service, building upon work of his predecessor, Henry Blanford, noticed a strange periodic reversal of sea-level atmospheric pressure between weather stations in the west at Darwin, Australia and the east at Tahiti, Polynesia. Normally pressure at Darwin is low relative to Tahiti, but this is not constant. It fluctuates, sometimes so much so that it in effect reverses, and these periodic reversals, particularly when they are severe, are now known as El Niño events. The irregular, periodic oscillation of pressure between the west and eastern Pacific stations south of the equator was simply referred to as the Southern Oscillation by Walker. Today, the phenomenon is known as ENSO, or the El Niño Southern Oscillation.

Walker was attempting to discover why the Indian monsoon sometimes failed by constructing statistical correlations of the Southern Oscillation with temperature and rainfall patterns in the region and globally. His work, well in advance of any serious notion of climate science, was pilloried by his scientific peers in the establishment, primarily due to the phenomenological nature of his statistical studies – there were no oceanic or atmospheric mechanisms known at the time that could explain these irregular fluctuations across the tropical Pacific. He had also met with limited success in applying his findings in explaining the behaviour of the Indian monsoon. Nonetheless, he and his team had stumbled upon one of the most important circulatory systems on the planet, the equivalent of the mesenteric artery in the human body. They had also pioneered the development of important concepts in time-series analysis, such as cross-correlation and auto-regression that are now commonly used every day in efforts to examine behaviour in time-dependent phenomena, such as stock prices and human physiology, as well as weather. Due to the lack of support from his contemporaries, and despite the accolades lauding his achievements with time-series analysis, interest in the Southern Oscillation died with Walker's final publications on the topic in the late 1930s.

Fortunately, Walker's studies were not put to rest permanently. Their re-emergence coincided with oceanographic studies in the Pacific as part of a programme commissioned for the International Geophysical Year in 1957 and, by good fortune, an unusually strong El Niño event. The studies provided coupled measurements of atmospheric pressure, wind and sea surface temperature for the first time, revealing coincident changes (cross-correlations) in all of these across the tropical Pacific. These and later studies, combined with the earlier work of Walker and his team, led a Norwegian-American meteorological professor at UCLA, Jacob Bjerknes, to conclude that the atmospheric pressure changes observed by Walker were driven by changes along a gradient of sea surface temperature extending from the coast of Peru to the area of the western Pacific surrounding the islands of Papua New Guinea, Indonesia and northern Australia. Bjerknes concluded that it was the variation in wind strength of the tropical easterlies that impacted sea surface temperature differences. In 1969, he organized his earlier findings into a mechanistic model. In it he described how descending, high-pressure masses of dry air over the cold waters upwelling along the coast of Peru were pushed across the equatorial Pacific waters by the easterlies. This air, picking up moisture through evaporation over the warm water of the tropical Pacific, would arrive in the west and form low-pressure masses of convective clouds that would rise, condense and then release large volumes of rain across the region. The air, having released its latent heat through rainfall, then travels at high altitude back across the Pacific, only to subside over the colder tropical waters off the coast of Peru. The easterly surface winds push warm waters westward, shoaling these along the Asian rim of the Pacific and deepening the thermocline.[15] Shoaling in the west as surface water is pushed away from the coast of South America encourages further upwelling of cold water and a shallowing of the thermocline in the eastern Pacific. Bjerknes described this coupled relationship between air pressure, wind strength and sea temperatures across the tropical Pacific as the Walker Circulation in posthumous recognition of Gilbert Walker's pioneering work. Wind strength and its coupled relationship with sea surface temperature differences across the Pacific was the mechanism that had eluded Walker, mainly due to a lack of good sea temperature data. The missing mechanism linking the statistical correlations discovered by Walker with the behaviour of the oceans and atmosphere had finally been revealed.

These further studies also revealed that El Niño is in effect one phase in a continuously changing Walker Circulation that also sees the western Pacific at times become even warmer than normal. This occurs when easterly winds are unusually strong and coincide with cold waters off the coast of Peru extending much further towards the central Pacific. When this "tongue" of cold water lengthens its extent westward, rainfall across the western Pacific intensifies, and the opposite conditions to the warm-phase of El Niño develop. This cold-phase is now referred to as La Niña. When the circulation is transitioning between these two extreme states, ENSO is in a normal, or neutral, condition. But there is no "standard" condition for the Walker Circulation, since our timeframe of direct monitoring extends only a few decades, and we have yet to see whether it can achieve even greater variability. Severe El Niño phases can bring monumental flooding to the central west coast of South America while delivering devastating drought and forest fires to islands of the western Pacific, particularly Indonesia. These are precisely the outcomes during the most severe events to take place over the past 30 years, occurring in 1983–1984, 1997–1998 and 2015–2016. During a severe La Niña phase, these weather systems are often reversed. Critically, strong ENSO phases have also been linked to drastic changes in rainfall across many regions of the planet beyond those adjoining the ENSO belt of tropical Pacific waters (Ropelewski and Halpert 1987; Mason and Goddard 2001). During these periods, there appear to be two latitudinal belts – one within 12 degrees and another centred at 25–35 degrees north and south of the equator – that appear most sensitive to severe phases of ENSO. The impact on precipitation in these belts is not uniform across the planet with certain areas in the Americas and South-East Asia most affected. The impact of severe El Niño and La Niña events also appears to have the opposite effect on rainfall at different latitudes. Northern South America, including Colombia, Panama, Venezuela and the Guianas, experience significant drought during most severe El Niño phases, but receive excess rainfall during severe La Niña phases. A similar pattern plays out in the western tropical Pacific, centred on Indonesia, but including the Philippines, Papua New Guinea, the Solomon Islands and northern Australia. Areas of northern Mexico, the south-western United States, Texas and Florida show a positive rainfall relationship with the arrival of El Niño conditions and a negative relationship with severe La Niña events. The same is true for areas of southernmost Brazil, northern Argentina and central Chile, and to a lesser extent, southern Australia. The relationships, like those established by Walker, are statistical in nature and a full understanding of the cause–effect relationship between ENSO and precipitation remains elusive, particularly in the case of the extra-tropics.

It is clear from accumulated evidence that the Walker Circulation is an incredibly important gear regulating the global transfer of energy across oceanic basins, linked to the behaviour of the North and South Pacific

Gyres, and the behaviour of the Hadley Circulation – the predominant pattern of air movement that transfers energy from the equatorial regions to the mid-latitudes. Knowledge of the Hadley Cell long pre-dated Bjerknes' work on ENSO and no doubt helped to shape his characterization of the Walker Circulation. Both Hadley and Walker Circulations transfer energy through a similar atmospheric mechanism linking a high energy, ascending air mass of low pressure and high moisture at one end to a low moisture zone where air with low energy descends in a high-pressure system. The difference between the two is that in the normal state of the Hadley, there is a movement towards the edge of the tropics from high to low moisture conditions at north-westward angles due to the effect of our planet's spin. The Walker returns its dry air eastward and within the same swathe of the tropics just south of the equator. We know that the Hadley is driven by differential heating of the Earth's surface from equator to poles and that, to paraphrase Aristotle, nature always seeks to fill a low-energy space. But why there are changes in wind strength and sea surface temperature across the tropical Pacific, an area receiving roughly equal amounts of solar radiation and temperature, remains unclear, as does the order of precedence in conditions. Is it sea temperature that drives the winds, or winds that drive changes in sea temperature? The two tropical circulations are invariably linked, but oceanographers and climate scientists have only started to better understand oceanic behaviour and its determinants over the last 40 years. Much of this work has focused on the wave dynamics of the more fluid components of our planetary surface, their propagation and role in "triggering" events.[16] Consequently, long-term outcomes remain fairly difficult to predict with any accuracy. We are, however, getting a much better grip on the behaviour of tropical oceans and their role in driving global weather variability. What we do know with clear certainty, however, is that the powerful heating of the vast, tropical oceans by the Sun fundamentally drives these motions, in the sea, in the air and over land. Thus the tropics are at the heart of this global system, playing a paramount role in the absorption and export of energy globally. But the factors shaping the distribution of bioproductivity afforded by the planet's powerhouse are more complex, as we will see in the next chapter.

Notes

1 This maintains the liquid-to-viscous state of the mantle material. The fluid state of the mantle creates convection currents circulating between the hotter, high-pressure lower zone and cooler, low-pressure upper zone. These currents drive tectonic and volcanic activity through the overlying lithosphere. Crudely stated, mantle material consists of a highly impure mixture of heated glass (SiO_2) and milk of magnesia (MgO).

2 A vast number of components can be considered in tidal range models, but seven to nine of these account for 90 per cent plus of tidal amplitude in most locations. These are grouped by the source of gravitational force, namely lunar

(M) or solar (S), and their periodicity, being semi-diurnal (2), diurnal (1), quarti-diurnal (4), harmonical overtides, or longer semi-annual, annual or multi-annual. A catalogue describing a spectrum with over 500 tide harmonics shaping ocean tides has been published by Hartmann and Wenzel (1995).

3 Estimation of full dissipation rates advanced rapidly with the advent of satellite-based altimetry; see Kantha (1998), Cartwright and Ray (1991) and Egbert and Ray (2000). The main satellite mission advancing our knowledge of changes in the ocean's elevation, known as the TOPEX/Poseidon, was launched in 1992 through French–US collaboration. Its successor, Jason-1, was launched in 2001.

4 These localized areas with no or very little tide are referred to as *amphidromic points*. For the M2 component there are 13 of these areas, including the Mediterranean and Caribbean Seas, that experience very low tidal movements due to their isolation from the larger oceans.

5 US, Canadian and Icelandic polar scientists have presented evidence that suggests a series of massive volcanic eruptions triggered the Little Ice Age through the growth of sea ice. Feedback effects of expanding sea ice led to prolonged cooling and harsher winters. See Miller *et al.* (2012).

6 Eccentricity, combined with precession and obliquity, are collectively referred to as Milankovitch cycles, after the early-twentieth-century Serbian mathematician credited with their first formal description. See Hays *et al.* (1976).

7 The intermittent nature of absorption across the short-wave infrared spectrum creates "windows" through which long-wave radiation can escape to space. If these windows are occluded or closed through increased absorption, most commonly via cloud cover, then less energy escapes.

8 These aerosols, along with other suspended matter, typically originate from volcanic eruptions, forest fires, dust-storms and fossil fuel combustion.

9 The UV and visible portion of solar radiation with wavelengths less than 0.7 micrometres.

10 Diatoms, cyanobacteria and dinoflagellates.

11 Absorption peaks at the blue and red wavelengths in plants and reflect radiation in other spectral bands.

12 The amount of incoming radiation received less the amount reflected, mainly by clouds, ice, snow and sand, plus that emitted as long-wave infrared, mainly from land, water and clouds.

13 Aboard NASA's Terra and Aqua satellites.

14 Sichuan has the highest winter-forming stratiform cloud cover on the planet and this strongly influences the average cloud cover for the area. These are the low-lying, grey-blanket formations that typify many other regions affected by seasonal land–sea temperature differences, such as the Pacific Northwest region of the United States and Canada. See Li and Gu (2006) and Klein and Hartmann (1993).

15 The thermocline is a distinct layer in water bodies where temperature declines precipitously as the absorption of solar heat by the surrounding water reaches its limit. In the case of the tropical oceans, absorption occurs usually down through the first 100-metre layer only.

16 A good starting point to understanding oceanic adjustment processes related to ENSO (for the mathematically minded) is Philander (1990).

3 The blue and the green

Blue seas

A flight from the north-west coast of North America to South-East Asia is arguably one of the most boring. For someone that enjoys the spectacle of changing features as you fly over our planet's surface, the flight seems an eternity. The feeling that time has been momentarily suspended is placed in hyper-drive every time I stare out of the cabin window. None of the visually stunning geometries that are revealed during high-altitude flights over the continents are to be seen – none of the great erosional patterns created by the forces of wind and water that appear in fractal repetition across the landscape. There are no mountain chains and snow-capped peaks, no agricultural fields with their varied shapes, sizes and colours, no forests or savanna, meandering rivers or cities with their radiating motorways and sprawling infrastructure. You cannot get a feel for the pace and direction of change, be it driven by natural or human forces. In fact, there is very little below to push your thoughts one way or another. The view along a trans-Pacific flight is a raging fire at night – although strangely calming, it does not appeal to our more demanding empirical tendencies.

There is one interesting feature though. Across this sea-plain, I am struck by the change in water colour, from the murky green-brown hues along the coast of the Pacific Northwest to the azure, deep blue and then aqua-marine tints as the plane proceeds from the central Pacific into the regional waters of the South Pacific islands of Tonga, Fiji and New Caledonia. Clusters of small tropical islands dot the surface, surrounded by bright blue rings, the type that only come with the clear, crystalline water that seems to forever feature in the "tropical island paradise" marketing campaigns of national tourism boards. The change in colour across the Pacific is subtle, and seems inconsequential. But this change indicates a geographic transition, a pattern in the growth of one of the most important building blocks of life – phytoplankton. These algae, the marine equivalent of grasses, are the primary source of biological production in our planet's oceans. Harvesting sunlight using chlorophyll, they transform solar energy into chemical energy that is then used to grow, very much in the same way that grasses and other land plants fix carbon from CO_2 and convert this into more complex biomolecules, such as starch. Photosynthesis runs most efficiently on the blue and red wavelengths of the visible spectrum (called photosynthetically active radiation, or PAR) and areas with high concentrations of chlorophyll, either on land or at sea, tend to appear greener than areas that do not contain an abundance of autotrophic life. The tropical seas are more blue than green precisely due to their relative lack of chlorophyll-containing, PAR-absorbing phytoplankton. Phytoplankton's ability to create "something from the Sun" is at the heart of oceanic life, sitting firmly at the base of the marine trophic pyramid – a network of prey and predator that sustains the dense soup of microscopic plankton that "graze" on the phytoplankton and in turn, are grazed upon by many larger invertebrates, such as shrimp and krill, oysters, jellyfish, sponges and anemones. Indirectly, phytoplankton sustain virtually all of the larger marine vertebrates, fish and mammals through the chemical energy that they continuously inject into the oceanic food web. They are of such critical importance to marine life, that it is safe to say that our seas would be empty pools without them.

Phytoplankton production is highly dependent on the availability of visible light. Revisiting the energetics of tropical oceans discussed in Chapter 2, we recall that the amount of incoming solar energy reaches its maximum over the tropical oceans. Given the immense potential of the energy available to drive photosynthesis in the tropical oceans, it is not unreasonable to then expect that this part of our planet should also be at the pinnacle of biological productivity. Roughly estimated to be between 1.7 and 2.5 petajoules per square kilometre per year, the total solar radiation available annually at the surface of the tropical oceans amounts to around 300 exajoules (10^{18} joules) – an amount difficult to comprehend. Put in a more tangible quantity, it is enough energy to meet the electricity need of every house in the United States at an annual average consumption rate of 41 million kilojoules per home[1] – not for a year, but for a period of 55 years.

Tropical oceans then should be ridiculously productive. It takes around 40 kilojoules of energy to fix a single gram of carbon through the most common type of photosynthesis and if we apply this to the average amount of energy available across the tropical oceans, we can roughly estimate the amount of carbon that would be expected to be fixed biologically each year as a consequence. These estimates range between 6.8 and 9.8 exagrams of carbon.[2] This theoretical ceiling of production is a nearly immeasurable weight, only rivalled in magnitude by the mass of rock that forms the Earth's crust (in 10^{20} kilograms), or that of the air that constitutes our atmosphere (in 10^{19} kilograms). But despite the Brobdingnagian surface area covered by the tropical oceans, and the Herculean energy engine that is formed from this enormous solar collector, the tropical oceans are surprisingly impish when it comes to biological productivity.

In fact, the size of the gap between the amount of carbon that could be fixed theoretically each year based on the availability of solar radiation alone and the amount actually observed via satellite-borne instrumentation is staggering. Taking a look at the average annual net primary productivity, or NPP, for the tropical oceans as depicted in Figure 3.1, it is clear that they lack the kind of biological firepower that we would have anticipated from calculations based exclusively on their solar energy supply. These data, derived from a series of productivity models constructed by the University of Montana's Numerical Terradynamic Simulation Group,[3] show an average annual NPP of the planet's tropical oceans of 3.3+/–0.5 petagrams for the period 2000 to 2006 – a thousand-fold plus shortfall in relation to the estimated theoretical capacity available based on solar energy alone. Moreover, the model results reveal that tropical oceans host only around 43 per cent of the global marine NPP, equal to just under a quarter of the average annual NPP for the entire planet, but in an area that covers 42 per cent of the planet and 75 per cent of the surface area of the tropics.

This odd underperformance is important since it appears that much of the energy available within the tropical powerhouse is not biologically productive. NPP is a measure of total growth after accounting for energy used in the operational support of photosynthesis, mainly through gas exchange processes (transpiration). To use a parallel drawn from the financial world, this is the operating profit accrued after all of the costs of production have been taken into account. This first step in biological

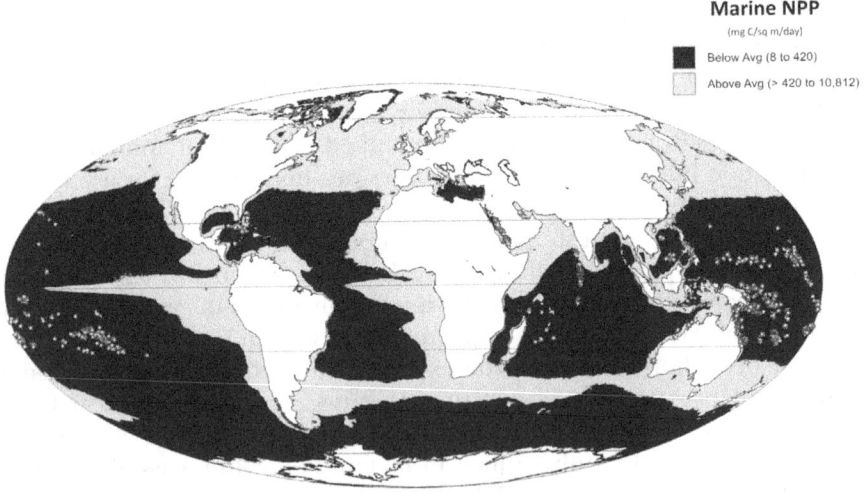

Marine NPP

(mg C/sq m/day)

■ Below Avg (8 to 420)

☐ Above Avg (> 420 to 10,812)

Figure 3.1 Distribution of marine bioproductivity as depicted by areas registering values above and below the global average.

Note

The distribution of the world's major coral reef formations indicated by ★.

productivity is immeasurably important to modern society not only through the food we eat, the wood and fibre we use and its regulating control on the composition of our atmosphere, but due to the unique bridge it represents between the largest "free" supply of energy available to the biosphere and life's own inexhaustible demand for it. There is no known alternative to chlorophyll-based primary production that can autonomously sustain the scale and pace of chemical activity as we see here on Earth. This makes the chronic underperformance of tropical oceans even more paradoxical since there appears to be a great disconnect between the distribution of an insatiable consumer – biological life – and a virtually limitless supply of its most coveted product.

What is keeping the two apart and why are tropical oceans, on average, less productive than we would expect? The reason behind the low average NPP has to do primarily with another input equally necessary for phytoplankton growth – nutrients. Nutrients have been shown to wield almost as much, if not more, control over the abundance and diversity of phytoplankton as does the availability of sunlight. These are common minerals such as iron, nitrate, phosphate and silicic acid, as well as the biomolecules thiamine, riboflavin and others that make up the Vitamin B complex. This is the same Vitamin B group that is found in meat, beans, bananas and other wholefoods. It is the same complex as that sold in tablet form at pharmacies everywhere and considered equally critical to human health and function.

Laboratory studies have suggested that these nutrients play a fundamental role in determining the abundance and community composition of phytoplankton by co-limiting enzymatic functions that are critical to cellular growth (Sañudo-Wilhelmy *et al.* 2012). Field experiments conducted as early as the 1990s, such as those pioneered by Professor Peter Liss' marine sciences group at the University of East Anglia in the UK, have shown that artificially increasing the availability of some of these nutrients, such as iron, at a large scale will spark rapid increases in phytoplankton production. These studies, the marine equivalent to the great fertilizer trials conducted by agricultural scientists of the nineteenth century (Rossiter 1975), revealed that marine primary productivity is structured more or less around the same set of variables known for centuries to control terrestrial plant growth – light, temperature, moisture and nutrients. In the case of oceans, it is a lack of nutrients that witnesses the greatest decline in productivity. Recently, some oceanic regions characterized by particularly low concentrations of Vitamin B have been described as "vitamin deserts" due to the extreme effect this scarcity has on phytoplankton abundance (Giovannoni 2012). What gives rise to vitamin deserts and why are the tropical oceans filled with them? The poor nutrient status of tropical oceans taken as a whole is a product of global currents and the forces that drive them. At the heart of this system of currents are the same five massive, ponderous rotating wheels that carried the early European explorers and traders around the world, intrigued the polymath Benjamin Franklin and science-fiction writer Jules Verne and continuously convey energy poleward from the tropics – the subtropical gyres.

The eye of the gyre

The geographic centre of each gyre is located several degrees poleward of the Tropics of Cancer and Capricorn in what are commonly referred to as the Horse Latitudes. These outer reaches of the celestial tropics are also the same latitudes at the centre of the atmospheric "subsidence belt". Fed by the convergence of the poleward flow of the tropical Hadley Cell and the equatorward extent of the mid-latitude Ferrell Cell, the Horse Latitudes are bathed in extremely dry air as it descends towards the surface. If you recall from Chapter 1, it is this movement of air that bifurcates the tropical belt into two moisture zones – a central wet zone and peripheral dry zone. Over the continents, this depleted air mass provides little in the way of moisture and at these latitudes the greatest dry deserts have formed. The Sahara, Atacama, Namib and Great Australian are largely products of the convergent descent of dry air at these latitudes interacting with other factors that discourage rainfall. The descent extends across both land and sea and if we run a line on a map within the same belt of latitude away from any of the great continental deserts, we eventually arrive near the eye of a subtropical gyre. The same belt that brings blue skies, resplendent sunshine, but little rainfall over North Africa, Baja California and Western Australia also structures the subtropical gyres by creating similar zones of stable high pressure in the atmosphere above them. The differences between these stable high-pressure zones and adjoining low-pressure systems create the winds that drive the major boundary currents forming the gyres. The spin and curved surface of our planet interact with the geographic distribution of the major land masses in complex ways to bend these currents, creating the circular motion of their surface waters. Some of these waters are captured, or entrained, through these processes and become trapped in the slow-moving vortex, spiralling at a decreasing pace inwards, eventually arriving at the eye. By the time these waters reach the centre, they have been stripped of their kinetic energy and isolated from potential sources of nutrients. Bathed in the solar radiation of the tropical belt, the conditions near the centre of the subtropical gyres are calm and relatively warm, but the right measure of nutrients, temperature and solar energy that supported the heightened biological productivity of these waters as they passed through other regions has been lost. The "sea within an ocean" is a pit-stop in the perpetual conveyance of water around the globe – deprived of its energy of motion and the materials that support and sustain dynamic change, water in the eye of the gyre waits quietly to re-engage with the global oceanic conveyance system. Some studies estimate the residence time of water trapped in the gyre at more than a decade.[4]

The central vortices of these great systems are also some of the most remote points that can be reached from any continent. Before the steam engine, crews aboard sailing ships spoke with dread at being captured by these "waters without wind" that lengthened their trans-oceanic journeys and increased the chance of running short of food and freshwater. In many ways – calm, warm and biologically unproductive – they are forlorn spaces where the dynamic processes attached to the continents scarcely register. Here the effect of our planet's mighty land masses on environmental conditions is arguably at its very weakest. Little of the massive sediment, organic matter and freshwater that is disgorged into the oceans each year from the Amazon, Mississippi, Congo, Yangtze and other rivers, successfully reaches these remote destinations. Airborne dust and aerosols emanating from deserts, fires and volcanoes, cities and factories are entrained by the very same winds that drive the boundary currents of the gyre. Consequently, most of these potential sources of limiting nutrients such as iron, silica or phosphorus arrive sporadically, if at all, in the eye.

Blue seas turn green

The size and position of the gyres dominate the tropical oceans and make them, as a whole, relatively unproductive. But there are pockets of exceptionally high productivity where the right conditions of nutrients and temperature co-occur. These pockets take on three forms: coastal upwelling, riverine estuaries and coral reefs. Combined, they account for the majority of the NPP attributable to tropical oceans.

The first of the three, upwelling, develops in the wake of strong surface currents. The displacement of warmer surface water creates a complex, vertical hydrodynamic gradient that pulls deep seawater upward where it is entrained by the wind-driven surface current. Displacement changes the pressure gradient, like a pump. In theory, upwelling zones could occur anywhere in the ocean where the strength and persistence of the wind is sufficient to displace surface water with a force necessary to generate an upward current. In practice, they propagate most frequently along coastlines where incoming currents are deflected and winds move parallel to the shoreline. Almost all coastal zones experience some measure of upwelling, but for most regions these are episodic or seasonal, the former arising when special atmospheric conditions allow short-lived, ephemeral displacement of surface water to occur and the latter forming and dissipating in time with the seasonal shift in the magnitude of prevailing winds. Most of these upwelling currents are too transitory to be recognized as individual features of the global marine system. There are special situations, however, where surface displacement occurs continuously throughout the year and these permanent upwelling zones are recognized as individual currents. Four permanent upwelling currents along the western margins of the Atlantic and Pacific basins continuously shunt cold water to the surface. The Humboldt and California Currents ascend in the eastern Pacific and the Benguela and Canary Currents – the same Canary that played such an important part in the colonization of the Americas – rise along the western margin of Africa. By feeding cold water to the surface and then towards the equator, each forms a critical leg in the circular movement of water that gives rise to the great subtropical gyres. The ascending cold water also carries large quantities of the nutrients and vitamins from below. These are the products of dead organisms, sinking to the ocean floor where they are locked in the refrigerated depths and combine with minerals to form a "super-food" solution of natural fertilizer that is readily available for uptake by phytoplankton as they bask in the copious solar radiation available at tropical latitudes, or in seasonal "blooms" at higher latitudes as the Sun tracks poleward during the long boreal and austral summers.

A less concentrated form of permanent upwelling also occurs in the equatorial waters of the Pacific and Atlantic. These zones arise respectively on the western margins of South America and Africa and extend much further offshore than the five major zones of permanent coastal upwelling. On a map, they appear as tongue-like projections of colder surface water, reaching towards the deep water of the open ocean. Wedged between the equatorial limits of the great subtropical gyres, these extension of cooler water are formed by the displacement of warm water westward towards the eastern continental margins. Like coastal upwelling, the shoaling of warm water westward generates a space that is quickly filled through ascending cold water, spiking the surface zone with a massive influx of nutrients. The westward winds driving the formation of these highly productive equatorial zones of upwelling are formed as part of the Walker Circulation – the equatorial pressure gradient force that drives the El Niño phenomenon (ENSO) through fluctuations in the difference between its east and west limits. When ENSO fluctuations are extreme, the zone of upwelling associated with the Walker Circulation can expand and contract dramatically. Severe El Niño events coincide with a rapid contraction of the upwelling zone and a precipitous collapse in NPP. During extreme La Niña events, where the pressure gradient of the Walker Circulation sets up conditions that are opposite to those of an El Niño phase, equatorial upwelling expands much further sea-ward than its average, boosting the contribution of this zone to global NPP.

The second type of high tropical marine NPP is created where some of the mightiest rivers on the planet meet the ocean. The Amazon and Congo inject enormous quantities of nutrient-laden sediment into the ocean and this land-borne influx supports massive phytoplankton blooms tens of kilometres offshore. Some of the highest NPP values recorded for any tropical marine location are associated with NPP hotspots arising at the mouths of these rivers. The productivity attached to these zones varies as it does for upwelling areas, but this variation does not reflect changes in wind strength. Instead, the NPP of river estuaries reflects changes in the volume of sediment and its nutrient content that is delivered via freshwater surface flow.

In many ways, the upwelling currents and estuaries that give rise to some of the highest levels of NPP in the tropics are simple supply–demand systems. Their productivity is almost entirely due to a concentrated influx of nutrients either from the ocean floor or from the adjoining land. Productivity increases or decreases with a rise or decline in available nutrients. The third zone of high tropical marine productivity, however, has nothing to do with an influx of vitamins or nutrients. In fact, these areas of high productivity are more commonly associated with the highly unproductive regions of subtropical gyres than with eastern boundary upwelling currents or large, tropical river estuaries.

The zone of productivity I am describing is, of course, the coral reef. The spatial relationship between areas of relatively low marine NPP and coral reef locations can be seen in Figure 3.1. Using these coverages to calculate the average NPP within 200 kilometres of the nearest coral reef yields an estimated 540 milligrams of carbon per square metre per day. This is less than a quarter of the average value estimated for permanent upwelling zones and tropical estuaries in the same coverage.[5] Studies measuring the amount of carbon fixed by coral reefs vary, but an average NPP on tropical Pacific reef formations has been roughly estimated to be around 3,000 milligrams of carbon per square metre per day, nearly a six-fold increase in productivity relative to the surrounding waters (Crossland *et al.* 1991). Why are coral reefs so incredibly productive when they are located in such unproductive areas of tropical ocean? They have evolved a bio-engineered work-around – a form of symbiosis that brings together phytoplankton and coral, each with strengths that together allow them to overcome their individual physiological deficiencies. The phytoplankton, a type of dinoflagellate often referred to as a zooxanthella, provides most of the energy, carbon and nitrogen that the host coral requires in return for nutrients not readily available in the warm, blue waters inside the subtropical gyre. As is the case with the ecology of all organisms, there are exceptions and not all corals form this relationship. The ones that do not are more often found at greater depths where light is scarce and nutrients are more readily available. It remains unclear which partner in this mutualistic relationship made the first move, since many zooxanthellae are also capable of living without corals. But the overwhelming majority of coral reefs are of the shallow marine type and these almost always have developed a partnership with zooxanthellae. This symbiotic relationship, a true marvel of biological evolution, boosts productivity in nutrient-poor conditions that otherwise would not support abundant autotrophic life. The local impact of this eco-engineered primary productivity on the abundance and diversity of marine life cannot be overstated. Coral reef communities – some of the most biologically diverse and productive assemblages on our planet – reside in the middle of some of the most nutrient-poor waters on the planet, defying the odds of survival through evolutionary co-existence.

Hyper-blooms

Land and airborne inputs can fundamentally boost the productivity of marine environments on the shallower margins of the continental shelves by injecting nutrients into the offshore environment. But unlike nutrients arriving near the surface on upwelling currents, the phytoplankton blooms that sometimes accompany unnatural spikes in land-borne nutrients, often due to agro-chemical run-off or untreated sewage outfall, can be "counter-productive". In some locations, it has been shown that these spikes lead to rapid oxygen depletion through the water column under certain conditions. This occurs immediately following a hyper-bloom event, as mass mortality and decomposition of phytoplankton quickly consumes all of the dissolved oxygen needed to sustain higher submarine life. A temporary death zone emerges in the affected areas, characterized by a precipitous decline in available oxygen through the water column. This is followed by a rapid and massive die-off of most creatures that cannot move quickly enough to escape the growing zone of hypoxia as it spreads. This includes many crustaceans, starfish, shellfish and other less-mobile life that inhabits the seafloor and lower parts of the offshore environment. Relative to the normal marine life cycle, the death knell of a hypoxia event can ring suddenly and with catastrophic efficacy.

The effects of a warming global climate on the wind fields that drive upwelling cycles have been implicated. Where polluted freshwater discharges into areas experiencing strong upwelling, the post-bloom hypoxia can quickly reach a state of complete oxygen depletion, or anoxia, and widespread collapse of the marine trophic structure ensues. The coastal areas susceptible to these events are multiplying and now impact many of the most productive coastal areas surrounding the North Atlantic and the Yellow Sea region between China, South Korea and Japan (Diaz and Rosenberg 2008). Events of this nature have also been documented at an increasing frequency off the west coast of North America in more recent years. With approximately 20 per cent of the annual marine fishery off-take generated in the coastal upwelling and river estuarine zones, hyper-blooms and the consequent collapse of benthic productivity are deep causes for concern. A convergence of increasing land-borne discharge of chemically spiked run-off and more erratic upwelling behaviour increases the likelihood of boom-and-bust productivity with coastal fisheries absorbing the costs of this increased volatility.

The environment in the gyre is far less susceptible to these impacts due to its remote location, but this remoteness also brings with it less natural productivity to be lost. It also means that between the two – the artificial impacts on coastal marine biomass caused by expanding dead zones and the naturally low productivity of the gyres – there is not much ocean where the marine biomass necessary to sustain a growing human population can develop. In theory, the eye of the subtropical gyre should be the purest marine environment on our planet, free from the pollution that is increasingly a characteristic of many coastal marine systems, except for the recent emergence of one unyielding contaminant – plastic.

While many coastal marine environments are becoming habituated to the boom-and-bust cycle of contaminant discharge caused by an expanding use of agro-industrial inputs on land, the tale of the gyre as dustbin began in the 1980s when several scientists conducting surveys in the North Pacific Gyre region just north of the Hawaiian Islands noticed very high concentrations of strange particulates in their water samples. They were surprised to discover that these particles were not a natural substance new to science, but in fact the product of a wide range of degraded polymers used in the production of everyday plastic containers. First evidence of widespread contamination of the open ocean with abandoned plastics was collected as early as the 1960s and 1970s when an alarming increase was observed in the number of plastic items found in the stomach contents of sea-birds in Hawaii and in the plankton-rich waters of the North Atlantic (Kenyon and Kridler 1969). More recent studies have shown that the accumulation of plastic has proliferated rapidly in the eyes of the subtropical gyres (Cózar *et al.* 2014). In these zones, recorded densities of plastic approached an average level of around 300–600 grams per square kilometre. Remote areas of ocean sampled outside the subtropical gyres as part of the same research voyage contained less than one-tenth of these levels, although this appears to be rising. With an average annual production of organic carbon fluctuating around 84 kilograms per square kilometre, plastic densities at one-half kilogram are not about to dominate the ecology of the gyres anytime soon. But the same spiralling currents that have deprived the most remote subtropical oceans of their biological productivity for millennia have within a mere half-century funnelled an endless stream of plastic into one of the largest combined repositories of waste synthetic carbon molecules on our planet.

While the interactions and dependencies are not fully understood, nor the relationships with changes in sunlight, ocean currents and sea temperature fully resolved, it is reasonably clear that the enormous "seas within seas" that are created by the oceanic gyres produce conditions of temperature and nutrient availability that are not able to support high levels of NPP, despite the solar resources available. As a consequence, marine NPP in the tropical oceans is highly concentrated in areas where temperature, and especially nutrient availability, are more favourable – areas such as the coast of Peru, where coastal upwelling supports very high NPP. Coastal zones can also experience elevated NPP, particularly where tropical rivers, such as the Amazon or Congo, disgorge large amounts of dissolved, land-borne nutrients. Where tropical coral reefs dominate the local mari-nescape – along with the mangroves and seagrass beds that are often asso-ciated with these formations in shallow tropical atolls, bays and inlets – NPP rises dramatically. In some regions, such as the Indonesian archi-pelago or the Great Australian Barrier Reef, NPP can reach close to its global maximum. But taken on a global scale, the total area of these hyper-productive NPP zones is a mere bucket in the vast tropical seas and their contribution to global oceanic NPP is infinitely small – a thimbleful in the unproductive volume of tropical surface waters ensnared by the great oceanic gyres. To discover the true mammoth of biological productivity in the tropics, we have to turn our attention to the land and the forests.

Green trees

Travelling across the equator from Caracas, Venezuela to Lima, Peru involves a flight equally as tedious, if not monotonous, as a trans-Pacific journey. A flight of a mere seven and half hours, it hardly compares to the seemingly unending journey from North America to South-East Asia, but it does require a lengthy trip across an ocean – an ocean of trees. The monotony of the Pacific is replaced with the immense lowland basin of Amazonia that at its heart – situated about 4 degrees north of the geo-graphic equator – remains an impenetrable sea of green. This land-borne sea is broken by the sinuous slide of freshwater from the Andes as it col-lects repeatedly into larger and larger branches, finally arriving at the main stem of the Amazon River where it is shunted eastward to the trop-ical Atlantic. This great basin has no environmental comparison in any part of the terrestrial tropics, be it calculated in terms of forest area, volume of water discharged to the ocean, solar radiation absorbed, total number of species present or any of the many other metrics we might select. It is in myriad ways, the culmination of the conditions and factors that epitomize the unbridled biological capacity of our planet – a Gold-ilocks effect of temperature, moisture and sunshine born from the gravita-tional interactions that govern our relationship with the Sun. These "just right" conditions are assisted by the geographic happenstance of a large land area situated along the western margin of a tectonic plate that has produced a mountain chain, the Andes, that is ideally suited to catch (north-)westward bound moisture. A good portion of this moisture is then re-circulated eastward through the streams and rivers that feed the Amazon. It is its own hydrological feedback system and there really is no parallel on our planet. Of course the factors that have produced the Amazon, like those that have created upwelling zones on the ocean peri-meter, are present on all three continents that slice through the tropical belt, it is just that the Amazon basin is so much larger.

Figure 3.2 shows the average NPP of land (solid line) and ocean (hatched line) at each degree of latitude across the entire planet, based

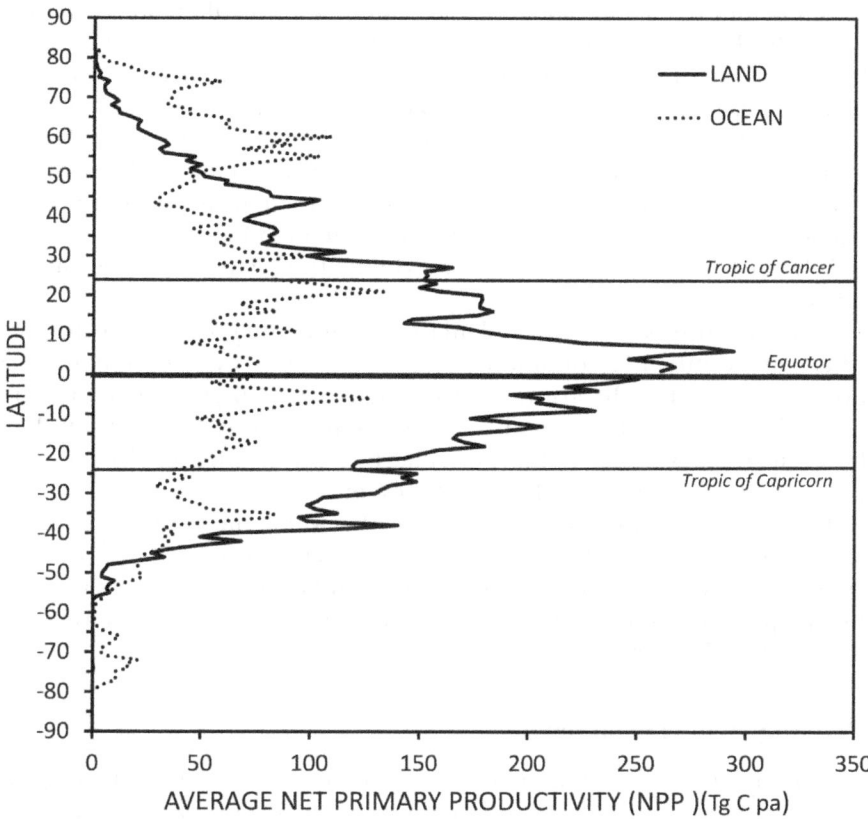

Figure 3.2 Distribution of bioproductivity expressed as latitudinal averages for land
and ocean.

on data collected via the MODIS device, on board NASA's Terra satellite.[6]
These values represent "yearly averages of latitudinal sums of pixel averages" and, as a result, reduce the effect of smaller-scale variation. They
give a broader view of the north–south trend in NPP by sacrificing detail
of the east–west variation across the planet and how this varies from month
to month and year to year. The prodigious biological capacity of the terrestrial tropics and their Amazonian flagship is evident in the very high
NPP that we observe for the region. But when we compare this NPP with
that of the extra-tropical regions and the tropical oceans, it is clear that it
also acts as a cornerstone to global bioproductivity. Figure 3.2 shows that
the highest biological productivity on the planet is found in the terrestrial
tropics closest to the equator where, on aggregate, NPP is two to three
times the average amount estimated for the tropical oceans. This equates
to about 42 per cent of global NPP each year. This sky-high level of

productivity is achieved across a mere 18 per cent of the planet's surface, with much of this amount concentrated in an even smaller swathe of land within 10 degrees latitude north and south of the equator. Separating the latitudinal distribution of global NPP into terrestrial and oceanic components reveals how much better tropical terrestrial ecosystems are, on average, at fixing carbon than the tropical oceans. It also reveals how much more productive temperate oceans are, particularly those of the North Atlantic and Pacific and along the continental shelf margins, than tropical waters when compared on an area-weighted basis.

Why are the terrestrial tropics so productive? The answer is simple: forests.

Tropical forests, like phytoplankton in upwelling zones, are formed where environmental conditions for growth are at their optimum. This doesn't mean all of the critical factors engendering high primary productivity – light, nutrients and moisture – are simultaneously or always at their best, but that their combined potential to limit growth relative to other parts of the planet is normally at or near a minimum.

It is the best expression, in terms of the wholesale variety and abundance of life, that evolution on our planet currently has to offer. The result is a very deep and broad assembly of organisms interacting through myriad pathways to maximize production in an environment experiencing a perpetual glut of solar radiation.

Diversity begets productivity

This rich biodiversity is arguably the most important factor in tropical forests achieving some of the highest levels of primary productivity recorded on our planet.[7] The reason for this has to do with biological evolution. With an increasing pool of species follows a larger number of subtle differences in adaptations – variations on a theme that allow each variant to exploit the measure of available resources it needs to support the successful reproduction of the next generation. Species that cannot achieve this measure of resource exploitation fall out of the pool, perhaps to survive in another region with a more amenable composition or, in the case of wholesale changes to conditions across their range, perhaps to go universally extinct. Professor Dan Janzen, an eminent tropical ecologist, thinker and conservationist, once described the species that were in the process of losing a grip on their place in the pool as "the living dead" or "[ecological] anachronisms" (Janzen and Martin 1982) – they were still represented by a few, old individuals, but clearly had lost the ecological wherewithal to maintain their populations under changing (often human-induced) conditions (Janzen 2001). These ecological zombies are a natural component of the evolutionary process, but one whose ranks have been swollen by accelerating human consumption. The abundance and turnover of species in tropical forests makes the system as a whole very efficient in the way that primary production occurs with each individual plant exploiting every last amount of available resource in a way that garners it some sort of advantage via its set of morphological and physiological attributes. For the majority of plant species, this involves vertical growth and the need to lay down the structural support necessary to move upward and compete for the abundant tropical light. Observing the plants and insects that make up the bulk of biodiversity in tropical forests, I have always been dumbfounded by the "efficiency" of evolutionary adaptation – how successful organisms go about their business in a way that seems equally inspired in both grace and compulsion.

As a consequence, forests are, like coral reefs, carbon-accumulating ecosystems and, like their bio-diverse, marine equivalent, have over time evolved the means to overcome deficiencies that would otherwise impose significant limits on their long-term productivity, despite the abundant sunshine. Many of these means are adaptations to the crucible of resource competition that is created by such a large pool of species – the attributes of tropical forest plants, such as shallow and dense root mats, that allow them to conserve and efficiently recycle nutrients released through the decomposition of dead and decaying material, or deep tap roots that allow some large trees to cope with variation in water availability due to changes in soil and topography. The plasticity of plant foliage that can undergo changes in leaf size and chlorophyll density in response to its vertical position within the forest works to maintain bio-efficiencies while the production of potent secondary chemicals aims to repel persistent pests constantly challenging a plant's defences in an attempt to purloin its hard-earned productivity towards their own biological success. While many attributes are adaptations to a competitive environment, many tropical forest plants are also known to form special biological relationships, such as those with root-infecting mycorrhizal fungi and rhizobial bacteria (Perreijn 2002; Allen 1991). These function much like the zooxanthellae in coral reefs – boosting the growth and reproduction of both participants through an exchange of food and nutrients that each participant on their own would be unable to obtain in similar quantities. Most tropical forest plants also produce a combination of flowers and fruits with characteristics that attract a specific set of pollinators and seed-dispersers capable of supporting their reproductive success – loosely bound by the promise of a nectar reward, but nonetheless effective partnerships that unknowingly advantage the future generation of forest plants during the most hazardous stage of their life. Of course, all of these ecological mechanisms operate in natural habitats outside the tropics as well, but not with the breadth and cadence of interactions that characterize tropical forests.

The inner workings of novel ecological interactions are being teased apart each year by scientists, adding to the impression that the tropical forest is, more than anything else, a co-evolutionary cauldron where organisms continuously battle it out for their slice of the productivity pie. Yes, a few successfully go it alone, and some perish in an attempt to do so, but the majority of forest organisms enhance their respective shares over their lifetime by engaging in some measure of ecological *quid pro quo*. Viewing forest plants and animals in this way may seem anthropomorphic, but the ultimate driving force for an individual is survival, reproduction and successfully passing on its genetic material to offspring that are equally or more successful at reproducing. In a highly competitive environment where survival and reproduction rely heavily on a capacity to repel or avoid mortality from a wide range of potential predators and pathogens, capturing the productivity potential of the wet tropics often results in non-conspicuous evolutionary outcomes. The mutualisms formed between organisms – often separated by considerable taxonomic distances – offer a path for individuals to enhance their productivity through an evolutionary short-cut. Forming mutual relationships is less costly in terms of the inevitable trade-offs that must otherwise be struck through a process of classic evolutionary adaptation whereby the characteristics of an organism evolve under a battery of selective pressures. These work-arounds are evolutionarily efficient and commonplace – for example, more than 80 per cent of plant species on the planet are believed to form some sort of relationship with symbiotic bacteria or fungi (Wang and Qiu 2006). But it is arguably the unsurpassed ecological complexity of relationships formed between organisms in tropical forests that make this biome the most prolific primary production system on the planet.

The consequence of this productivity is reflected in the size of the resident biomass – both living and dead – of tropical forests. Most evergreen tropical forests – those formations that naturally occur in the wet tropics – are thought to generally store between 125 and 250 metric tonnes of carbon per hectare above ground. Most other regions outside the wet tropics by comparison are thought to store less than 75 metric tonnes per hectare (e.g. Saatchi *et al.* 2011). In most upland tropical forests, the living biomass accounts for upwards of 90 per cent of this storage. Necromass, the sum of all dead and decaying material, including logs, branches and leaf litter, accounts for the second largest share. Relatively little carbon is stored in the soils of these forests since most is exported through the groundwater and surface flow of rivers and streams or broken down and absorbed by decomposers. That is, except in areas that are permanently inundated, such as swamps. Up until recently, the extent of these formations was not considered meaningful at a global scale. Tropical swamp forests have been known for some time to be potent carbon accumulators when conditions of drainage, topography and rainfall favour the formation and accumulation of peat layers. We also knew that these areas are found in all three tropical regions. But recent work has now established that these are much more extensive than previously thought, particularly in the Congo Basin and the Amazon (Page *et al.* 2011; Draper *et al.* 2014; Dargie *et al.* 2017). Where they occur, they contribute significantly to the carbon storage function of tropical forests, in some instances exceeding the amounts stored in the above-ground biomass.

So, tropical forests are without doubt highly productive due to the favourable environmental conditions that prevail at the waist of our planet. We know that these conditions support higher productivity where the seasonality of rainfall and temperature is not too high or too low, but somewhere in-between. We also know that these conditions foster a large, dynamic pool of species that, in turn, are interacting continuously in ways that overcome potential nutrient limitations and make the most of the over-abundant solar radiation and moisture. But the crowning contribution of tropical forests to global productivity is not solely a consequence of these characteristics. Other tropical and subtropical communities, such as coral reefs, estuaries and wetlands, often display NPP levels that equal or exceed those commonly associated with tropical forests. These systems also benefit from relatively favourable conditions, are built around a backbone of complex ecological relationships and can efficiently capture nutrients. They also can store large quantities of carbon. But these highly productive habitats are relatively rare in the global scheme and tropical forests are not. It is the areal extent of tropical forests that make them a potent force of global bioproductivity.

This may seem odd when we recall that the tropics accounts for a minor fraction of land area on the planet. Surely the vast lands of the mid-latitudes would contribute a larger amount to the total productivity of our planet each year than the relatively land-scarce tropics? Such a proposition would indeed have proven correct some 5,000 years ago when native forests dominated the entire northern hemisphere and humanity had only begun to claim a stake on the agricultural potential of these lands. But today, in the aftermath of a multi-millennial agricultural revolution that has secured our place as the dominant species on the planet, the situation has changed. It has left tropical forests as the last, great bastion of natural productivity.

Green to brown

We can take a closer look at the distribution of global land cover and NPP to better gauge exactly how this transformation has positioned tropical forests at the apex of global carbon assimilation. To examine how our planet's vegetation might have looked before the spread of agrarian societies, we need an estimate of the global distribution of different vegetation types, such as forests, grassland, shrubland, desert and tundra, among others. These broadly shadow levels of productivity we can expect in relation to the amount of vegetation cover. A comprehensive classification of natural vegetation types on our planet conducted by the international conservation organization, the World Wide Fund for Nature (WWF), proved very useful in this regard. This initiative was undertaken by a large team of scientists scouring a vast amount of information over several years in order to classify the planet's natural vegetative condition into hundreds of "ecoregions" contained within 14 larger biomes and classified areas dominated separately by rock, ice and open water (Olson *et al.* 2001). I simplified this vast amount of information by collapsing the WWF biomes into four main categories of land cover in order of declining productivity: broadleaved forests (highest NPP), coniferous forests, grassland and shrubland, and a final category containing areas with little or no vegetation that I call DIRTS, that includes desert, ice, bare rock, tundra, and snow (lowest NPP). The total land area covered at each degree of latitude by each of these four categories can be seen in Figure 3.3. To develop a general sense of the extent that land cover has changed over the past five or so millennia, I also needed to find a suitable modern-day coverage to compare with the "pre-agriculture" scenario generated using WWF's ecoregional system. The European Space Agency (ESA) conducted a worldwide classification of land cover in 2009 through its GLOBCOVER programme and this appeared ideal for use in the comparison. This effort produced a coverage consisting of 22 high-resolution land cover types resolved from the raw multi-spectral sensor data collected on board the ENVISAT satellite.[8]

Five of the cover classes describe human-dominated landscapes. Two of these, intensive irrigated and rain-fed agriculture, were combined into a single class. These were consistent with systems predominating across much of Europe, many parts of North America or Asia and, increasingly, across the Brazilian *cerrado* just south of the Amazon and across the Indonesian archipelago. Two other cover types described areas of less intensive agriculture consistent with more traditional, small-hold farming techniques that rotate between crops and natural vegetation – the sort you would encounter along the upper reaches of the Amazon and Congo rivers or in places such as Papua New Guinea and parts of Central America. These too were merged into a single class. The final cover type recognized by the ESA system delimits areas transformed by urban growth. This would include any area covered by an artificial, impermeable surface, and would include cities, towns, roads and other forms of infrastructure put in place to support urban society. At global scales, urban area remains nominal in terms of cover. These three amalgamated cover types provided a base signature for the changes that have taken place to the pre-agricultural, natural cover of the planet as depicted using WWF's ecoregional cover. A sixth pathway of conversion, forest to pasture, was not explicitly recognized in the GLOBCOVER system. I had to produce a "synthetic" land cover class ("Pasture") using a spatial subtraction routine to roughly estimate the amount of land converted from forest in the WWF cover to grassland under the GLOBCOVER system and assume the bulk of this transition has taken place as a consequence of human activities.

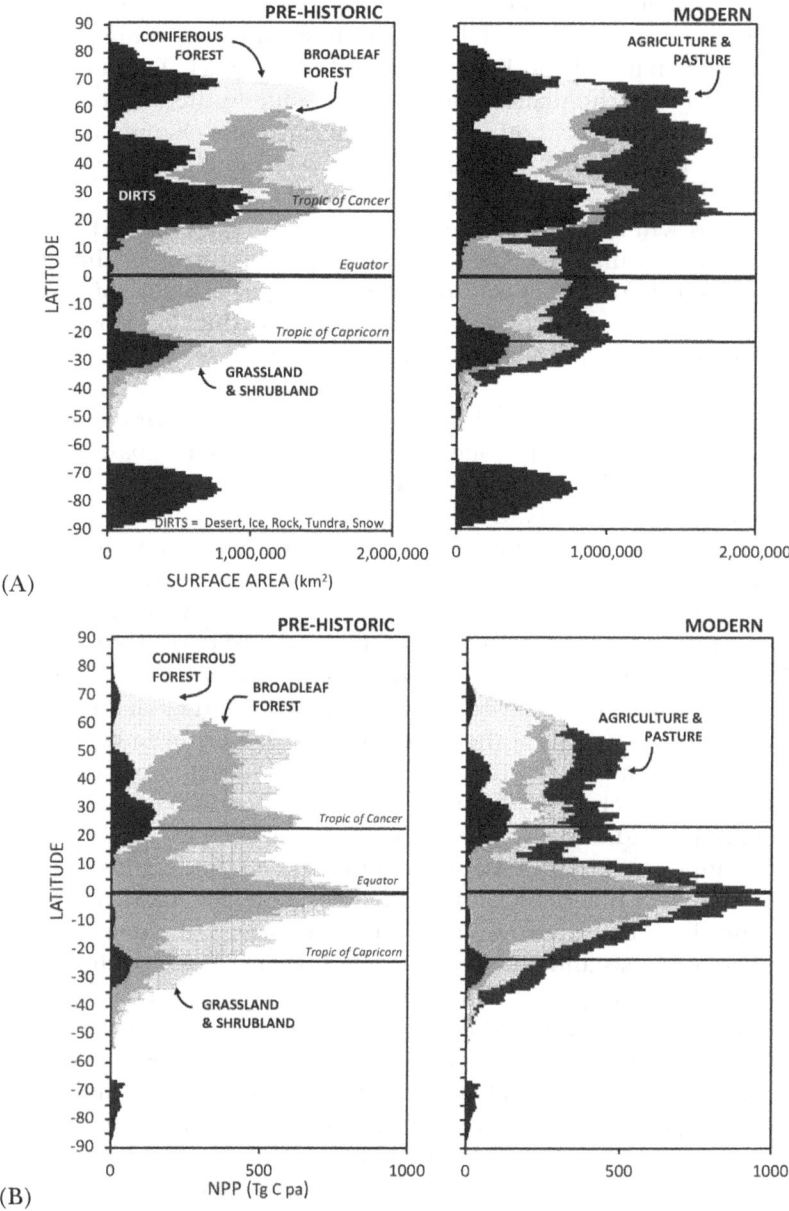

Figure 3.3 (A) The latitudinal distribution of major land cover types prior to agriculture and the current change in their areal extent caused by appropriation of these for agriculture and pasture. DIRTS is a composite of surface types with little or no bioproductivity. (B) A crude estimate of total NPP distribution by major land cover type prior to onset of agriculture and in relation to current distribution of types.

This may not be the case in all instances. In projects of such magnitude, there are invariably areas that have been erroneously classified, a point that has been clearly indicated by research scientists working on both WWF's eco-regional and the ESA GLOBCOVER systems. The land cover across some areas of the planet is less uniform, due to the effects of topography, soils, hydrology and other factors that can cause changes in vegetation at scales that are difficult to resolve using satellite-borne sensors. However, both products embody "state-of-the-art" capabilities and as a consequence represent some of the best available depictions of land cover at the global scale. Comparing these classifications with areas of the planet that I know well both within and outside the tropics – a sort of random ground-truthing – instils me with confidence that the ESA coverage at least reasonably mirrors the land cover you would observe on the ground in most places.

Comparing the WWF and ESA cover classifications reveals several fundamental features of our collective terrestrial resource base and the important position of tropical forests as part of this foundation. It shows that broadleaved forest, a land cover class that is distributed between 60 degrees north and south of the equator, was naturally as common outside the tropics as it was within at the time of our ancestors' drive to achieve greater food security. We can see this in Figure 3.3. Comparing the two diagrams depicting the latitudinal distribution of land cover types prior to agricultural expansion and in its current state clearly shows the near extinction of natural broadleaved forests between 20 and 60 degrees north latitude. Most of the forests at the edges of this belt have been replaced by some form of pasture or scrub grassland, perhaps due to the fact that at these latitudes, the climate is too dry or too cold to sustain widespread crop systems. Instead, these systems appear to have been established across the more temperate latitudes that prevail in-between these limits. Concentrated in a belt between 35 and 60 degrees, intensive agriculture has become the dominant land cover type on our planet at these latitudes. The temperate forests and native grasslands that once covered this part of our planet have been squeezed into a few residual spaces – an even smaller fraction of these reflecting the composition and structure of the stands that dominated the landscape across much of North America, Europe, Eurasia, Australia and New Zealand. A large portion of the remaining area of forest, as depicted at these latitudes by a thin sliver in Figure 3.3, is in fact of very recent origin, having re-established in the wake of modern changes to the economics of farming. As food production has become more industrialized and subject to market economies of scale and consumer preferences, financial considerations have reduced the amount of land that is viable to cultivate. Where these lands are no longer needed, some have been sold off and converted to urban use. In other cases, government and conservation societies have provided financial incentives that favour forest re-growth or a switch to biomass production and carbon sequestration. But these increases amount to a drop in the ocean of losses that have accrued over the past several millennia.

The story of how ancestral, old-growth or native temperate forests have become a pale reflection of their former grandeur is not merely about farming and livestock production, although ultimately these are the land uses that led to their disappearance. Often agriculture followed in the natural wake of industrial timber cutting. Approved by governments, monarchs and land-owners and carried out by shrewd industrialists, large-scale forest cutting was pitched as social progress and improvement. The reality was that timber was the structural material of choice for virtually all of the transportation economy up until very recently. An unending demand for ship-building timbers through to the end of the nineteenth century led to great swathes of temperate hardwood forests being cleared, particularly across Europe. The advent of railroads in the late nineteenth century further fuelled demand for rail ties and carriages just as the demand from ship-yards waned. Massive expansion of housing developments, particularly across the Americas where wood-framed building became the norm due in large part to the immense size of the forest resource, added to the demand for timber, particularly from coniferous forests. These early forests also became an easily obtainable source of fuel, driving much of the early industrial production in the New World before the adoption of coke and coal. Many of the remaining areas of these forests reflect this demand and have lost most of the characteristics that define old-growth stands: a mix of uneven-aged trees of various sizes with a diverse community of understorey plants and an abundance of deadfalls. They have become more tree farms than forests. As the Oxford historian Michael Williams wrote about the early North American use of timber: "Wood [in America] was abundant, it was ubiquitous, and consequently it was used prodigally in a multitude of ways. It entered into every aspect of life, quite literally from cradle to coffin …" (Williams 1989).

This could just as easily have been applied to other parts of the extra-tropics that still had abundant forests at the time, such as Australia, New Zealand and Canada. The differences between the two land cover systems in Figure 3.3 suggest that by the end of the twentieth century these widespread changes amounted net losses of between 12 and 13 million square kilometres of broadleaved forest, a staggering 68 per cent of the original cover, and three to four million square kilometres, or nearly 20 per cent, of coniferous forests across the northern and southern extra-tropics. The agricultural and timber production systems that shadowed an expanding and increasingly migrant, global population have left only a small collection of scattered vestiges of the once magnificent temperate hardwood forests to remind us of their appearance before our ancestors collectively re-designed the land cover at these latitudes. Amazingly, many of these remain under threat from the same industries that see them as unexploited resources, although their meagre size offers little in the way of long-term financial sustainability. "Creaming" of the long-term natural productivity, free of any financial investment, remains the primary objective. Over the same period, between 3 and 3.5 million square kilometres of tropical forests, around 16 per cent of the original cover based on the WWF eco-regional system, have disappeared. As a consequence, although temperate broadleaved forests were once more widespread and contributed more to global productivity, tropical forests now account for three-quarters of the remaining area of this broad forest type – one dominated by the more diverse flowering (angiospermous) plant group compared to coniferous (gymnospermous) types – and contribute at least 2.5 times more to global NPP than temperate forests (e.g. Beer *et al.* 2010).

This pattern of land cover change presents a growing dilemma for tropical forest countries that have been left with their stewardship, as well as those extra-tropical countries that have re-purposed much of their former forest cover towards agriculture and industry. Perhaps most striking is just how little of our planet remains naturally productive outside tropical forest regions. Again, we can turn to Figure 3.3, but this time focusing on the relationship between annual NPP as it varies with latitude and how this has changed with the transformation of land cover.[9] By comparing the graphs depicting estimates of NPP for a pre-agricultural planet and current land cover conditions, we can see the tremendous changes in the relative contributions of the various land cover types to global productivity. The drastic reductions in broadleaved forest in the northern subtropics between 20 and 30 degrees is particularly disconcerting if one considers the relatively large fraction of the Earth's surface covered by highly unproductive landscapes within this belt of latitude. At a broader level, there is a large amount of global land area – around 35 per cent – that is occupied by biomes with little or no natural vegetation cover. These are areas dominated by desert, tundra, polar and high alpine ecosystems that experience temperature and moisture conditions that fall well outside the envelope of climate that generally sustains more vigorous plant growth elsewhere. Where some plants have found the path to a positive carbon balance under these difficult conditions, they are chronically starved of light, water or both. Growth is compressed to a few favourable weeks of the year. The prospects of a high NPP in these biomes are as remote as the desolate lands they occupy – a coincidence of geographic position and the forces that drive the distribution of rain and sunshine across the planet. The discrepancy between the global contribution of these unproductive biomes to land cover (35 per cent) and annual NPP (10 per cent) is immense. Again, we can see this enormous change in Figure 3.3. A series of large black peaks appearing to dominate the surface area distribution of cover types across the planet precipitously shrink when scaled on the basis of their productivity. Comparing the NPP of these areas with those under other land cover – particularly broadleaved forests – further illustrates the extreme conditions constraining growth across more than a third of the land area of our planet. Technology – the great game-changer of human advancement – has much to overcome in making a third of the planet's terrestrial area productive enough to sustain our growing global population.

Combined with the realization that the vast oceanic gyres occupy more than a fifth of the global oceanic surface and are equally unproductive, a very large portion of the Earth's surface is simply incapable of the sort of natural growth that we might otherwise assume to be available in the future. It requires immense external subsidies of energy (to increase temperature) or water (to increase moisture), or both, to make tundra, desert, polar and alpine environments productive. Equally, to fill vast parts of the tropical and subtropical seas with phytoplankton would necessitate a colossal subsidy, but in this instance of vitamins and minerals that are naturally scarce at the centre of the subtropical gyres and may come with perverse consequences, such as anoxia. The economic costs make these sorts of conversions improbable, except where the needed input is already readily abundant, such as is the case with Iceland's geothermal greenhouses, or the mountain-born waters of the Nile that create a vast agricultural space in the midst of a desert. Perhaps technological advances will relegate the prospect of a shrinking reserve of productive space in the face of a growing human population to a footnote of past concern. Perhaps not. Until then, the remaining vast spaces appear for the most part to be unproductive ones since we have already altered and appropriated the bulk of productive landscapes and marinescapes outside these areas.

Much of the land area available to humans in the northern extra-tropics is either too unproductive to support our growing populations or has already been appropriated for agricultural or urban purposes. There appears to be very little room left to run. In the context of modern society, a preponderance of low terrestrial bioproductivity in the extra-tropics, high conversion rates of native forest and grasslands for agriculture and low bioproductivity in the tropical oceans should sound the alarm bell. But it hasn't. Most extra-tropical countries remain well fed and prosperous despite the challenges of their relatively unproductive landscapes. The key to this paradox – a lack of natural productivity and a surfeit of economic prosperity – finds its root explanation in the subsidizing role that the terrestrial tropics has played in bringing a measure of prosperity that western European nations could not and would not have known without it. Appropriating the bioproductivity of the tropics was the singular greatest force for global change, sparking revolutions and revolutionary thinking about trade, governance, finance and human rights. It altered what the world eats, how it is clothed and what it believes, and it pioneered the global spread of goods, services and ideas. The legacies are vast and deep and began with a simple bioproductive blip that made Europe biologically isolated and the surrounding seas a testing-ground that would in the waning years of the fifteenth century allow Europeans to glimpse the promise of tropical productivity.

Notes

1 These calculations are based on a monthly average household energy use figure of 957 kWh (US EIA 2015) and a count of 133 million for the number of US households (US Census Bureau 2015).

2 Based on the natural variability in the amount of solar radiation available at the surface due to factors other than latitude, such as marine boundary layer cloud cover, smoke, industrial pollution and volcanic ash, among others.

3 The models are based on satellite-borne observations of chlorophyll concentrations, sea temperature, light availability and photosynthetic efficiency. For further details of NPP data coverage see www.science.oregonstate.edu/ocean.productivity and Zhao *et al.* (2005), Behrenfeld and Falkowski (1997a) and Behrenfeld and Falkowski (1997b).

4 In the South Atlantic subtropical gyre, residence time has been estimated at 4–10 years, based on differences in surface salinity concentrations from centre to edge. In the North Pacific Gyre, bomb-produced tritium in surface seawater was used to estimate a residence time of 9–15 years. See Gordon and Bosley (1991) and Michel and Suess (1975).

5 The NPP of waters surrounding reef formations was calculated by extracting values from the raster coverage using a nearest-neighbour algorithm and averaging pixel values across the total set of extracted values. These were compared to averages of pixel values extracted for permanent upwelling and estuarine areas in the tropics. The coral reef coverage was provided by Reef Base/UNEP-WCMC.

6 The terrestrial NPP data are a 0.05 degree aggregated improved Collection 5.1 annual MOD17A3 NPP output from the *Numerical Terradynamic Simulation Group* (NTSG) at the University of Montana.

7 The relationship between species diversity and productivity has been widely debated and examined across a wide range of biomes, in both controlled experiments and natural systems. While results are not universally in agreement, the majority of studies support the view that increased species diversity and variety of plant attributes is associated with increased productivity. See Loreau *et al.* (2001) and Erskine *et al.* (2006).

8 Further information on the GLOBCOVER project can be found at: http://due.esrin.esa.int/page_globcover.php.

9 NPP values presented in Figure 3.3 were derived from the OSU combined global NPP model for the years 2000–2006. A new, annual average coverage was then intersected with GLOBCOVR assignments. The pool of NPP values for each biome was then applied to the WWF biome classifications to arrive at an estimate of NPP distributions prior to the onset of widespread human land use changes. This allows to some extent for the natural variation in NPP associated with latitude that would occur within the broad categories of land cover.

Part II
Exchange and globalization

4 An Icelandic banana

The view south-east along Route 1 from Reykjavik is punctuated by displays of birch and rowan huddled among a mix of bare rock hills and impoverished grasslands. The landscape along the southern coast of Iceland pays homage to its volcanic origin, the relentless struggle that plant life experiences at these high latitudes and importantly, the shaping hand of its inhabitants in configuring the present ecology. The volcanic plain that runs beneath Route 1 was once, from most historic accounts, more forest than grass. The agricultural ambitions of the early Norse settlers, like those of colonists everywhere, drove the expeditious removal of the few naturally forested areas in an attempt to gain a foothold on an island with little luxuriance on offer. Their ambition, wagered primarily on barley, was a matter of survival at a time when consuming locally produced food was not a preference, but a necessity. Despite this life-and-death imperative, by the mid-twentieth century traditional agriculture in Iceland had been virtually abandoned, after generations of farmers struggled to make it a permanent feature of the national landscape. The history and culture of Iceland is thickly woven with tales of their ancestors repeatedly gambling, and losing, while trying to feed their people. Nearly 900 years of punctuated crop failure, disease and famine levied a long series of cataclysmic losses on their isolated community, driven by periods of extreme wind, frost and cold. Severe soil erosion following the early clearance of the native forests hastened the decline of the already small arable estate. Only a modest production of hay and barley has survived today, grown to feed the uniquely Icelandic breeds of sheep, horse and cows. Livestock and hay have become Iceland's "default" traditional agriculture.

Iceland's latitude and topography leave few frost-free days in the year – ranging from 90 days inland and at elevations exceeding 100 metres up to 150 days along the south coast – a fact that relegates traditional agriculture to an incredibly small area over a very short, summer window. The immediately arable land – area not in need of significant amelioration – counts less than 5,000 hectares. The darkest periods in Iceland's history coincided with the abrupt closing of the summer window over this small area – when climate deviated from its long-term seasonal averages and delivered a summer-time frost precisely when struggling barley, potatoes and other staples needed it least. Total crop loss and famine ensued. Summer frost, like a visit from the *jötunn*, ancient giants of Norse legend, became the Icelanders' bane. It is in this historical context that the landscape along Route 1 takes on a new visage: one of quiescent acceptance. Until we arrive in Hveragerdi.

Here the contrast between the frost and fire that dominate Iceland's landscape is visceral. The continuous, billowing streams of white vapour spiralling upwards from the hot springs at the edge of the village convey the sense of pent-up energy waiting below. This small town of 2,300 people sits directly above the Mid-Atlantic Ridge, a segment of the 65,000-kilometre network of seafloor ridges, or seams, where tectonic plates emerge from the viscous mantle, cooling as they do so, to form the planet's crust. Most of these plate edges rest deep within the oceans, but where they rise to the surface, as in the rare case of Iceland, the release of energy can be spectacular. Hveragerdi rests on the edge of the extinct Grensdalur volcano in one of the most seismically active areas of the country. It was purposefully built at this site to capture the local upwelling of this massive geothermal energy, and the plentiful supply of naturally heated water it produces. This resource has been put to good use, as elsewhere in Iceland, to heat homes and water and as an attraction to the growing number of visiting tourists. But Hveragerdi is of particular interest due to the series of clustered greenhouses in the heart of the town, run entirely on geothermal power. Hveragerdi in fact translates as "Hot Garden", the odd transparent boxes of green foliage appearing almost as an extra-terrestrial outpost in the otherwise subdued, frost-stricken landscape. The most alien aspect of these greenhouses, however, is not their location, but what grew inside: bananas.

Banana production in the small town began in the 1940s, soon after the first geothermal greenhouses were constructed. At the time, the cost of importing tropical fruit was prohibitive and the power necessary to continuously heat the greenhouses to the ideal temperature for growing bananas, 25° Celsius, was cheap. For a decade or two, bananas were grown at a commercial scale to feed demand in Reykjavik. This then dwindled to several dozen or so plants that remain today after the price of bananas from tropical Central and South America declined precipitously in the 1970s, sparking a surge in imports. Economies of scale always rendered the commercial prospects unsustainable, even with an endless supply of cheap energy. Icelandic bananas take two years to mature, compared with three to five months in the tropics, requiring up to eight times as many plants, and greenhouses, to maintain the same rate of production. The fate of the geothermal banana of Iceland was inevitable. Yet, for a period it was thought that this island nation, paradoxically, produced more bananas than any other in Europe. This belief, however, appears upon inspection to be only partially true. Bananas have been produced in much greater quantities in the Canary Islands, a province of Spain, since the late nineteenth century, but the islands are geographically part of Africa rather than Europe. Portugal, Greece and Cyprus currently produce a small amount each year, around 30,000 tonnes, but supply was devastated during the war years. In this context, the greenhouses of Hveragerdi, Iceland were probably the largest European producer of bananas during the post-war period from 1945 to 1955.

Why did Icelanders decide to grow bananas? There are many other traditional European staples that would have benefited from the regulated climate only greenhouses can deliver. Flower and fruit production in many of these crops, such as pears, apples and oranges, benefits from some measure of chilling, called vernalization. Light chilling improves the appearance, and marketability, of oranges as long as they are not exposed to temperatures that are too low. Although vernalization normally forms part of the seasonal cycle in less inclement parts of Europe, Hveragerdi greenhouses provided the best environment in Iceland to ensure a period of controlled vernalization did not decay into unwanted frost.

Prohibitive import duties, difficulties with post-war shipping networks, and the harnessing of cheap geothermal energy are the factors that drove the production of the banana, but Hveragerdi could have just as equally chosen to grow a traditional European food. People do not have to eat bananas to live a perfectly healthy life. Perhaps it was the challenge or desire to showcase the capabilities of the newly established geothermal greenhouses that drove their decision to produce an archetypal tropical fruit in the middle of a frosted landscape. Maybe a visit to the tropical luxuriance inside the greenhouse temporarily alleviated Icelanders' wintertime burden of living on the Arctic's edge? Over the past decade, sweet peppers, cucumbers and tomatoes have been the only agricultural produce in Iceland registering significant growth in production, entirely within greenhouses. They take less time to produce than the banana, but these fruits also originated as tropical crop plants in India (cucumber) and Mexico (pepper, tomato), albeit grown now in almost every country in Europe. Today, a visit to Hveragerdi reveals that the greenhouses are more typically filled with tomatoes, roses and other ornamental house plants or have simply been abandoned. The rise and fall of the banana on an Arctic island, so intrinsically removed from its natural, tropical homeland, tells us a great deal about the complex interchange that has taken place between tropical nations and the rest of the world over the past 500 years. But the Icelandic banana is a terminus in the history of this tropical fruit and its spread across the planet. To further understand the complexity of this exchange, we have to travel south to the Canary Islands, to examine how the banana became a food icon and the events that led to its bizarre circumnavigation of the planet, eventually landing in Iceland.

In the global pantheon of domesticated fruit, the banana, and its close relative, the plantain, reign supreme. More than 76 million tonnes of its nearest rival, the apple (all varieties), were produced in 2012 alone, but this amounts to barely over one-half of the weight of bananas and plantains grown during the same period (FAO 2016b). Production is so prolific that one out of every five kilos of domesticated fruit produced globally each year consists of bananas and plantains. Perhaps even more astounding is the recognition that despite the world's population growing by 70 to 90 million people each year – more than doubling from around 3 billion to 7 billion in the 50 years spanning 1962 to 2015 – the production of bananas has not only kept pace with this staggering growth, but expanded four-fold from around 35 billion to just short of 140 billion kilograms over the same period (FAO 2016b). That translates to nearly a doubling of fruit for every person on the planet from 11 kg in 1962 to 20 kg (44 lb, ~ 40 bananas) in 2015.

The Canary Islands today are the largest supplier to Europe of bananas originating within the European Union, but the historical role of these islands in the universal spread of the king of fruit far exceeds their very modest contribution to a global retail trade now worth over US$29 billion. The importance of this small archipelago of 13 islands in the story of the banana rests not in its contribution to production, but in its rather strategic position at the start of the westward arc of the North Atlantic Gyre. This gyre, one of five immense systems of rotating ocean current on our planet, dominates the North Atlantic. Bound by the opposing continents, it is driven by gradients of heat and salt that are formed across the basin and form part of the global oceanic conveyor. These gradients are formed primarily by differences between the amounts of heat received in the tropics and the Arctic. Evaporation and rainfall, amended by freshwater inflows from rivers and melting ice-pack, alter salinities, combining with temperature to shunt seawater in waves, or fronts, in a series of source–sink movements. The Gulf Stream is the primary pump of this current in the North Atlantic, conveying the warm, salty waters of the tropical Caribbean northwards into the much colder, fresher region of the Scandinavian North Atlantic. Cooling, these waters sink, moving southward along the British Isles and Iberian Peninsula towards the North African coast.

The North Atlantic Gyre is formed of currents driven in part by the temperature and salinity, or "thermohaline", conveyor that produces the Gulf Stream and in another part by the relentless push of two major wind systems in the region: the trade winds in the south and westerlies in the north. The main factor creating the loop in the current is the Earth's rotation and the control this exerts on these winds – the Coriolis Effect. Winds and sea currents are the main planetary mode of conveying energy from high to low states – they work to keep the planetary energy budget in a state of dynamic equilibrium. Large differences between the annual amounts of solar energy received at high latitudes and in the tropics should dictate that the winds travel directly along the shortest path in a south–north direction from continuously high (tropical) to seasonally low (arctic) energy states. The Coriolis effect, the effect of rotation, is one of continuously deflecting this south–north movement at right angles, creating a "curl" in the direction of wind and current. This effect, the same force that manifests in the movement of tropical hurricanes, constantly deflects the major (air and sea) currents in the North Atlantic in a clockwise fashion, creating the gyre. The scale of the North Atlantic Gyre, although not the largest on the planet, is vast. Some estimates place the amount of seawater conveyed along the weaker, south-eastern sector, the Canary Current, at $6 \times 10^6 \, \mathrm{m^3 s^{-1}}$ (Hernández-Guerra *et al.* 2005) and the stronger north-western flow, the Gulf Stream, at a staggering $93 \times 10^6 \, \mathrm{m^3 s^{-1}}$ (Richardson 1985), a volume of just over 37,000 Olympic-size swimming pools shunted through a stationary vertical plane every second. At the heart of the gyre, similar to the still air at the eye of a hurricane, is a massive body of calm water, the Sargasso Sea. A sea within an ocean, the 3.5 million square kilometre Sargasso Sea is bounded by the four currents forming the gyre. While these currents subtly alter mean position, velocity and depth in response to seasonal and multi-annual oscillations in wind, temperature and salinity, they have remained surprisingly consistent since first recognized over 500 years ago – an important condition if one is to rely solely on the wind and ocean current to arrive in the New World.

It is the eponymous Canary Current that sped the first voyages of Christopher Columbus and his fellow explorers in the late fifteenth century to the southern run of the North Atlantic Gyre and then on to the Caribbean islands of Cuba and Hispaniola. The Canary Islands thereafter quickly became the Spanish staging ground for the largest and most rapid invasion in history. Columbus, Vazquez, Cortés, Pizarro, del Soto and the thousands of Spaniards that settled and subdued the lands of Central and South America commenced their journey from these islands and took with them the agricultural tools they believed were best suited to garnering rapid economic advantage from the vast tropical landscape set out before them. Journals and accounts written at the time suggest that the overwhelming bulk of these arrived with the friars from the main mendicant orders, as they built a new network of missions and convents through the Spanish Crown's policy of *encomienda*. The banana and plantain were some of the earliest of these tools to arrive in the New World and were soon spread throughout the Caribbean and Central and South America. An account of the early years of Spanish conquest, occupation and settlement by Captain Fernandez de Oviedo y Valdes described the banana's arrival in Santo Domingo (Dominican Republic) from the Canary Islands in 1520 (Libro VIII, Capitulo I, Parte X, p. 292 in Fernández de Oviedo y Valdés 1851).

The banana and plantain proved well suited to the tropical climate and soils of Hispaniola. The plant was easily transported, grew quickly with little nurturing, produced nutritious fruit within a relatively short time from planting and quickly became a product of trade, producing income for growers. It is not surprising then that *Musa* root cuttings became an indispensable part of the toolkit that accompanied each group of colonists as the Spanish began quickly to expand their network of settlements outward from the new regional capital at Santo Domingo. The pace of expansion and development was unprecedented for the time – more than 20 new cities established over the course of a mere 50 years following Columbus' final voyage, each with a coterie of *encomenderos*, and a religious and administrative network to ensure that both the Church's and the Crown's economic interests were protected and expanded. From the beginning, the banana and plantain became an integral, if not pivotal, part of these developments. From their experimental home on Hispaniola, they quickly made their way to the neighbouring islands of Puerto Rico, Jamaica and Cuba. They were carried by Fray de Berlanga to the newly established Panama City when he was ordained as bishop in 1534. Twenty years later, his ecclesiastical colleague, Fray de Quiroga dispersed cuttings to the main settlements across his diocese of Michoacán in Mexico. After the Treaty of Tordesillas, the Portuguese claimed a large swathe of eastern Brazil, and the first of a series of captaincies was established at Pernambuco in 1534. Plantains and bananas followed the growing stream of Portuguese colonists from the island territories of Madeira and Cape Verde, invariably spreading along the Brazilian coastline with each new settlement. By the end of the sixteenth century, bananas and plantains had become commonplace features in the agricultural landscape across much of the Caribbean, southern Mexico and Central America. Some historians have suggested that the banana and plantain were brought to the New World by de Berlanga to provide sustenance for African slaves and that their presence offers a time stamp to the introduction of plantation slavery in the Americas. I tend to disagree with this view. As far as we know, the banana was introduced to the Canary Islands from the Guinea coast by Portuguese traders, only a couple of decades prior to Spain's annexation of the islands. Valde's description of the banana plant and its introduction by Friar de Berlanga does not evoke a sense of familiarity, as do related descriptions of efforts to introduce citrus or olive trees, two ancient stalwarts of the Iberian agro-economy. Instead, the banana is discussed in light of its performance and prospect as a useful crop plant. This is particularly important when placed in the context of the neotropical landscape. Spanish expeditions and early efforts to establish colonies across the Americas were frequently besieged by indigenous communities or poorly provisioned and showed a grave unfamiliarity with the food plants of the region and how to prepare them. The banana and plantain offered an easily transportable plant that could be sown in most lowland soils and would produce fruit within a year. The fruits are high in starch and the caloric content – a third of that found in maize, nearly equal to that of cassava and higher than the sweet potato – rivalled the three main staples of indigenous agriculture at the time of the Spanish invasion. The greatest attraction of banana and plantain cultivation, however, resided with their outstanding productivity. Over six times as much weight in bananas could be produced relative to the indigenous maize varieties. Plantains could yield 40 times the weight of potatoes in an acre of land. Compared to the time and effort required to sow, grow, harvest and process the main indigenous crops, the banana and plantain would quickly become the "fast food" of neotropical cuisine. The famous nineteenth-century Prussian explorer and naturalist, Alexander von Humboldt, described it thus:

> ...that as thirty-three pounds of wheat and ninety-nine pounds of potatoes require the same space as that in which four thousand pounds of bananas are grown, the produce of banana is, consequently, to that of wheat as 133:1, and to that of potatoes as 44:1.
>
> (Humboldt 1814)

There are only a couple of dozen first-hand accounts of those early decades of Spanish and Portuguese colonial expansion across the neotropics. Those that have survived are often embellished with wild descriptions of unrivalled potential. Examining documents archived at the Archivo General de Indias in Sevilla, suggests that many of these accounts, or *cartas de relaciones*, clearly were filed simply as a contractual obligation to patrons back home – an ebullient corporate report if you will – designed to impress, while hoping to instil the prospect of even greater return among potential patrons remaining on the sidelines to such risky, unquantified adventures. Other reports were filed as a matter of duty to the Spanish Crown by commissioned military officers. These letters, however, often reflect a pre-occupation with remuneration rather than process or events. Letters from Cortés, Tapia, Alvarado and Mena catalogue complaints about inadequate rewards for their services or difficulties with competing factions of *conquistadores* seeking patronage and permission, a dynamic that would later unravel the Spanish viceroyalties. Historical accounts written around the middle of the sixteenth century, but often published much later, shine some light on the processes and events that characterized the sweeping transformations accompanying the early phase of Spanish colonization. Yet very few of these discuss the mechanics of the agricultural revolution taking place as bananas and plantains, along with sugar, rice, oranges and other Old World staples, spread throughout the new colonies. Consequently, precise details of how and when the banana and plantain spread throughout the New World remain open to a creative historical interpretation. What we do know is that these plants quietly played a massive role in transforming the social, cultural and physical landscape of the Pre-Columbian Americas beyond recognition.

There are thousands of varieties of bananas in existence. Most of these are found today in the core ancestral homeland of the species, the Philippines, and the South-East Asian region between Papua New Guinea and Vietnam (Perrier *et al.* 2011; Li *et al.* 2013). A large number of these are "wild-type" varieties, still contain seeds, do not look much like the modern market banana, and are not particularly nice to eat. The key stage in the banana's rapid movement around the planet was the emergence of a small pool of domesticated seedless hybrids that are much closer to the standard fruit grown and shipped around the world today. A study by Perrier and colleagues in 2011 utilized genetic analyses, linguistic affiliations and archaeological evidence to establish these centres of early domestication and the most likely prehistoric migration routes that led to a handful of early varieties arriving in India, Africa and the Middle East. Palaeobotanical evidence places the emergence of early domesticated varieties in South-East Asia at around 6,500 years BP. The path travelled by the banana to Africa is less clear, but routes via the Indian subcontinent and across the Indian Ocean to East Africa are believed to have been used by migrants or traders around 2,500–3,000 years ago, based on evidence from Cameroon and Pakistan. At approximately the same time, another variety of banana, the Fe'i (*Musa* variety *troglodytarum*), was migrating eastward from Papua New Guinea to the Solomon Islands, Vanuatu, New Caledonia, finally arriving in Hawaii and modern-day French Polynesia around 1,200 years ago. Other varieties later followed this path across the Pacific. When Captain Cook became the first European to set foot on Hawai'i during his epic third, and last, voyage of 1778 (he was killed while trying to kidnap the Hawaiian king, Kalani'ōpu'u, to hasten the return of his stolen out boat), the banana and plantain had already been on the island for at least a half-millennium.

The first Portuguese traders introduced the banana and plantain to the Canary Islands from Equatorial Guinea in the mid-1400s, but by that time the fruit was being cultivated by nearly everyone in the tropics outside the Americas. The corms (underground stem) of these bananas, subsequently transported to the New World, were not the fruit that we know in the markets of Europe and North America, or that was grown in the greenhouses of Hveragerdi. These were other varieties, still produced today but for local consumption in the household gardens of the Caribbean and West Africa. To trace the origins of the mass-produced banana, we must travel back to the ancestral homeland of the banana in South-East Asia again and a variety eventually known as "*Gros Michel*". The great age of European exploration was nearly 300 years old by the time that Nicolas Baudin, a French navigator, explorer and naturalist, began a series of expeditions to South-East Asia in the latter half of the eighteenth century. These were commissioned by the Hapsburgs through Baudin's fortuitous acquaintance with the Austrian monarch's head gardener, Franz Märter, on board a merchant vessel transporting passengers from France to its colony at New Orleans in Louisiana (the United States would not annex this territory until 1803, under the Louisiana Purchase). Märter was organizing voyages of scientific exploration for the monarch at the time, Emperor Joseph II, a fastidious student of "enlightened absolutism" and great patron of scientific advancement. Märter recruited Baudin to transport a voluminous catalogue of natural specimens along with their collector, Franz Boos,[1] Austrian naturalist and, later, director of the Schönbrunn Palace gardens, from Mauritius (Ile de France, a French colony at the time) and Réunion (Ile Bourbon, still a French possession). Baudin learned from Boos how to preserve and maintain biological specimens during the exceedingly long and weather-exposed voyages often endured as part of these scientific explorations. It was these skills that led Baudin to collect and successfully transport live specimens of the *Gros Michel* plant from Canton in China to the island of Martinique around 1790. Within 50 years, the *Gros Michel* had been dispersed throughout the Caribbean and Central America from its base in Martinique[2] and became the mainstay variety of a small, opportunistic import business that was about to explode into a global industry.

The first regular shipment of bananas was bought and loaded aboard a steamer hired by an enterprising German-American immigrant, Carl Frank, and taken from the Panama Canal Zone (then part of Colombia) to New York City in 1866 (Adams 1914). Shortly after the US Civil War, Frank had recognized an opportunity while serving on a postal mail ship. He observed the productivity of banana growth during the time he had between trips and was able to quickly come to agreeable terms with a number of farmers in the area for a small supply of bananas. He then arranged a steamer to transport the inaugural supply to New York, where he sold them for a profit of 100 per cent. This opportunity arose from the dramatic developments in mechanical engineering that would come to characterize the Industrial Revolution at the turn of the century. Advances in steam powered engines combined with the replacement of the paddle wheel – perfectly suitable for short river transport but not open oceanic voyages – with the more efficient hydrodynamic screw propeller, which allowed cargo ships to travel further and faster than ever before. This change shortened trip times from the Caribbean to the United States considerably while increasing the fuel efficiency of larger ships, a key factor in reducing the cost of transport. The shift from wind to steam paddle wheel to propeller transformed oceanic transport more than any other development at the time. It pried open a can of new opportunities in the exchange of goods between the tropics and the extra-tropics, just as the discovery of the North Atlantic Gyre did for the Spanish in the fifteenth century. Bananas and other perishable fruit up to this time simply would not survive oceanic voyages, arriving blackened and mouldy, unfit for consumption. This made the banana a relative late-comer to the growing trade in tropical foodstuffs. Dry bulk goods that could survive extended periods at sea, such as coffee, black tea, cane sugar, cocoa, rice and dried fruits, had been the primary objective of most oceanic trade up to this time and these products contributed much earlier – at least 250 years in most cases – to the growing diversity of food items characterizing the European and North American diets.

Prior to the steam engine, bananas were known in North America and Europe, but to most of the population as an exotic object in the far-away lands of adventure books, along with elephants, tigers and monkeys. Occasionally small quantities would outlast the time at sea and haphazardly show up at ports in the United States, making them an unusual luxury food item consumed exclusively by society's upper classes. They were more akin to truffles than apples, and few people at the time would have imagined the banana as a common component of their kitchen pantries. Prices were initially paid per fruit, wrapped in tin-foil, then by the bunch; they were always sold at enormous profit in the eastern ports of Boston, New York and Philadelphia. Priced in today's dollar (adjusted for inflation), some consignments could attract nearly $4.80 per *fruit*. Frank saw the confluence of potential demand and mechanical innovation and the enormous profit to be made at their meeting. But he was not alone in his epiphany, and other, more deliberate, entrepreneurs soon thereafter gathered their financial resources. The consequence would be a handful of super-sized commercial enterprises that would rapidly transform the banana from culinary oddity to dietary staple across most of the planet.

A long list of books has been written about the events surrounding the commercialization of bananas.[3] The saga is a classic tale of entrepreneurial spirit, market competition and the financial rewards that can arrive by recognizing an opportunity to grow a demand and then supply it. It is also, sadly, filled with tragic tales of abuse, exploitation, and corruption in a time when corporate enterprise rarely took care to consider its social contract with those that laboured towards its financial reward or the environmental consequences of its operational choices. In this regard, the banana trade lost an opportunity to pave new pathways in commercial production distant from those developed by sugar, tea and coffee growers at the time. Instead, banana suppliers often chose to perpetuate many of the same traditional plantation practices and attitudes that had been the labourer's bane in the tropics since the earliest days of European colonization. The consequence was chronic political instability, retarded economic growth and the malaise attached to these powerful social inhibitors in many countries where they operated, the memetic "banana republics".

After a rapid succession of mergers, acquisitions and bankruptcies, three giants of the banana industry emerged by the 1920s: the United Fruit Company (UFC, later Chiquita) – the largest – with properties in Jamaica, Panama, Costa Rica, Cuba, San Domingo (Haiti) and Honduras; the Standard Fruit Company (SFC, later Dole Foods), based in New Orleans with most production originating in Honduras; and Elder and

Fyffes (later, just Fyffes), a UK and European fruit importer, under 45 per cent UFC ownership, trading primarily in bananas from the Canary Islands and the Caribbean. There were many shipping lines, such as the Royal Mail Steam Packet Company, that took on consignments of bananas as an ancillary service to improve the profitability of their voyages. Other small suppliers, particularly in Jamaica and the Windward Islands, continued to export fruit directly to the ballooning markets in Europe and North America, but it was the three big companies, more than any other, that eventually fixed the banana as an everyday food item in the collective minds of people living in the extra-tropical latitudes.

The number of bananas being shipped to overseas markets had grown from around three million in the 1890s to five billion by the 1920s, with little end of demand in sight (Adams 1914).[4] At the same time, this enormous increase in volume sprouted vast economies of scale and, as a direct consequence, a steady decline in retail prices, stimulating yet further demand. By the 1950s, the banana industry had survived the economic scare of the Great Depression and severe disruption of shipping routes during the Second World War to overtake Europe's and North America's prior favourite fruit, the apple, in terms of both greater supply and lower price. Hundreds of thousands of hectares of tropical forest land had been planted with the *Gros Michel* variety since the 1900s in countries throughout Latin America and the Caribbean, almost entirely to serve the export industry, and these were now producing at full capacity. Yet, despite the superlatives that had come to define the global banana trade – largest area planted, greatest volume exported or highest revenue achieved – one aspect remained singular: the plant itself. The millions of tonnes of fruit arriving at the markets each year were produced from an identical genetic copy – a clone – of the same plant brought to Martinique from southern China by the Frenchman, Mssr. Baudin, more than 150 years prior. Virtually the entire banana crop of the New World, covering thousands of square kilometres in more than 15 countries, was derived from a single genetic type. The banana is in fact part of a larger group, or order, of plants that includes the ginger, cardamom and turmeric family (Zingiberaceae), the arrowroot (Marantaceae), the South African bird-of-paradise and Madagascan Traveller's palm (not actually a palm) (Strelitziaceae) and Amazonian Heliconias, all of which can propagate vegetatively by throwing off underground shoots to form new plants. In the case of the banana and plantain, the above-ground plant we see is merely a tightly bound whorl of leaves unfurling from the real stem underground. Consequently, the banana can produce new pseudo-stems over and over again, but they are all genetically identical, more akin to grass in a lawn than trees in the forest. This vegetative propagation was a great asset in the rapid expansion of *Gros Michel* plantations as with many other tropical crops, for instance cassava or sugar cane, since it by-passes the critically vulnerable stages of seed germination and early growth that account for the overwhelming majority of losses in sexually reproducing populations. Clonal reproduction at this scale reduces the hazard (the likelihood of death, having survived to any given life stage), but at a cost. The lack of genetic diversity across the plantations means that should a pathogen evolve the chemical key to unlock the plant's defences, no further adaptation is required to successfully invade the next suitable host and the entire stock becomes open to infection. Humans have fought, and survived, epic battles against pathogens mainly thanks to our genetic diversity and distribution across large, often isolated, areas. One of the great theories explaining diversity in tropical forests is based on the need of most, but not all, species to continuously escape harmful pathogens through dispersal and low population densities. This leads to a few individuals of many species in relatively small areas of forest. Our agrarian pursuit of food security, however, has taken crop plants in the opposite direction – less genetic diversity purposefully collected in highly concentrated areas. Export bananas are one of the least diverse and most highly concentrated crops grown on the planet and by the 1950s, a new strain of the ubiquitous fungus *Fusarium oxysporum*, had found the key to *Gros Michel* and made its way from Asia to the plantations in the Americas. The impact on the banana trade was devastating, particularly to the interests of the largest plantation operator, the United Fruit Company. Their profits dropped precipitously. In a single decade, profits at the UFC went from US$66 million in 1950 to just under US$2 million in 1960.[5] After a half-century of building up and then dominating the banana trade in the Americas, the company had become powerful, moribund, institutionalized and myopic. The middle of the twentieth century proved to be a low point for the mighty company. Its increasingly erratic behaviour towards its plantation workers, their elected governments and civil society would throw many Central American countries into disarray. "McCarthyism" and the "red menace" were reaching the height of their influence in Washington and any threat to the unfettered interests of American companies was seen as another front in the Cold War. The consequence for countries such as Guatemala and Honduras was decades of domestic turmoil, war and stalled development.[6]

The pre-occupation with maintaining absolute control over their properties in Central America only furthered institutional resistance to change at the UFC. The result was a significant delay in the vital search for a replacement variety that could resist the *Fusarium* onslaught. Ultimately, the change that was needed arrived from work done by the much smaller Standard Fruit Company. They had quickly initiated trials to identify an alternative to the *Gros Michel*. After various attempts to establish a viable alternative from the original gene pool in Asia, they came across a successful variety growing in Brazil – the *Cavendish*. By the mid-1960s, with profits on the wane and the old guard passing on, the UFC brought in new management. They quickly moved to adopt the *Cavendish* as their new variety and the final *Gros Michel* bananas were shipped to the United States in 1965.

The *Cavendish*'s journey, however, began nearly 125 years before the final collapse of the *Gros Michel* trade in the Americas. Like the *Gros Michel* and earlier varieties introduced by the Spanish and Portuguese friars, the *Cavendish* originated from the banana's ancestral homeland in South-East Asia. Like its predecessors, the *Cavendish*'s journey from the tropical forests of Asia to plantations in America was also precipitated by the combination of individual ambition, ruling class patronage and a hyperactive desire to acquire and collect everything that the tropics, still mysterious and exotic to most outside the region, had to offer. The tropics yielded their secrets openly and freely – a global commons with little sense of the natural capital it held within its landscape. When Charles Telfair received a new shipment from Canton, China filled with specimens to add to a growing collection housed at the famous Pamplemousses tropical garden in Mauritius (by then, a British colony), one of the living specimens was the *Cavendish* banana.[7] The year was 1826 – 35 years on from his French predecessor, Nicolas Baudin's arrival in Mauritius, and the onward transport of the *Gros Michel* to Martinique. From its new home in Mauritius, however, the next leg in the westward journey of the *Cavendish* – London, England – seems improbable. Telfair authorized the export of a few specimens to a collection near London, thought to be at Bury Hill, a sweeping estate in the Surrey countryside owned by a brewing magnate of the time, Robert Barclay. Barclay, like many of his social rank, had a strong interest in horticulture and gardening and this invariably brought a small group with both the interest and the means into frequent correspondence regarding their acquisitions. It appears that Barclay fell gravely ill shortly after the arrival of his new acquisition and it is in this context that specimens of the *Cavendish* banana were sold to the Chatsworth estate, the hereditary seat and home of the 6th Duke of Devonshire, William Cavendish. At this juncture, the banana variety that arrived in the northern English county of Derby, having travelled from China via Mauritius and London, was referred to simply as the Dwarf Chinese.

The Swede, Carl Linnaeus, had unveiled his taxonomic system of binomial nomenclature nearly a century earlier and the new system of classifying organisms sparked a frenzied passion for specimen collecting among the wealthier members of society. Interest in live plant specimens peaked in nineteenth-century Great Britain, transforming the traditional garden and forest through the introduction of woody trees and shrubs from every region of the planet. The more exotic the specimen, the greater was society's accolade. The Duke of Devonshire counted himself among the greatest patrons of this cutting-edge horticultural practice. To advance his ambitions, the Duke took on a highly intelligent and capable gardener to head his efforts to transform Chatsworth grounds into a site of world-class splendour. His name was Joseph Paxton and he would rise under

Devonshire's patronage to become one of England's greatest gardeners and horticulturalists, culminating in his design of the astounding 39-metre-high glasshouse built in Hyde Park, London for the Great Exhibition of 1851 – The Crystal Palace.

It was Paxton that received the Dwarf Chinese banana from Barclay's estate in 1829 and nurtured it towards producing fruit for the first time in Great Britain. The news quickly spread amongst his fellow horticulturalists through articles in various magazines, including his own periodical, Paxton's *Magazine of Botany*. It was here that he stated that the Dwarf Chinese banana shipped to Chatsworth via Mauritius had been welcomed into Linnaeus' system with a new Latin binomial: *Musa cavendishii* – named in honour of the Duke of Devonshire.[8] Paxton's success with the *Cavendish* spurred him and the eponymous Duke to create a hot-house – named the Great Stove due to its appearance and tropical temperature – on the Chatsworth estate where their new celebrity could be multiplied. Greenhouses of this size were unknown for the time; for example, the great palm house at Kew Gardens would not be installed for another decade or so.[9] By 1836, Paxton had seen the Great Stove through to completion and set about filling the new hot-house with tropical specimens, among these being a large number of productive *Cavendish* plants. He and the Duke took immense pride in their "mini-plantation", perhaps not fully divulging that the exotic fruits formed through parthenocarpy – did not require pollination – and therefore eliminated the most critical barrier to tropical fruit production in greenhouses. Demand for specimens was great and Paxton eagerly set about distributing them among his fellow horticulturalists and admirers. In 1848, missionaries headed for Samoa in the South Pacific carried specimens of the *Cavendish* with them. They were well received by islanders already familiar with the Fe'i and other varieties of bananas and plantains that had arrived with their prehistoric ancestors. It quickly spread to Fiji and Tahiti and then throughout the region, transported and traded among the growing list of colonial settlements, and by the 1850s had arrived in Hawaii. Within a mere 35 years, the *Cavendish* had circumnavigated the globe. Crucially, or perhaps unfortunately, it was overlooked by Carl Frank and his contemporaries as the commercial banana trade in the Americas began its ascent towards global primacy through cultivation of the *Gros Michel*.

Around the time that Paxton's banana was making its way across the Pacific, it also, inevitably, arrived at that strategic cross-road of the global maritime super-highway – the Canary Islands. The precise details surrounding the arrival of the *Cavendish* in the Canary Islands are unclear, but it is generally believed that it formed part of the general diaspora from Great Britain in the 1850s that coincided with Paxton's efforts at Chatsworth. It is also plausible, however, that the *Cavendish* arrived from Mauritius directly. The islands had been intimately involved in trade with the Americas since its very start and were emerging as a convenient location in the development of tropical cash crops to supply the European markets. They also remained a stop-over for most ships travelling to the Americas, West Africa and Asia via the South Cape (the Suez Canal, by-passing the South Cape, would not be operational until 1870). As the commercial supply of *Gros Michel* developed in the Caribbean, a parallel trade in *Cavendish* was established from plantations in the Canary Islands. The development of deep-water ports at Puerto de la Luz and Santa Cruz de Tenerife in 1884 sparked the beginning of a commercial trade to supply the UK, and later, European markets. The Canarians were delighted to see the banana trade develop, after experiencing two centuries of boom-and-bust development, first in wine, then sugar, cochineal and, finally, tobacco. The economic strain by then had taken its toll, leading to mass emigration to the Caribbean in the late 1800s, in particular to Puerto Rico, Cuba and Venezuela. Akin to the nascent trade in the Americas, the Canary *Cavendish* began as a marginal trade as shippers explored opportunities to fill their vessel's hold on the return leg of its voyage. One company, the Liverpool-based Elder Dempster & Company regularly ran vessels to West Africa and the Canaries were a frequent stop-over along the journey. Their initial interest in the deep-water harbour developments in 1884 was the establishment of a coal bunker facility. Steam-ships at the time were fuelled by coal. Coal is a voluminous fuel and bunkering of the material in Tenerife reduced the area in the hold needed to store fuel for the voyage. This meant more space for cargo, and more profit per voyage. Soon after opening an office in Tenerife, the head of Elder Dempster, Alfred Jones, recognized the potential of a fully committed banana import business to complement the coal bunker facility and began to acquire land to grow the *Cavendish* at a commercial scale. Within a few years, the trade was exporting more than 130 million bananas per annum from the Canaries, almost exclusively to the UK. By the turn of the century, this had increased to over 170 million (Davies 1990). The volume of bananas exported from the Canaries to Europe was a small fraction of the trade that was developing in the Americas. Despite the advantage of distance, the islands ultimately lacked sufficient land and labour needed to expand production, unlike tropical America. Ultimately, Elder Dempster needed to seek partnerships to expand the trade across the UK and Europe and this meant finding a source in the Americas. Elder & Fyffes, partially owned by the United Fruit Company, was the product of this necessity, but the development of the Canary *Cavendish* would, unbeknownst at the time, become the standard for the modern banana business. Within 50 years, the *Gros Michel* would be devastated by Panama Disease and the *Cavendish*, born Chinese, adopted as British, and reared as Spanish would ultimately become American … and Icelandic.

For many reasons, both celebrated and maligned, the banana illustrates how much tropical nations have affected, and been affected by, their northern neighbours. From its humble origins in the kitchen gardens of tropical Indomalaya to its position as a globally traded commodity, the odd journey of the modern plantation banana illustrates the profound influence the tropics have on our planet, and the complex, often uneasy, sometimes merciless, biological, economic and social exchange that has characterized relationships between tropical and extra-tropical nations for more than five centuries. The Icelandic banana – perhaps a folly; as much to show how we can re-define and control the natural order of life as it was to establish on the edge of the Arctic a viable source of a tropical food that by the time had become commonplace throughout the extra-tropics. The spirits of Paxton and his patron, the Duke of Devonshire, were alive and well in Hveragerdi, continuing to play their part in accelerating the pace of what has evolved into a worldwide melee of shifting biological, economic and social geographies that began with Iberian naval adventurism and the great agricultural exchange that rapidly followed.

Notes

1 Franz Boos worked with Jean-Nicolas Ceré, director of the Jardin du Roi at Pamplemousses, shortly after Baudin's voyages in the 1780s. The Jardin du Roi is the world's first and oldest tropical botanical garden and is still in operation in Port Louis, re-named the Sir Seewoosagur Ramgoolam Botanic Garden after the first prime minister of independent Mauritius.
2 Baudin's banana was originally named *Martinique*, and only *Gros Michel* after another Frenchman, Jean Francois Pouyat, transported it to his estate at St. Andrew, Jamaica in 1836. It is most likely that the bulk of *Gros Michel* bananas that spread across the Caribbean and Central America originated from this stock.
3 Some of the best include Koeppel (2008), Davies (1990), Jenkins (2000) and Wiley (2008).
4 Estimate based on 300,000 to 5,000,000 bunches with 100 bananas per bunch.
5 United Fruit Company, *Annual Report to the Stockholders* (Boston: Various years, 1950 to 1960).
6 The machinations of the UFC, more than any other, led to decades of political strife and war in these countries. See Striffler and Moberg (2003).
7 This is, in part, speculation on my behalf. Actual events that led to the *Cavendish* arriving in Mauritius are not recorded, only that it arrived in 1826 from China. It is very likely, however, that it arrived from Canton (Guangdong) Province. This is the historical centre of banana cultivation in China and the largest producer today. Given that climate suitable for bananas is limited in China, alternatives would include neighbouring provinces of Guangxi or Fujian. Telfair would have known of Baudin's collecting foray in Canton that brought the *Gros Michel* to Mauritius and this makes it the most probable source of the *Cavendish* as well.
8 *Musa cavendishii* was described on page 51 of Paxton's *Magazine of Botany and Register of Flowering Plants*, Volume III, 1836. The *Cavendish* is now considered a variety – *Musa acuminata* (AAA Group) "Cavendish" – rather than a species.
9 The Palm House is the largest survivor of the great Victorian greenhouses. The Great Stove at Chatsworth fell into disrepair during the First World War and was demolished in 1920. Fifteen years later, The Crystal Palace at Hyde Park met its end in a ravaging fire that consumed the massive structure within hours on a night of high winds.

5 The great crop swap

While living in Guyana for much of the 1990s, we would travel at least once a month to Georgetown, the capital, from our house located some 200 kilometres inland. Guyana, a small, former British colony is wedged between Venezuela and Brazil on the north-east coast of South America, along with its equally small neighbours – Suriname, a former Dutch colony, and Guyane, a French overseas territory and home to the European Space Agency's rocket launch-site at Kourou. The interior of Guyana (and the surrounding region as a whole, the Guiana Shield) is one of the most forested, and least populated, on the planet (Hammond 2005).

These trips would take us along a road more pot-holed than paved, dusted and muddied than filled with tropical greenery, speckled with the objects and activities characteristic of unregulated, wild frontiers in the throes of haphazard development. In fact, this was the only road to the deep interior – buzzing with fuel tankers and heavy machinery heading to gold and timber concessions. Large 4×4s over-filled with passengers and goods travelling between the coast and the southern border with Brazil pass by swathes of cleared forest, planted with cassava and pineapple. Periodically we would stop along the way at road-side huts peddling drinks and meals of bush-meat, fish and rice. Trucks with spot-lights would arrive at dusk, looking for animals emerging from the forest to graze or dry off after a heavy rain. The verge was spotted with wreckage, small crosses and plastic flowers – monuments to the disproportionate loss of life along an extremely dangerous, unpatrolled and poorly maintained road.

The monthly trips were a necessity. Our residence at a timber concession was poorly equipped to provide food for us, let alone the 400–500 staff working and living at the site. The white sandy soils, one legacy of the region's ancient geological past, barely nurture the stress-tolerant chilli pepper, let alone crops that have more elevated nutrient requirements. What was available, appropriately referred to as "ration", re-defined basic living. When it reached the township store on pay-day, there was a stampede of people much less fortunate than those of us with relatively well-maintained vehicles and the convenience of choosing when we travelled to Georgetown. The food available in Georgetown, by contrast, consisted of the fresh fruit and vegetables that one would expect from a tropical market: a wide variety of bananas, plantains, mangoes, papaya, sour-sop, passion-fruit ... and apples.

Why precisely one would encounter a tray of Red or Golden Delicious, Gala or Fuji apples amongst the splendour of tropical produce in the open-air markets of Guyana is ultimately the same question asked of bananas growing in Iceland. People in Guyana do not need to eat apples to live a perfectly healthy life. Apples in the tropics may not have reached the status of bananas in the extra-tropics, but every month we spent going to the market in Georgetown, apples were available in some variety and abundance. The apple is also strangely ever-present in other tropical markets I have visited, including: the magnificent Adolpho Lisboa municipal market in Manaus, capital of the Brazilian Amazon, the covered Juan Sabines market in central Tuxtla Gutierrez, capital of the southernmost Mexican state of Chiapas, and the markets of Port Louis, on the island of Mauritius near Madagascar, among many others.

The apple is the world's second most consumed fruit – as a group, more than 75 billion tonnes were produced in 2012. Hundreds of commercial apple varieties exist, the product of continuously refining attributes towards their intended purpose – eating, cooking, juicing or distilling – over millennia. The bulk of the global crop, over 55 per cent, is produced in China, on the edge of the apple's ancestral home. Current genetic studies indicate that the wild-type ancestors of the modern apple were most likely growing across a belt of land sweeping from the Caspian Sea in the west to the westernmost province of China, Xinjiang, in the east. The fruit has been grown in Europe, however, for more than 2,000 years, since the armies of Alexander the Great are reputed to have returned to Europe with apples from their expeditions into present-day Tajikistan and the frontiers of northern India. Poland, Italy and France are major producers, and exporters, of the apple today. Yet, the apple that we see in the markets of many tropical countries does not come from China, nor Europe. Apples are being imported from producers in the New World, such as Chile, the USA and the temperate southern states of Paraní and Santa Catarina in Brazil. Akin to the history of the banana and plantain, the apple has become commercialized in the New World and is consumed throughout the region in a manner that is strangely insensitive to local abundance or the variety of native fruit available. Apples are certainly not consumed by people in the tropics with the same daily ritual as bananas are eaten in North America or Europe, but apples have become inculcated into local urban cultures of tropical cities in a very similar way. Often they are purchased as a luxuriant alternative to the staple tropical fruits – almost aspirational, a legacy of former colonial connections that continue to influence the social norm in most tropical countries. Adoption of far-flung foods seems to be an inherent part of our shared strategy to diversify our resource base. While the apple was becoming part of the tropical market in cities, the banana and plantain had long been fused to the traditional mix of crops managed by indigenous people living in remote rural areas. In some Amazonian tribes, such as the Awa people living along the frontier between lowland Colombia and Ecuador, the plantain has become the staple crop in their "traditional" system of agriculture (Orejuela 1992), despite available evidence suggesting that this plant arrived in the upper Amazon region no earlier than the late sixteenth century. Virtually every indigenous community in the Amazon recognizes the plantain or banana as a component of its traditional agricultural system and has done so since at least the nineteenth century (Brochado 1977).

Apples and bananas are not alone in their march across regions where biology, climate and geography tell us they do not naturally belong. The displacement of tropical food and fibre production from their ancestral homes is now more rule than exception. A great, irreversible crop swap has taken place between even the most disparate climatic and cultural regions on the planet. The effect is so pervasive that most people do not know where their everyday food items originate. What we do know is that the early European explorers had a penchant for identifying potential money-spinners and then quickly transplanting them to "safe zones" where competing empires could not "steal" their newly acquired blockbuster. As was the case with the banana and plantain, this story played out over and over for centuries as (mainly) European nations utilized every means at their disposal to ensure that their hard-fought monopolies over the global supply of any given food or fibre source were protected while exerting an equal amount of effort in finding a way around their competitors' own monopolies.

The greatest tales of European discovery and conquest are actually stories about the acquisition of new agricultural products and the effort to form a monopoly over their supply. Christopher Columbus was one of the first players, in the first act of this very long play. He was tasked with finding a new route to Asia's spice and textile markets first accessed along the Silk Road by the armies of Alexander the Great nearly 2,000 years previously. By the fifteenth century, however, the Silk Road was no longer safe to European traders. The Ottoman Empire wrested control of the western reach of the road running from Serbia to eastern Turkey under Mehmed II and he did not favour the unfettered trade between Western Europe and Asia. This, foremost among several factors, including maritime discoveries made by the Portuguese, spurred the Spanish Crown to invest in Columbus' voyages. Columbus was not a strident empiricist in the shape of von Humboldt, Bates, Darwin or other great nineteenth-century naturalists. His accounts of those first voyages to the Americas are filled with vague and repetitive descriptions of the lands and people he encountered for the very reason that he had little contextual knowledge in which to place what he saw and wished to dispel the obvious controversy attached to what he and his crew had found. It must have been painfully clear that this was not the Indian subcontinent, nor Japan or any part of Asia. He did, however, take a keen interest in anything that he believed might have material value, particularly in the absence of any significant discoveries of gold and silver that clearly dominated his initial interactions with indigenous societies. This was, after all, the very reason he had been commissioned to find a new route to Asia – to establish an alternative to the insecure overland approach and the secretive and difficult voyage around the South African Cape dominating the trade with the world's great Asian spice producers, and the considerable wealth it afforded at the time. His description of food and materials that the indigenous people were utilizing received more attention than any other topic and represents the earliest historical account of many products that today have an annual net market value in the hundreds of billions of dollars.

For most modern entrepreneurs, the process of product discovery, production and growth into a network of worldwide distribution and sale should be seen as a great success story. It is fair to say, however, that the citizens of the neotropics have not, on balance, reaped the benefit of efforts to globalize the natural crop diversity born from their indigenous predecessors. This diversity includes plants central to global food security, such as maize (corn), cassava and potatoes and everyday fruit, vegetables, nuts and spices, such as chilli and green peppers, pineapple, cocoa, cashews, vanilla and tobacco. Out of the 22 most important food and fibre crops to emerge from the neotropics since the arrival of Columbus, only eight have been predominantly sourced from the Americas since the 1960s (Figure 5.1).

Today, the global production of neotropical crops is a 160-billion-dollar market. Apart from maize, more than half of the production of the most important indigenous, neotropical crops has been transferred to "Old World" regions of Africa, Asia and Europe – cassava, cashews and cocoa to Africa, potatoes to Europe, peppers, tomatoes, groundnuts, papaya and vanilla, among others, predominantly to Asia. Among globally traded products, only avocados and Brazil nuts remain firmly planted in their "home" ranges in Central and South America. This massive transfer of plant genetic resources has resulted in an average yearly loss of around 110 to 120 billion dollars, three-quarters of the global net value of crops originating from the neotropics. Over the last half-century, the cumulative value attached to the transfer of production value from the region to areas outside the Americas is conservatively estimated at 6.2 trillion dollars (FAO 2016b).[1] If the contribution of the United States to corn, potato, tomato and other vegetable production is taken out, the loss grows substantively greater. This conservative estimate would put the annual loss to the Latin American and Caribbean (LAC) economies at around 1.2 per cent of regional GDP growth (all countries in the Americas, excluding USA and Canada). In comparison, the total value of gold and silver estimated to

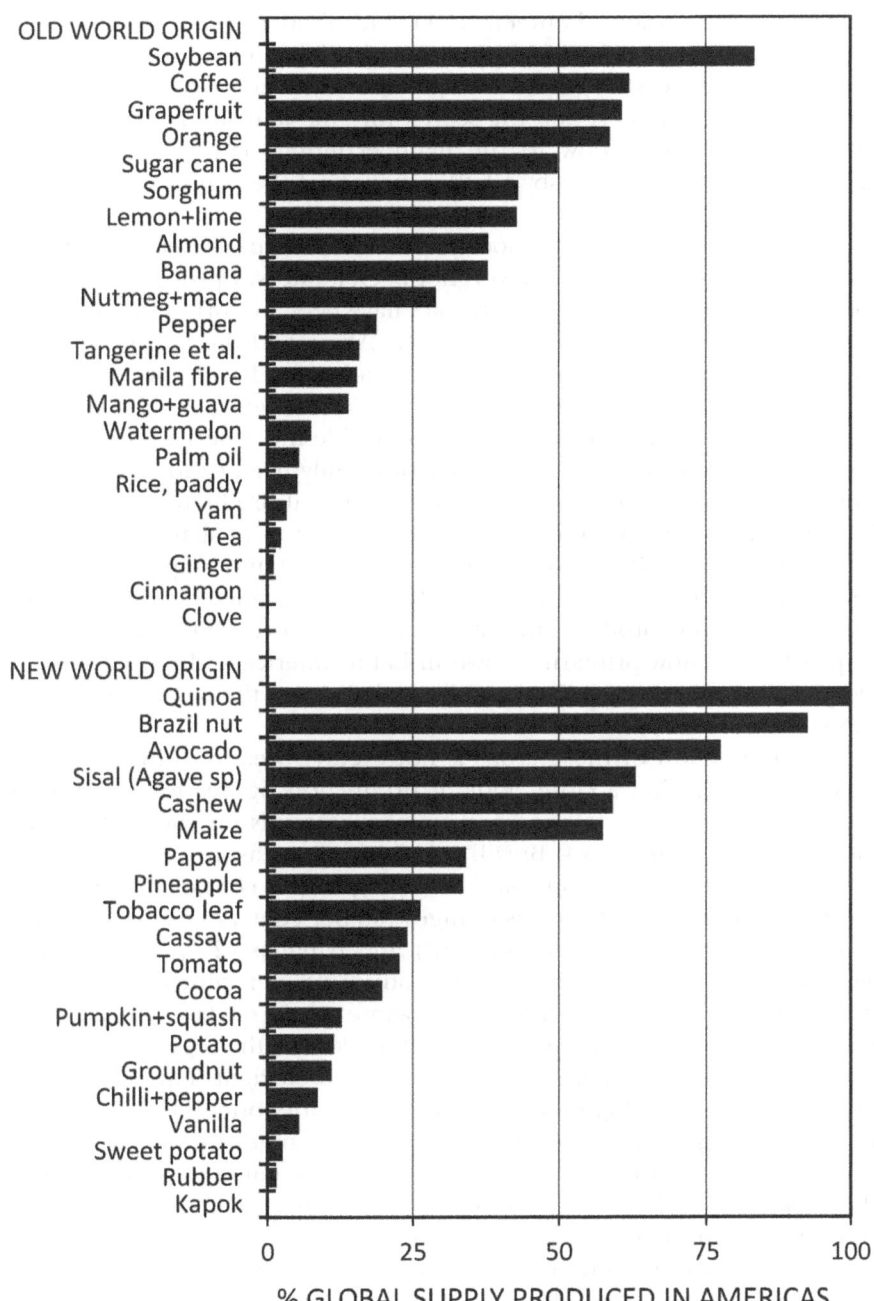

OLD WORLD ORIGIN
Soybean
Coffee
Grapefruit
Orange
Sugar cane
Sorghum
Lemon+lime
Almond
Banana
Nutmeg+mace
Pepper
Tangerine et al.
Manila fibre
Mango+guava
Watermelon
Palm oil
Rice, paddy
Yam
Tea
Ginger
Cinnamon
Clove

NEW WORLD ORIGIN
Quinoa
Brazil nut
Avocado
Sisal (Agave sp)
Cashew
Maize
Papaya
Pineapple
Tobacco leaf
Cassava
Tomato
Cocoa
Pumpkin+squash
Potato
Groundnut
Chilli+pepper
Vanilla
Sweet potato
Rubber
Kapok

0   25   50   75   100

% GLOBAL SUPPLY PRODUCED IN AMERICAS

Figure 5.1 Crosby's Columbian Exchange and its agricultural legacy in the New World Tropics.

have been extracted from the New World through colonial Spanish and Portuguese possessions between 1500 and 1800 can be conservatively estimated to be around US$36 billion (at 2014–2015 gold market prices) (Select Committee 1810; Cowan, unpublished[2]). The story of New World encounter and conquest and the search for gold and silver ultimately has proven to be a side-show to the main economic consequence of the Columbian Exchange (Crosby 1972) – the loss of the region's natural biological patrimony.

The loss of indigenous food plants to industrial-scale production outside the Americas has, however, not been a one-way process. Dozens of food plants from outside the neotropics have been imported to the region from their native agro-ecosystems in Asia, Africa and the Pacific. Many of these are prominent components of the global food and fibre trade, but very few have emerged as world-class market-makers as is the case for numerous neotropical crops now grown outside the Americas. Among the most important crops from the Old World, only five of those now grown in the Americas account for the majority of global production (see Figure 5.1). Subtract production in the USA from this account again and soybean (56 per cent of production in Americas from USA) and grapefruit (73 per cent) quickly drop out. This leaves coffee, oranges and sugar cane as the three global food products that began as crops in the Old World (sub-) tropics but are now primarily grown in Latin America and the Caribbean. Despite the region's pivotal role in the global rise of the banana, the lion's share of production remains in Asia.

The production of these crops is not evenly spread within the region either. Again, taking a closer look at production figures for these three world-leading crops, it is clear one country accounts for the bulk of production in these imports – Brazil. It is by some measure the largest producer of coffee in the neotropics, averaging 47 per cent of total regional production each year, as well as oranges (68 per cent), sugar cane (57 per cent) and even soybean (58 per cent) and bananas (27 per cent). This agricultural titan even manages to produce 18 per cent of the apples grown in LAC, exclusively from its temperate southern states – a source of many apples that show up in the tropical markets in the region. Only a few other countries have managed to capture a double-digit share of regional production in any adopted crop – Colombia with coffee (17.5 per cent), Ecuador through bananas (20 per cent) and oranges in Mexico (14 per cent). Thirty-two of the 41 countries in LAC have failed to produce more than 5 per cent of regional output in any major crop transplant.[3] When this is placed in the much larger pool of global producers, the production of many Old World crops in the Americas hardly merits mention.

It is not solely a story of loss and disappointment to tropical America though. There are several instances where LAC countries have successfully adopted transplanted crop varieties, beyond coffee, sugar cane and citrus fruit, to challenge on the global stage. Guatemala has become the world's largest producer of cardamom (a plant in the ginger family, Zingiber-aceae), a high-value product originating in southern India that attracts prices in the spice market only exceeded by vanilla and saffron. Grenada, a long-standing producer of nutmeg and mace,[4] is the regional leader in the production of this Indonesian native, although production on the island only accounts for a quarter of a global supply that otherwise remains firmly tied to its ancestral homeland. Spices are not major commodities, such as sugar or maize, and there global markets, while broad, are not deep and they generate only a fraction of the revenue of staple food and fibre plants. Consequently, while the process of exchanging agricultural products has led to some economic cross-fertilization, the import of Old World crop production to the New World has not balanced the opposing flow of native products out of the region. Analyzing estimates of net crop value made by the United Nations' Food and Agriculture Organization (FAO) shows that global production of crops native to the Old World attained an average net value of around 370 billion dollars annually between 1961 and 2014. But of this total, only 86 billion, or less than a quarter, was attached to production in the Americas. The cumulative value of Old World crop production over this same period, adjusted to 2014–2015 constant international dollars, amounted to just over 13 trillion. Production from the Americas accounted for less than a fifth of this total. If we then pool the cumulative value of 44 major crops originating from both Old and New World sources, the Americas have accrued a net direct benefit of close to 4.6 trillion dollars since 1961. Compare this with the nearly 17 trillion in value that has been generated by Africa, Asia, Europe and Oceania over the same period and we can see how poorly the Americas have benefited from the exchange of agricultural legacies. Of the benefit that has accrued, the lion's share has gone overwhelmingly to Brazil, the world's largest producer of coffee, sugar cane and oranges and second largest of soya bean, after the United States.

Now imagine an alternative history where indigenous crops only expanded along the frontiers of their ancestral ranges, and the same, modern-day supply of food and fibre largely coincided with the areas they occupied during prehistory. In other words, Old World crops remained in Asia, Africa and Europe and New World crops remained in the Americas. The net value of producing cocoa, potatoes, maize and cassava, among others, within the neotropics, would be just over 41 per cent of the total global net value of crops. The actual fraction of global agricultural value captured by the Americas as a result of the great swap is closer to 21 per cent, a loss of a fifth of global crop value during the process of crop exchange. Again, remove the significant weight of the US agricultural behemoth and the gap grows greater – to around a quarter. To most businesses, losing this sort of market share would be a death knell. It is easy enough to argue that neotropical countries have had the same opportunities to make the most of the great swap as those in Asia, Africa and

Europe, but failed to walk through the unlocked doors. We could argue that Brazil and a few other countries have done just that. So why the great disparity in economic benefits?

There are a number of factors that led crops expatriated from the Americas to perform better than those imported from (mainly) Asia and Africa. A number of these: land area, labour force size, distance to market, are simple constraints on the factors of production needed to deliver the quantity necessary to create a "cash-crop" system. Others are related to the ecology of the crop itself, such as the impact of natural enemies – pests and disease – on production in mono-cultured plantations. Certainly the land area available for tropical crop production is greater in Asia, Africa and Australia than in the neotropics and the population of these regions in the sixteenth century was nearly 20 times the estimated population of the Americas (Mcevedy and Jones 1978). This difference has steadily diminished to just over five-fold by 2015, but still provides a compelling advantage when it comes to farm labour costs. In addition, many of the mono-cultured crop systems, such as rice paddy, were well established in the palaeo-tropics by the time Columbus arrived in the New World. This would have made the introduction of more intensive mono-culturing of New World crops much easier than in the Americas where much of the traditional food production was carried out through complex, rotational cropping systems that combined many different food plants. This approach is thought to have improved the nutrient status of the impover-ished tropical soils and subdued losses to natural pests and disease but at a cost to the quantity of each crop produced. Transplanting crops to geo-graphically isolated regions left many of the endemic pests and diseases behind, allowing the exotic cash crop to flourish. But it is not unreason-able to say that all of these "capacity" factors were simply co-variants of an underlying root cause attached to the history of social and economic com-petition among European nations driven by an inability to adequately resource their own economic and demographic expansion. The rapid rise of market competition, driven by a vibrant merchant class, catapulted exploration and discovery towards business and enterprise and unlocked a new source of globalized growth.

Such is the case with the history of the great crop swap. Wild tropical plants that had been transformed into crops through local selective breed-ing over millennia by indigenous agronomists were within centuries growing across every region of the planet except the poles, even Iceland. The last biological exchange of this magnitude prior to the great crop swap is believed to have taken place between North and South America sometime in the Miocene, around 20 million years ago, when shifting tec-tonic plates gave birth to the Caribbean Sea and an arc of volcanoes began to form the Central American isthmus. The isthmus as we know it today was formed over a span of 15–18 million years through a number of phases balancing sea-level and tectonic changes (Coates and Obando 1996).

Once complete, the connection between the two continents led to a biological exchange of countless animals and plants (Stehli and Webb 1985). Prior to this event, we have to look back to the early Jurassic, the Age of the Reptiles, around 200 million years ago, when the ancient continent of Pangaea rifted into two continents – Laurasia (North America–Europe–Asia) to the north and Gondwana (South America–Africa–Australia–Antarctica–India) to the south, splitting flora that had been distributed contiguously across a single, super-continent for approximately 125 million years (Pitman *et al.* 1993). In the context of these natural events, the spread of crops by humans over the past 500 years is a staggering transformation of our planet's biogeography. It has also created a new, synthetic *trophogeography* (from Greek τροφῆς, trophé = nourishment) where traditional crop systems and dietary staples do not reflect natural provenance, land suitability and custom, but historic interactions between (sometimes) unwitting indigenous suppliers and foreign traders and market-makers that understood the broader commercial potential of new crop varieties. The new trophogeography has led to a transformation in the history of human survival, allowing populations to prosper in regions where providence was previously scarce.

The history of this transformation was born from a series of individual actions, as we have glimpsed in the story of the banana – de Berlanga carrying the banana from the Canaries to Santo Domingo; Baudin shipping the *Gros Michel* variety to Martinique from his expeditions to China; the Duke of Devonshire's and Paxton's desire to be the horticultural celebrities of Victorian Britain; and the formation and growth of the global banana trade companies from small-scale side ventures of a few mercurial shipping agents servicing ports in Latin America, the Caribbean and West Africa. The events leading to the modern distribution of many other tropical crops are surprisingly similar to that of the banana and plantain. From the 1500s through to the mid-1800s, a very real game of "purloin the plant" was played out across the planet, involving armed forces, diplomats, companies and horticulturalists with the aim of making or breaking germplasm monopolies. The stakes at the time were astronomical since their countries and economies were in the ascendant and individual fortunes were often achieved directly or indirectly through connections to overseas resources. The swap of cultigens was not restricted to resources from the Americas alone, although Columbus' discovery, and the Columbian Exchange that followed, was undeniably a critical catalyst (Crosby 1972). Vast plantations of crops native to Asia and Africa were interchanged throughout the tropics. Some plants exhibiting greater environmental tolerances were exported to the farmlands of southern Europe, the United States, South Africa, and Australia to supply and grow demand for (sub)tropical products in these regions. The machinations of the period did more than any other to change the trophogeography of our planet forever.

A perfect example is coffee. Despite some lingering misconceptions of a Latin American (Colombia) or Indonesian (Java) origin, coffee (*Coffea arabica* L.) is an Ethiopian plant, grown for millennia in the highlands south-west of Addis Ababa and the region surrounding the mouth of the Red Sea (Hattox 1985). The infusion that we recognize today as black coffee emerged as a part of formal culture in the South Arabian region no later than the middle of the fifteenth century, around the time that the young apprentice Columbus was beginning to learn his way about trading ships and ports in Genoa. Coffee plants were shipped across the Red Sea to the Arabian Peninsula, where the first centre of commercial-scale production was developed in the Yemeni mountains near the coastal port city of Mocha. Arab traders saw an expanding market across the more affluent parts of the region in Egypt, Syria, Persia and Turkey and perfected a way of altering the chemical composition of the coffee seed while extending its storage life through roasting (Clarence-Smith and Topik 2003). Coffee houses, able to store sufficient stock to satisfy their growing number of daily consumers, sprang up in the Azhar precincts of Cairo and within 50 years were present in most cities across the region. Coffee became an integral part of everyday social life in the Middle East, including Mecca, where thousands of Muslim pilgrims encountered the "Arab wine" for the first time, returning home with tales of a revolutionary drink. Italian traders also encountered this new social phenomenon during their visits across the Mediterranean and carried roasted beans back to their home ports. The first registered coffeehouse opened in Venice in 1615 (Hattox 1985).

Coffee was celebrated wherever it spread, but not by everyone. Clerics and religious scholars – both Muslim and Christian – argued for decades over the morality of such a "stimulating" infusion (Crawford 1852). Makeshift coffeehouses were frequently opened only to be promptly closed following the issuance of religious edicts. The case for rejection or acceptance was muddied by the fact that the infusion was not known when the Biblical and Koranic texts were written – they could provide no clear, direct guidance to those given the power and responsibility to determine acceptable social behaviour during the Middle Ages. By the middle of the seventeenth century, coffee had been deemed a moral beverage by the Church, perhaps in part in response to its meteoric rise in popularity among the masses, and the way was cleared for one of the greatest global expansions of any tropical cash crop. A coffee-mania had taken hold in the Western world, but indulgence did not arrive cheaply.

At the time, prices paid for the few, rare spices and condiments that had already reached Europe, such as cinnamon, nutmeg, clove and black pepper, were astronomical and only afforded by the wealthiest social classes. In sixteenth-century London, an ounce of spice would typically cost the equivalent of 50 eggs, two pints of ale or a dozen rabbits and consume nearly a full day's wages of a common labourer. Faced with a broader, regional diet composed largely of meat, fish, bread, cheese, cabbage and gruel, accompanied by mustards, honey and garlic, the prospects of financial gain through the supply of new, revolutionary foods to a growing Europe were nonetheless staggering. The risk of losing these prospects to competing crown interests, however, was even greater. This may seem like hyperbole, but it is important to remember that a vast number of products that we commonly consider as mainstays of "traditional" European cuisine originated as imports from the tropics during this period: Italian espresso (coffee) and pomodero sauces (tomato), Greek tzatziki (cucumber) and French cassoulet (haricot beans), to name just a few. The impact of their arrival was of such great effect that they quickly became part of the cooking vernacular of Europe and North America: apple with cinnamon, olive with tomato, milk with tea. The cuisines of modern Europe may be traditional by the measure of time, but not by provenance.

A desperate scramble to retain supply monopolies emerged with virtually every instance of a new food crop discovery. During the sixteenth century, Arab traders would boil or roast green coffee beans to prevent viable seeds falling into European hands keen to cultivate their own supply (Hattox 1985). The Spanish experimented and expanded their interests in cocoa production, preparation and refinement for nearly a century before introducing the product to the broader European market, out of fear that their discovery would be purloined by other financially minded interests. The Portuguese travelled in secrecy along their new route to the Spice Islands to prevent other nations breaking their stranglehold on the supply of exotic, tropical spices from China, India and Indonesia. Insider information gleaned from Dutch sailors on board Portuguese ships ultimately revealed the maritime secret.

The seventeenth and eighteenth centuries witnessed both the full voracity and the ultimate futility of these various efforts to make and break the emerging global monopolies in tropical crops. The Enlightenment had brought a new-found entrepreneurial and engineering spirit to European society and this quickly translated into significant technical improvements in the way that tropical crops were transported and processed. The time-savings attached to these improvements meant that larger quantities could be supplied in their final form to markets across the continent. Economies of scale ensued, driving prices lower and stimulating demand from the ever-expanding working classes that could now afford small quantities of tropical products that were previously rare and beyond their economic reach. The problem now was to overcome supply constraints.

European spice supplies under the Portuguese were unreliable and erratic. They were sourced from semi-wild stands through third-party producers and predicting the size of the spice consignment arriving aboard a Portuguese trading vessel was more gambling guesswork than mercantile trade. This deeply dented market confidence and hindered the expansion of demand across Europe. Having successfully dethroned the Portuguese, the Dutch East India Company, or VOC (Vereenigde Oost-Indische Compagnie), set about subduing indigenous resistance in the Spice Islands, most often violently, in an effort to not only re-create the Portuguese network of trading outposts, but also expand and establish the means to increase spice production and improve the reliability of supply through mono-cultured plantations. The native population was brutally eliminated from the Banda Islands and strict controls were placed on visitation. Trees outside the newly formed plantations were chopped down. Live-specimen smugglers were dealt with swiftly and severely. The seeds of the nutmeg tree, the source of the spice, were boiled or washed in lime to ensure that no viable germplasm left the islands. These extreme measures undertaken to corner the market in nutmeg, mace, and clove supplied from the Banda and Malaku Islands commenced an age of immense wealth creation for the Dutch, principally through the activities of the VOC. They would replicate these successes with increased cinnamon and pepper production in Ceylon (Sri Lanka), on the islands of Java and Sumatra and along the Malabar Coast of India. At its height, the expanding spice empire reached from Taiwan and southern Japan in the north to Papua New Guinea in the east and Sri Lanka and southern India in the west. The network was administratively controlled from Batavia, later re-named Jakarta, and connected to Europe through intermediate ports and colonies established in Mauritius and Cape Town, South Africa. The VOC's activities would transform tropical Asia from a region ruled by a network of autonomous, trading kingdoms to a centralized constellation of colonial outposts coordinated efficiently towards the production of tropical goods for transport to European markets.

When we consider the long-term prospects for sustainable development in the lower latitudes it is worth remembering that the overwhelming majority of tropical nations have been, at the very best, passive participants in the spectacular period of wealth generation rooted in the European Age of Discovery and reaching its full velocity during the Industrial Revolution. The scramble to acquire germplasm for the most profitable crops immutably re-shaped tropical regions, their landscapes and the people living in them in ways that continue to fundamentally dictate how wealth is created and where it flows across the globe. If anything, what the great crop swap and the scramble to make and break crop monopolies tells us is that where lasting wealth flowed, lasting economic development settled. Tropical colonies played a particularly important role in the planetary flow of this wealth through the re-direction of their bioproductive capacity towards the cultivation of cash crops, particularly the crystalline, salt-like sweetener known as cane sugar.

Notes

1 All values are in constant international dollars. This smooths out some of the distortions in value created by using market exchange rates that can be volatile, often due to speculative forces.
2 Dr. Richard Cowen, Chapter 8 – New World Silver. (mygeologypage.ucdavis. edu/cowen).
3 Data from FAO (2016b).
4 Mace is the red, fleshy coat, or aril, that surrounds the seed of the nutmeg, *Myristica fragrans*.

6 Sweet salt and the tropical trade

Sweet salt

Having landed again after a long flight, we jump in the 4x4 and head to our lodgings in the heart of the city. Leaving the airport entrance, we are soon driving along a green barrier of sugar cane. Rows and rows move by, punctuated by the occasional plume of smoke from fires set to prepare the cane for hand-cutting and help rid the fields of unwanted weeds, snakes and rats. Flakes of black ash fall on my lap. As our vehicle sweeps by the large stretches of mono-cultured fields, I watch rows of cane-cutters in their straw hats deftly swipe their razor-sharp machetes into the wall of vegetation. They make it look easy, but I had a go once, and can attest to the fact that it is not – there is a real skill in handling these ten-foot tall, densely packed stalks in a timely, systematic manner without cutting yourself or another in the process. The remnants of very old mills, some dating to the seventeenth century, inter-mix with their modern replacements and the enormous complex of fields. Trucks laden with cut lengths of the yellow-greenish cane pass by. The operations have an old-hand feel about them – a process that has been carried out over and over across generations – and it is clear that cane has been grown and processed here for a very long time. The odd point about this trip is that it mirrors two journeys in two countries, not one. Two places where sugar cane did not naturally exist, even in its wild form. Two countries separated by 13,000 kilometres, an ocean and 27 degrees of latitude: Guyana and Mauritius. Why these journeys are so similar has everything to do with the shared history of these two nations and the rise of sugar as the first truly global agricultural commodity. The story of sugar is a well-told tale. Legions of authors and historians have examined its evolution from a curious medicinal to an epidemic component of our modern diet, its role in slavery, empire and shaping trade and the global economy. But it also reflects the tectonic shift that occurred in the aftermath of the Iberian discoveries and the under-appreciated role that tropical plants and productivity have fundamentally played in the political, social and environmental changes shaping our world, both within and outside the tropics. The nexus of greatest impact ultimately shadowed the early Iberian routes linking Europe with West Africa and the Americas. But it did not start out this way.

Crystalline sugar – the granulated, sucralose sugar as we commonly know it – was not part of the European diet until the mid-1500s. Fruit, such as apples, pears, plums, grapes and figs, constituted the "desert" portion of a meal. These fruit, unlike other imported species that are also commonly consumed in Europe today, such as peaches, pineapples and bananas, contain mainly fructose and only small quantities of sucrose. The diets of most people in the Middle Ages included virtually no sucrose beyond the small amounts contained in their native fruit and vegetables. Honey also contains very little sucrose, consisting primarily of fructose and glucose. It was used in the kitchens across Europe prior to the arrival of cane sugar and evidence suggests that people in Europe were eating honey as early as the seventh millennium BC.

Literate Europeans knew of sugar from their Greek and Roman predecessors, through accounts of Alexander the Great, and the growing trade linkages between their empires and the various polities that controlled the Indian subcontinent during the first century (Reed 1866). Indians had for some time extracted a raw sugary paste from cane and Arab traders visiting the region began to return with small quantities in the form of pastilles. They had traded with producers in India for centuries and farmers in the Nile and the Tigris–Euphrates valleys eventually recognized the value of this tropical plant. Small quantities of sugar from the region were making their way via these connections to the massive spice markets of the Middle East. Invariably, the cultural cauldron that was the Mediterranean during the Middle Ages brought sugar from the East to the West. And when Islam sprouted during the fifth century, it sparked a cultural and intellectual revolution that saw the Muslim caliphate slowly expand outward from the Arabian Peninsula. When Muslims moved across the Mediterranean during the early eighth century, they brought cane sugar with them to islands such as Sicily and Cyprus and to southern Spain. Through the long period of the Crusades that followed, soldiers and traders across northern Europe began to return home from the Levant with samples. Crusaders had been exposed first-hand to the region's dynamic trade with Asia, introducing them to the many tropical products that flowed from the subcontinent. At the same time, the *emirates, caliphates* and *taifas* that came to control southern Spain and Portugal for nearly 500 years had exposed Iberians to new agricultural practices, such as irrigation, and the methods used to cultivate sugar cane. Europeans began to take greater notice of this super-sweet substance that could be shipped and stored as easily as honey, but produced in quantities that were simply impossible to achieve through apiculture.

Ultimately, it was medieval Venice – at the time a city-state of traders much like modern Singapore – that first introduced a steady, albeit small, supply of sugar to Europe through trade relations established with Arab proprietors since the eighth century. Sugar cane had been spread by Arabs across the Mediterranean and the plant must have been known to Europeans living in the region. The principal use of sugar at this time was not as a sweetener and large quantities of crystallized sugar would not have been making their way to distant markets. It was marketed for its supposed medicinal properties – to combat a plethora of ailments, induce vigour and combat fever – in a way that was typical of pharmacological practice at the time. It was also priced like many modern-day pharmaceuticals and thus remained a minor dietary source of sugar in Europe throughout much of the Middle Ages. As sugar evolved from a medieval medicinal paste to a crystalline, culinary ingredient during the twelfth century, Venice began stockpiling, and eventually refining, the substance. Monopolization of trade was the predominant business strategy throughout Europe at the time and Venetians were particularly adept at the practice, having grown a sophisticated ensemble of relationships with the Byzantine Empire and Muslim caliphates. Venice extended its control over Cyprus, Crete and Sicily, as well as much of the Greek Peloponnese – all important gateways to the Middle East – while establishing rights with suppliers in the region to exclusively trade salt. Salt, as today, was used mainly for food preservation and the northern European fisheries required large amounts. It was also one of the few flavourings available to medieval kitchens. Salt was also surprisingly expensive. During the thirteenth and fourteenth century, salt in England cost around six pence per hundredweight (=50.8 kg/112 lb) on average (Clark 2004). In today's prices, that amounts to about £600–£700 or around US$800–US$1,000 per metric tonne. This compares to modern wholesale prices (2016) for refined, iodized table salt at around £30–£50 per tonne.

The Venetians retained their exclusive rights to trade between the Middle East and Europe throughout most of the Middle Ages, and Venice's role as Europe's trade and financial centre swelled. However, in the middle of the fifteenth century, through a series of strategic mis-steps that saw it lose a number of altercations with an Ottoman Empire growing in strength (Inalcik 1989), the traditional Middle Eastern trade routes connecting Europe with South and East Asia were cut off. The decline of the Venetian pre-eminence in the import of tropical goods to Europe arrived swiftly as trade with the Mahometan regions became untenable and Venice found itself embroiled in a frontline defence of Christendom against the expanding Ottoman Empire. Venice's capacity to control the trade in salt, sugar and spices began to wane soon thereafter.

Perhaps the most important consequence of the change in Venice's trade relations with the Ottomans was the impact this had on trade in other parts of Europe. It became a major driving force behind the Iberian efforts to find an alternative means of supplying tropical commodities to the growing markets both at home and in northern Europe. Venice embroiled itself in a protracted fight against the Ottomans just as the

Iberians were completing the *Reconquista* and seeking alternative routes to trade with Asia. The Venetian trade with the Middle East had the knock-on effect of stimulating growth across regions further north. Merchant fleets in Western Europe slowly grew as they began to profit from the year-on-year increases in trade between the north and south. They also started to wield greater influence over the burgeoning market for exchange of commodities within the region. This trade, from the eastern Baltic Sea through the English Channel and to the Iberian ports, had succoured the growth of a new class of merchants that did not strictly rely on maintaining profitable relations with merchants in Cairo, Baghdad and Constantinople. The precipitous loss of trade with the Middle East had also cut off the flow of grain to southern Europe, creating a large void in supply and an opportunity for the nascent economies of the north (Braudel 1984). Northern Europe was soon delivering grain to the south in exchange for sugar, salt, as well as indigo and silk for the cloth industry and other products that could not be supplied regionally. As this trade grew, so too did their merchant fleets, port facilities and the skills and knowledge needed to profitably manage them. Antwerp, Bruges and Amsterdam grew to become important centres of intra-European trade. This north–south exchange would leave them ideally positioned to take advantage of the breathtaking opportunities arising in the aftermath of the Iberian navigational successes. By the time the Portuguese navigator Vasco de Gama sailed directly to India around the Cape in 1498 and returned with samples of crystalline sugar and a wealth of spices, Venetian control over the nascent trade in cane sugar had all but ended. Venice would miss this next great commodity wave and the city called *La Serenissima*, once home to the world's greatest merchant fleet, would, as a centre of trade, fall into relative obscurity and, eventually, disrepair.

When sugar did finally arrive en masse to the Western European markets, it first arrived in the holds of Portuguese and Spanish *carracks* and *urcas*, not Venetian galleys. Fleets of these ships could carry hundreds, sometimes thousands of metric tonnes of bulk commodities in a single voyage – a far larger quantity than any other merchant ship class operating at the time. They were initially built to service the relatively low-value, high-volume shipment of grain that dominated the early trade between northern and southern European markets (Braudel 1984). Prior to de Gama's successful voyage to India, the Portuguese – with financial capital from the Genoese – had developed some of the first sugar cane plantations on the islands of Madeira, the Azores, Cape Verde and Sao Tomé. Spanish effort concentrated on the Canary Islands (Kamen 2003). Despite some early failures, sugar as a cash crop began to take hold and when shipments of raw sugar arrived in marketable quantities from these fledgling enterprises, it was trans-shipped to northern and central Europe. Shipments from India via the new trade route around the South African Cape also began to arrive at Portuguese ports and the volume of sugar available to European markets increased rapidly. But these two sources, and in fact all other sources, would pale in comparison to the volume of sugar arriving from the newly minted European colonies in the Americas. Reign over the plantation in the New World tropics at the dawn of the eighteenth century was held without doubt by one product and one product alone, king cane.

Cane sugar was in part king due to its price and the profit to be had in its import, first to Europe and later to colonial North America. Like salt, sugar was unbelievably expensive up until the explosive expansion of maritime trade that occurred in the wake of Columbus' and de Gama's navigational discoveries. Again, various prices in medieval England compiled by Professor Gregory Clark provide some insight. Over the fifteenth century, sugar prices in England averaged around 16 pence per pound (Clark 2004). Inflating this price from this period to current value tells us that a pound of sugar would have cost a trader somewhere between £40 and £47 in today's prices. Scaling up to trade volumes, these figures equate to between £88,000 and £103,000 per metric tonne. Current prices (2016) for a tonne of highly refined sugar FOB run between £220 and £400, an extraordinary decline in real terms.[1] Paying over 400 times more than we do today, it is not surprising that the average person at the time considered cane sugar an unbelievable luxury. "Sweet salt" was also several magnitudes more expensive than the real salt and this made sugar a highly attractive alternative to merchants that did not already have a firm foothold in the existing trade in salt, grain and wool. As more and more supply reached the European market though, prices inevitably declined. By the middle of the seventeenth century, we know that a pound of sugar was sold at retail in London at 18 pence – between £9 and £10 per pound in today's terms – only a quarter of the price commanded 200 years prior (Robertson 1885). The decline began with the successful establishment of plantations in Madeira and the other Atlantic island possessions and then the flow of raw sugar from India. Once plantations in the New World tropics began to expand and mature into the eighteenth century, prices dropped even further. Prices by modern standards remained extravagant for such a common ingredient, but had dropped to between £6 and £7 per pound in current currency, a decline to one-sixth of the price prior to de Gama and Columbus.

Sugar prices declined further in real terms, halving to around £4 per pound in England by the mid-eighteenth century. By this time, the granulated form we commonly recognize as brown sugar had become a widely available staple. Bakers were increasingly infusing their products with sugar and concocting new ones. Importantly, many early pioneers in retail confection realized that sugar dramatically extended the shelf-life of their products, allowing producers and retailers the freedom to build inventory and avoid losses to spoilage in a way that was simply impossible for other readily consumed foods at the time. Long storage life equally advantaged the trade in other tropical products, such coffee, cocoa, tea and spices that could be transported and stockpiled with ease. Entire artisanal industries in silverware, porcelain and metalwork were expanding to supply new kitchenalia for the everyday presentation, storage and handling of sugar and other tropical food products as they became standard components of the household pantry. The confectionary industry and artisanal craftsmanship became a likeable expression of the domestic wealth being generated through the trade in sugar. But sugar, much like bananas, built wealth and privilege principally for a few at a heavy cost to many, particularly across the Atlantic.

Not all sweetness and light

From its first entry into the New World, sugar was destined to bring fortune to some and misery to others. The Spanish had sent cane with Columbus and the mendicant monks that would follow him shortly thereafter. These religious orders, along with military officers and wealthy, well-connected families, established some of the first sugar estates on Santo Domingo (Hispaniola). A decade or two later, estates expanded onto neighbouring Jamaica. These plantations were granted under *encomienda*, a word derived from the Spanish meaning "to entrust". It enshrined through royal commission the rights of *encomenderos* to receive tribute, but not land, in Spanish America. This tribute, taken from the indigenous peoples was similar in many ways to the medieval feudal system. Serfs worked fields to provide tribute to a feudal lord, and in this manner, *encomienda* granted a holder the licence to use indigenous labour in lieu of tribute. It was specifically modelled, however, on a system of tribute paid by peasants and Muslims to governors after the Spanish *Reconquista*. Few villages had the means, or likely the understanding, to provide tribute in the form that would satisfy their new Spanish masters and most were invariably put to work instead either in mining or in agriculture. In return, *encomenderos* were entrusted with protecting their wards against violence and providing for their housing, education and spiritual guidance. To do this, the scattered villages and families were coalesced into a series of larger settlements, called *reducciones* that made it easier to keep track of labour availability and the administration of work. It was this particular form of legal grant – *encomienda* establishing the boundaries and *reducciones* organizing the labour within these boundaries – that led to the early development of sugar plantations in the Caribbean. The primary purpose of this system of course was to facilitate the Spanish Crown's control over its new territories and catalyze economic activity, remembering that *encomienda* did not grant a land title to its holder.

Encomienda would not continue for long before the Iberians realized that the indigenous inhabitants were ill-suited to the arduous task of manual harvesting and planting of sugar cane at the scales desired for export back to Europe. Excessive manual labour devastated the local communities unfortunate enough to have found themselves caught within the net broadcast by an *encomienda*. Fray Bartolomé de las Casas spelled out the price being exacted from the indigenous peoples in his report to the Spanish Crown in 1542 and published a decade later as *Brevísima Relación de la Destrucción de las Indias*. His judicious plea to stop the abuses had its intended effect – to change the existing system that facilitated abuse and murder of the native inhabitants on the islands. *Reducciones* were seen as one of the primary policies facilitating this abuse. Reform was swift, arriving almost directly from de las Casas report reaching Madrid, and took the form of a new policy referred to as *repartimiento* (*mita* in Spain's viceroyalty of Peru). This was a new system of forced labour that eliminated many of the other obligations mandated under *encomienda*. It eliminated *reducciones*, allowed people greater freedom to live in their traditional settings and offered pay for work but under the constraint that a certain percentage of a community's working-age population were obligated by law to make themselves available for activities that advanced the interests of the Spanish grant-holder. Those most responsible for the abuses – clergy and government officials – saw their grants rescinded.

Viewed in the context of our modern system of justice and human rights, *repartimiento* hardly classifies as substantive reform. But to the Spanish conquerors of the Americas, it nullified everything they had established. Little had changed on the ground after the reforms. Compliance was slow and in some instances, provoked rebellion among *encomenderos* that wished to continue with a system that gave them unhindered rights to indigenous labour. Francisco Pizarro, conqueror of the Inca, was one of the powerful lords that refused to accept the new laws, ultimately deposing the crown-appointed Viceroy of Peru responsible for implementing the reform. Pizarro would later be arrested and executed for his resistance and thereafter there was a gradual acceptance of the changes. But like *encomienda*, the legal right to claim labour participation was hopelessly abused by many of the right-holders. A quarter of the population was conscripted under the legal precept of *repartimiento* and these labourers were meant to work no more than two or three days per week towards the grant-holder's interests. In reality, far more than a quarter of the indigenous labour force was coerced into working on a near-permanent basis, particularly on the plantations, where the cycle of planting and harvesting was unending. Confronted with the sweeping change thrust upon them over the half-century following Columbus, it is not surprising that many of the surviving indigenous inhabitants quickly made use of the reforms to escape these punishing demands and fled communities that remained under the jurisdiction of the Spanish. Forced into the most remote parts of the islands, they in effect became refugees in their own land.

In some parts of sixteenth-century Spanish America, *encomienda* and *repartimiento* would not have been entirely unfamiliar to the native inhabitants. It seems certain that many subordinate communities laboured as a form of paid tribute within the larger, more dominant social hierarchies that existed prior to Spanish arrival, such as the Aztec and Incan empires. In these instances, the Spaniards simply supplanted the top tier of the complex social hierarchies in existence at the time – a feat that the British would repeat a century and a half later in India (James 1997). But indigenous societies across the Americas were both diverse and complex and many communities were not accustomed to these organized forms of labour-based payment. Most early accounts – including those by Columbus and his crew – make it fairly clear that this was not the case with the indigenous Taíno communities inhabiting the island of Hispaniola. Taíno society was organized into chiefdoms, each headed by a *cacique*. As far as we know, *caciques* drew their power from their ability to grow their communities and often received tribute. But tribute was paid in material form by their community members at their own discretion – in feathers, food or handicraft – rather than through persistent demands for payment in a medium not of their choosing. Competition, conflict, violence, fear – these would have existed between tribal groups, particularly with the fierce Carib tribes inhabiting the eastern Caribbean, but the organizational structures that were created and the abuses that were carried out by the Spanish to subordinate labour and tribute within communities would have been anathema to them. *Repartimiento* would also expand to their possessions in Asia, such as the Philippines, and remain in place across some Spanish colonies as late as the early nineteenth century, but the laws governing tribute payment would largely disappear across much of the lowland neotropics in tandem with a dwindling indigenous labour pool. Perversely, the redundancy of the work laws governing indigenous inhabitants ushered in a new, inequitable reality of plantations worked by African slaves. The introduction of African slaves into the New World largely stemmed from the failure of *repartimiento*, particularly in the Caribbean and eastern Brazil, where lowland plantation agriculture reached its economic apogee. In the highland Andean and Sierra Madre regions of the Spanish viceroyalties of Peru and Mexico, deep-shaft mining prevailed as the main economic activity and indigenous populations continued to form the bulk of the labour force under various derivations of the *mita* system.

The ultimate success of sugar in the neotropics, driven by an almost insatiable demand in Europe, was almost entirely due to slave labour. The most profitable period of sugar production in the Americas, from the seventeenth through to the end of the nineteenth centuries, would coincide with the conversion of plantation agriculture from one that utilized the tribute-based model of the Spanish to one that exploited slave labour as pioneered by the Iberians but swiftly adopted by the Dutch, British and French. This switch relieved the chronic labour shortage. Though this shortage was almost certainly the initial driver in adopting slave labour, prejudice too became a powerful factor in discriminating roles and perpetuating the practice after this model proved to be profitable. The population of enslaved Africans grew and the social norms of a relatively small, isolated class of plutocratic plantation and mill owners began to view European labour in anything but supervisory roles as largely unacceptable. The development of a *plantocracy* – a system where European plantation owners were the ultimate authority over the lives of their workers – ensued. The Portuguese were the first to introduce cash-cropping in the Americas based purely on slave labour from Africa. Their *donatarios*, or captaincies, on the coast of Brazil had encountered the same difficulties as the Spanish *encomienda* in sustaining an adequate supply of labour from the pool of indigenous inhabitants. Like the Taíno, many Amerindians of the Brazilian coastal regions simply faded into the interior to escape the demands of the outsiders, often to find themselves beset by the Old World diseases that followed first contact. But the lack of adequate labour in the Americas only explains the development of the trans-Atlantic slave trade, it does not testify to its origins. A tradition of slavery across southern Europe, the Middle East and North Africa, provides a strong social context to its development, particularly by the Portuguese.

Under Arab rule between the eighth and thirteenth centuries, Moorish Portugal had grown accustomed to slavery, a commonly accepted practice throughout most of the Middle East and North Africa at the time. The caliphates raided and enslaved thousands of Christian Iberians. After the *Reconquista*, the new Christian rulers tolerated a continuation of slavery, particularly of *moriscos* – mainly Berbers from North Africa that had been forced to convert to Christianity. Morisco slaves are believed to have accounted for nearly 10 per cent of the population of Lisbon after the *Reconquista* and the practice of slavery, among many traditions built up over centuries of Muslim rule, indelibly structured Portuguese social norms after independence. Portuguese navigators discovered Cabo Verde, an archipelago of ten islands off the coast of present-day Senegal, in 1456. Uninhabited at the time, it soon became a critical conduit in the purchase of slaves from the Oyo, Dahomey and Ashanti in present-day Benin, Ghana, Senegal and Mali. Many were prizes taken through inter-tribal war or raiding, others were simply kidnapped or sold off to pay debts, but the advent of Portuguese interest in slave-trading along the West African coast inevitably stimulated the taking of slaves as a medium of exchange. This demand created a large, regional market where one may not have existed traditionally, but tribal rulers in Africa were keen to profit from this development. Ethnically distinct, they may have seen neighbouring tribes as threats, not allies. Raiding and slaving would have reduced these threats while allowing them to acquire European trade goods. At first, indigenous Africans thrown into slavery were put to work in Portuguese mines in West Africa. The Ashanti gold delivered through Ghanaian coastal ports became a particularly important source of finance for the trans-Atlantic slave trade. Slave labour was also sent from the African coast into the expanding sugar and coffee plantations of Portuguese Atlantic island territories, particularly Madeira and Sao Tome.

The activities of the Portuguese along the fifteenth-century West African coast acted as a proving ground for the use and transport of slaves. This meant that they had tested the profitability of slave-based cash-cropping and had already established solid trade ties with African tribal leaders and Arab slave-traders on the mainland when the Americas were discovered. These ties would invariably advantage the Portuguese, both in supplying their own plantations and those of the Spanish. When difficulties arose in the struggling Brazilian *donatarios*, the decision to send slaves must have been made with the lucid understanding of how to jump-start sugar production born from their previous experience. The first large group of slaves is thought to have arrived in 1538 at Pernambuco, a few decades after the somewhat hapless Portuguese colonization (Burns 1970). Approximately 100,000 people would be transported from Africa to the sugar plantations in Brazil over the remainder of the sixteenth century (Eltis 1999). A similar number would be shipped to Santo Domingo (Haiti) and Jamaica by the Spanish over this time. But it was Lisbon, not Seville and Cadiz, that had supplanted Venice as Europe's main sugar port due to its West African operations. The dawn of Brazilian sugar saw 2,600 tonnes transported to Lisbon but by 1600 this had increased seven-fold to a little above 17,500 (Simonsen 1944). Adjusted for inflation, the value of sugar arriving from Brazil at the time amounted to nearly £370,000,000 of purchasing power in today's terms,[2] an unbelievably large sum that would have made the Crown and the merchant classes extremely wealthy. Much of this sugar was quickly re-exported to other parts of Europe, particularly the Spanish Netherlands (Ebert 2008), and it was regional trade between Lisbon and northern Europe that soon revealed to the latter a growing void between the inflow of Portuguese sugar from their plantations and the demand across Europe.

Northern Europeans were invariably alarmed by the vast monetary wealth accruing to the Iberians through their trade in sugar and other tropical commodities. From the tropics, new and old, they were delivering nutmeg, coffee, indigo, cinnamon, black pepper, brazilwood, logwood, tobacco, cocoa, cotton and, of course, sugar. Gold and silver flowed from mines in the Americas and Africa, the latter proving a boon to the growing Iberian trade in Chinese tea, silk, spices and porcelain. Moreover, by the end of 1583 and after a relatively short-lived military conflict lasting a few years, the Spanish Crown under Philip II had subsumed Portugal and all of its overseas possessions into its empire through a familial right of succession. This consolidation of control over the tropical sugar and spice trade should have further strengthened the Spanish hand. Yet, the ultimate effect of this simple act – conjoining the Spanish and Portuguese empires – was to loosen the Iberian stranglehold on the tropics and its infinite supply of commodities and to free the tethered ambitions of the

northern European nations viewing enviously the lucrative trade in tropical goods. But the reasons for the Iberian decline and the sweeping
changes that would indelibly alter the fate of the tropics had as much to
do with sixteenth-century religion as they did with business and politics.

Non Sufficit Orbis

Philip II reigned as head of the first global superpower. A power that
extended its geographic reach far beyond that achieved by any of his European and Asian predecessors. After the war over Portuguese succession
was won, he controlled territories that extended across the Mediterranean
to North Africa and the Atlantic islands. Across the Atlantic, the Iberian
Union placed the entire New World under his singular control. Philip's
sugar plantations and trade interests extended to the eponymous island
archipelago in Asia, as well as the Spice Islands, Malacca and other trade
ports in present-day Indonesia, China, Japan, India and Sri Lanka. Virtually all of the Iberian "discoveries" made at the dawn of the sixteenth
century were now singularly consolidated under his throne. A possession
that circled the planet, prompting the king to commission coinage in the
year of the union stamped with the likeness of a large Spanish stallion
leaping over a diminutive Earth and emblazoned with a new royal motto,
Non Sufficit Orbis – The World Is Not Enough.

The sheer scope of resources available to Spain during Philip's reign
led to a rapid expansion of the tropical trade and Spanish presence in the
tropics increased markedly (Parker 1998). Franciscans in New Spain numbered fewer than 100 in 1536, but this had tripled by 1569, with a further
thousand taking up temporary residence across the Spanish Main to assist
in the re-education and conversion of *indigenas* (Rubial Garcia 1996).
Armed resistance from indigenous societies of course applied some
counter-force to Spain's expansionary plans, but disease played a significant role in quelling any real threat and the introduction of African
slaves provided the Iberians with the field labour that the indigenous
Americans could not or would not provide. None of these or the many
other challenges facing Philip's enterprise abroad proved insurmountable
as the accelerating growth of the plantations and mines fed his global
ambitions. Ironically, it was one of the smaller territories that he possessed
much closer to home – the Netherlands – that ultimately led to the end of
Philippine Spain's supremacy at the close of the sixteenth century and the
beginning of the end of Iberia's monopoly over the profitable tropical
cash-crop trade.

By modern norms, we might easily describe Philip II as a fanatic, based
on his actions and beliefs, but, within the context of his upbringing and the
age he lived in, this characterization would be misleading. His total com-
mitment to Catholicism led to an upsurge in deportations and forced conversions of "heretics" throughout the Spanish empire, but condemnation

of heretical beliefs had remained commonplace since the early days of the Inquisition under his grand-parents and was not unique to his reign. When the Protestant Reformation emerged a half-century after the publication of Martin Luther's treatises in 1517, it rapidly spread to the Spanish Netherlands. Catalyzed by reform leaders such as John Calvin and Dutch-born Menno Simons, Protestantism sat squarely in the spiritual mind-set of most northern lowlanders on the eve of a war that would indelibly alter Philip's imperial legacy, the sugar trade and the geopolitical face of the tropics.

At the time, the *Paises Bajos* consisted of 17 territories, the largest being Holland and Overijssel in the north and Flanders and Brabant in the south. Philip had inherited his Dutch possessions through the Hapsburg ancestry of his grandfather, Philip I, the son of the Holy Roman Emperor Maximilian I. When Philip I wed Queen Joanna of Spain – the eldest daughter of Ferdinand and Isabella – the Hapsburg possessions, including those of the Netherlands, were united under one crown with those of Spain. The northern provinces of the Netherlands were deeply affected by the Protestant proposition and quickly became a stronghold of reformational teachings. Philip's father, Charles V, had responded forcefully to the emergence of the Reformation early in his reign by deploying to the Netherlands many of the methods used in the Spanish Inquisition. But in later years, prior to his abdication in 1555, Charles' enthusiasm for prosecuting heresy appears to have dissipated, leading to a fragile, but relatively peaceful, détente with the Dutch reformists. This cessation of Charles' fight against heresy may have been driven by his leanings towards Humanism and its teachings, or an acceptance of the *realpolitik* in the Netherlands – that Protestantism was not going away anytime soon. But Charles also was a Burgundian who spoke Flemish and had married into the Spanish monarchy – his cultural loyalties would have been split, unlike those of his son Philip. Philip was a Catholic Spaniard through and through, an important factor in how he would ultimately address the religious schism taking place in the north.

Initially, Philip appeared to carry on his father's benign, if not conciliatory, stance towards the Dutch reformists. He appointed members of the Dutch and Flemish nobility as provincial magistrates, or *stadthouders*, offering the opportunity for a measure of home-rule through an assembly called the Estates General. Philip's appointment of his illegitimate half-sister Margaret, Duchess of Florence and Parma, as the Governor of the Netherlands in 1559 appeared to further support an intent to pursue a policy towards peace and stability in the region. But Margaret was being installed as the amiable face of the Crown, while another – Antoine Perrenot de Granvelle – was quietly being appointed as her prime minister and head of Council of State. It was in de Granvelle that the king invested the power to make decisions, although Dutch and Flemish *stadthouders* were also members of the Council.

As primary councillor to the regent of the Spanish Netherlands, de Granvelle quickly seized upon Philip's desire to rid the world of heresy. He organized the appointment of 3 new archbishops and 14 new bishops to the Spanish Netherlands. This move created an entirely new regional extension of the Catholic Church that vested considerable power with the clergy. They were appointed directly by Philip and would sit on the Estates General, effectively diluting any power the *stadthouders* would have in the new assembly. Moreover, any significant administrative action in the territory had to go through de Granvelle, and he sought approval from Philip rather than any of the *stadthouders*. This policy of direct rule thoroughly undermined the early promise of native-born participation in governing the provinces. The Estates General, diluted by the royal-appointed bishops, only existed on paper and the frustration of the betrayal quickly alienated many, particularly those from the northern Dutch provinces. Unrest in the region began to grow over the appointments, the fear of religious persecution that arrived with them, and the betrayal of the Estates General.

Although this pressure would ultimately see de Granvelle resign, Philip would commence a new campaign against the Dutch Protestants. He urged Jesuits to travel to the Netherlands and instructed that the edicts of the Catholic counter-reformation, embodied in the instructions formulated at the Council of Trent in 1563, should be put into effect. Again, the Calvinist leanings of many Dutch citizens catalyzed another protest that received no concrete response. By the summer of 1566, the populations of most major towns and cities became highly agitated. Without warning, Church cathedrals, offices and monasteries were attacked by mobs of common people. Religious property was destroyed and purloined; buildings were burned in what came to be known as the *Beeldenstorm*, or the Iconoclastic Fury. Despite Margaret's efforts to calm the situation, Philip had reached the end of his fuse and implemented a policy of military rule through his most trusted – perhaps most ruthless – military officer, Fernando Álvarez de Toledo, the Duke of Alba.

Upon his arrival with 9,000 loyal troops in August of 1567, Alba, nicknamed the Iron Duke, established the Council of Troubles. This aimed to identify the main agitators and ensure that they were summarily punished. Over 10,000 were called before the Council, including most of the *stadthouders* and minor noblemen, merchants and Protestant leaders. Thousands of these would be executed, imprisoned or banished, including several of the most prominent former *stadthouders*, in a reign of terror that would last five long years. William of Orange declined to attend and was declared an outlaw – his land and property confiscated and a bounty placed on his head. Alba forced the remaining Estates General to approve a 10 per cent sales tax, with the aim being to pay for his military intervention. The "Tenth Penny" was decidedly unpopular, adding further fuel to a simmering fire of rebellion. Despite the swelling discontent that arose as a consequence of the Spanish policies, William had remained loyal to the

Spanish Crown and the Church and measured every opportunity to avoid conflict and instability by adamantly refusing to side with the fanatical Calvinists, but equally not making overt gestures to condemn them. The arrival of the Iron Duke changed this and civil disobedience transformed into armed conflict. War had begun and William of Orange suddenly became the de facto leader of an organized, armed resistance to the Spanish Crown.

William, his brothers and their Protestant allies launched various attempts to gain control over the provinces in the subsequent years. None would prove successful and by 1571, one of his brothers was dead and his campaign short of money and men. Little progress had been made in securing territory until the unexpected capture of the Dutch port of Brielle in the spring of 1572. It was captured by a group of Dutch privateers, commissioned by William to disrupt the trade with the Netherlands that was critical in keeping Spain solvent. They found it undefended and the port quickly surrendered. This success, along with several that followed immediately in its aftermath, sparked an uprising that spread through the major ports and towns in the northern provinces of Holland, Friesland and Utrecht. In 1573, the same privateers won an important battle against a squadron of Spanish naval vessels off the coast near Hoorn in the Zuiderzee. The north had been secured through the most unexpected means – naval power. Philip recalled Alba, offered a full pardon to the rebels and cancelled the Tenth Penny tax, aiming to find common ground with the nobles and merchants thrown over to the other side under a common cause. But on the freedom to practice Protestantism, Philip refused to compromise.

Various political attempts to unite the 17 provinces ensued, but all failed. The South remained firmly aristocratic and Catholic, whereas the North had transformed into something more egalitarian and religiously tolerant. Many Calvinists were extremists and sought the end of other religious denominations, but most Dutch were simply keen to see the return of social stability and economic growth. This divide was too great to bridge and in January of 1579, the southernmost provinces of Artois, Hainaut and the Walloon Flanders signed the Treaty of Arras, committing themselves to the Catholic cause and the Spanish Crown, eliminating any prospect of a fully self-governing Netherlands. As a result, later that year the northern provinces of Holland, Zeeland and Utrecht along with four others signed the Union of Utrecht. In July of 1581, the northern Union officially released themselves from their oath of allegiance to Philip in the Act of Abjuration and declared their full independence from Spain, the Hapsburgs and the Catholic Church. The middle regions of Brabant and Flanders, mildly sympathetic to the South but unwilling to sign the Treaty of Arras, became the staging ground for the continuation of Philip's war under his new Governor, Alexander Farnese, the Duke of Parma. Ghent, Brussels, Bruges and Maastricht and the surrounding regions fell back under the fold of the Spanish Netherlands. Late in 1584, the great port of Antwerp, a repeated victim of Spanish looting and repression, was taken. The North desperately sought patronage from the surrounding monarchies – first the Duke of Anjou, but his attempt to re-take Antwerp in 1584 failed and he returned home only to pass away a year later. William of Orange too died that year after numerous attempts on his life – shot by an assassin seeking the bounty that had been placed on his head by Philip. Elizabethan England came to the aid of the North in 1585, sending several thousand troops under the agreement struck through the Treaty of Nonsuch. But this lasted only a few years. In 1588, the Union declared itself a confederated republic with no single ruling monarchy. Rule would come through the Estates General.

Philip deeply resented Elizabeth's meddling in the long-running saga of the Spanish Netherlands, let alone the usurpation of the throne from his deceased wife, Mary, and the elimination of the other Catholic line through the execution of her cousin, Mary Stuart, Queen of Scots, in 1587. The declaration of the Protestant Dutch Republic was yet another blow to his fight against heresy and he felt that Protestant England was complicit in the losses he suffered in the Netherlands. In the summer of 1588, Philip launched his *Grande y Felicisima Armada*, a fleet of 130 war ships gathered with the singular aim of invading England and deposing Elizabeth. Parma's troops stationed in the Spanish Netherlands gathered as the invading force, but naval action by England and the Dutch Republic harried the Spanish fleet and they failed to meet with Parma's army. The planned invasion scuttled, the Spanish fleet set sail for home, but was met by a large storm in the Atlantic with devastating losses. A third of the fleet did not make it back to Spain. Philip would continue half-heartedly in his attempts, but having lost the Dutch Republic and England to the Protestants, Spain was all but bankrupt, the Portuguese aristocracy was seeking to free itself from its overseer and news was returning from Philip's tropical territories of naval engagements, privateering and ransackings of his most prized possessions. Elizabeth's most accomplished naval commander, Sir Francis Drake, had looted Santo Domingo, captured Cartagena (de Indias) and occupied St. Augustine in Florida (Goldsmid 1890). Sir Thomas Cavendish attacked the Philippine port of Arevalo in the spring of 1587 and captured the treasure galleon *Santa Ana* along the Pacific coast of Mexico six months later before completing a circumnavigation of the globe and returning to England shortly after the defeat of the Spanish Armada. Philip II died in 1598 having failed to curtail the spread of Protestantism and severely weakened the Spanish empire in the process.

The rebellion in the Spanish Netherlands – a small conflict when measured in light of subsequent European conflicts – was nonetheless a critical turning point in the history of the cash-crop trade and defining who would ultimately shape the tropics. The Dutch were now able to pursue their own economic interests free of Spanish taxation and control and the English had come to realize their ability to carve out a piece of the tropics for themselves. Both now fully recognized the power of sea superiority and would direct the bulk of their military spending going forward towards the making of world-class navies. In December 1600, Elizabeth I would provide a royal charter to the formation of a global trading consortium – the East India Company. Two years later, the Dutch would commission the Dutch East India Company. Philip II, through his blinkered focus on defending religious dogma, had unwittingly unleashed two potent challengers to the century-long Iberian supremacy in the tropics and, perhaps most importantly, control of the lucrative sugar trade.

Mare clausum, mare liberum

The nascent Dutch Republic, having gained its tenuous independence from Spain, found itself in an economic predicament. Philip had acquired the Portuguese throne in 1580 and immediately cut off economic ties. The extremely profitable trade in tropical spices, sugar and salt was promptly moved south to Antwerp and the regional shipping of North Sea herring and Baltic grain was simply not enough to sustain the economy of a nation that had very little land area and few natural resources of its own. Moreover, the Spanish refused to acknowledge Dutch independence. Spain, Portugal, most of Italy and a large swathe of central Europe under Hapsburg control, as well as Catholic France, would now not openly trade with Holland. Bereft of the means to force a change in policy through military action against these overwhelming odds in Europe, the Dutch took the only option remaining to secure their economic survival – to take the fight against the Iberians to their tropical possessions and break their century-long monopoly on the spice and sugar trade.

Dutch privateers, raiding under the banner of the breakaway republic, had already captured Iberian carracks prior to Philip's death, and the value of their holdings clearly established the immense profitability of long-distance trade with the Indies. Along with intelligence gleaned from Dutch sailors that had sailed to Asia aboard Portuguese ships, the contents of the pirated vessels made it clear to the VOC and the Estates General that they had no other course but to turn to the tropics for economic salvation. This was easier said than done. The Dutch could not simply employ the VOC and their growing armada to plunder every Iberian trade ship they came across. They risked being seen by Europe as a breakaway "robber republic" devoid of the rule of law and contravening the existing etiquette that (loosely) governed relationships between states, even when they were at war. Being seen as pirates could hardly advance the Dutch desire to build a global empire based on legitimate trade and business. Regardless, the VOC needed to build up capital and the only means was to purloin that on offer through the steady stream of trade ships returning to Portugal from the Indies. The recently commissioned Dutch East India

Company began their campaign against the Portuguese in the Azores in 1602, capturing several trade ships laden with goods from Asia. Then in early 1603, a VOC ship captured the 1400-tonne *Santa Catarina* as it made its way through the Molucca Straits (Indonesia) on its return to Lisbon from China and Japan (Klooster 2016). The ship was filled with porcelain, silk and musk oil, and its contents would become the seed funding for a rapid expansion of the VOC. It would be the first in a series of seizures made by the Dutch in their fight against the Portuguese in Asia.

To dispel the notion that the prizes were produced through acts of piracy, a Dutch legal expert, Hugo Grotius, composed a thesis questioning the Church-promulgated division of the tropics between the Spanish and Portuguese based solely on rights of discovery. The Iberians had entered into a number of agreements in the wake of their navigational discoveries more than a hundred years earlier and these had established the boundaries, or spheres, of Spanish and Portuguese dominion in the tropics, both on land and at sea. The latter was based on the principle of *mare clausum*, or closed seas. Since the Iberians had sailed through all three major oceans by the time the three treaties had been signed, first in 1479 (Alcáçovas), then in 1494 (Tordesillas), and again in 1529 (Zaragoza), these agreements effectively prescribed exclusive navigation rights to the Iberians. All of the Atlantic south of the Canaries went to the Portuguese, including the Atlantic islands, the Gold Coast of West Africa and, importantly, the South Cape and sole oceanic portal to Asia. The Americas were given to the Spanish, less the Portuguese sugar plantations in easternmost Brazil, while the bulk of Asia and its trade established the reverse. The West was to go to the Spaniards, the East to the Portuguese. The powerful Vatican had already offered its blessing to the Iberians "owning the oceans", through a series of papal edicts that, along with other guidance, specified geographic boundaries to their influence based on "papal lines of demarcation". By arrogating to themselves ownership of the oceans, the Iberians were not only seeking to corner the global market in tropical commodities, but the globe itself.

The sweep of their claims seems incomprehensible, but none of the multi-lateral institutions, treaties or protocols that define global agreement today, existed then. At the time, there were also very few nation-states capable of or willing to undertake the circumnavigational exploits of the Iberians. The great societies of the New World were land-bound and lacking the technology for maritime travel. Most of the major tribal kingdoms of Africa did not utilize maritime travel beyond regional, coast-bound journeys. Only the Somalis proved to be masters of inter-continental maritime travel, sailing as far as China in pursuit of trade goods, and Somali ports were some of the most active in Africa prior to the Iberian discoveries. Arab, Chinese and Indian merchants were also very active in trade spanning the northern perimeter of the Indian Ocean, and the Portuguese invariably made use of this navigational expertise to find their way to the Moluccas (Indonesia). But the distances travelled by merchants from these advanced societies were relatively short compared to those taken by the Europeans. Among societies and people, only the Polynesians had mastered such great nautical distances in open waters. The unbelievable risks taken by the early Iberian explorers sailing into an oceanic void of such great measure cannot be overstated in this context. Their arrival in the New World was a consequence of the Ottoman trade embargo and the relative dearth of tradeable commodities available from continental Europe compared to those emanating from the tropics, China and Japan (Morris 2010).

But the spectre of inadequate bioproductivity at home also loomed as a driving force behind Europe's desire to take to the seas. Europe of course sits at extreme northern latitudes – the same lines that run across central Canada and Siberia – and the bulk of the region's bioproductivity was, at the turn of the sixteenth century, vested not in the land, but in the seas. If we examine the distribution of NPP on land and at sea using the same data coverage employed at a global scale in Chapter 3, it is apparent that nearly 80 per cent of NPP in northern Europe (50 to 70 degrees north) is attached to the surrounding seas and that this contribution declines southward to roughly 20 per cent in the Mediterranean region (Figure 6.1). The relative contributions of land and sea to overall bioproductivity converge on unity around 50 degrees but invert as one moves south or north of this inflection. Of course reconstructing NPP for Europe during

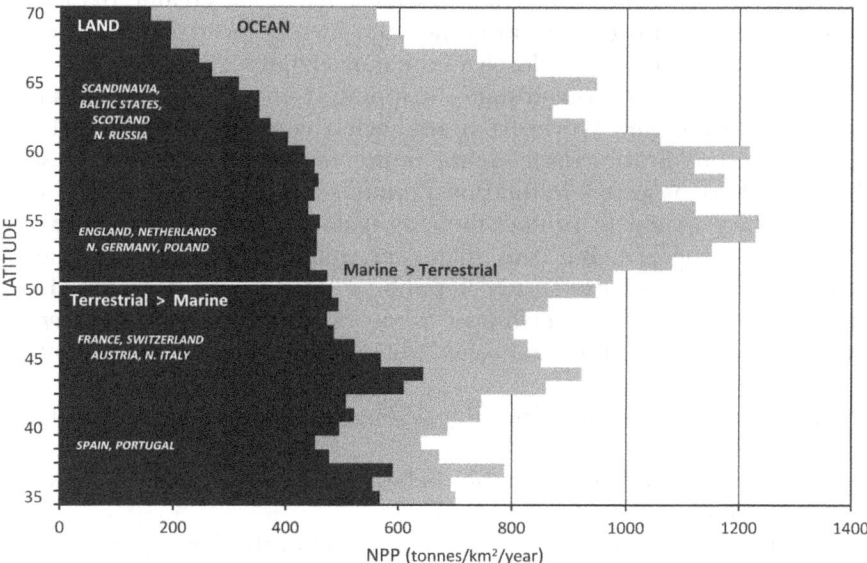

Figure 6.1 North–south changes in average terrestrial and marine Net Primary Productivity (NPP) across Europe and the latitude where they invert.

medieval times is difficult since detailed maps of land cover across the continent during the sixteenth to eighteenth centuries were patchy and imprecisely compiled without any common underlying methodology or purpose. Instrument-based records for European climate – the most extensive available for the planet – only extend back to the end of the seventeenth century and there are a plethora of limitations that question the accuracy of many techniques employed to reconstruct medieval climate (Jones and Bradley 1995). What we do know is that Europe at the time was experiencing sporadic periods of climate much cooler than any experienced today.

Although this period is often referred to as the "Little Ice Age", the moniker misleadingly suggests a continuous cooling that is not supported by the best available evidence. Records do suggest that the seventeenth century – a century defined by rapid European expansion into the Americas – was much cooler than present-day conditions, but the 1700s appear to rebound towards much warmer average temperatures (Jones and Bradley 1995). Cooler periods would have suppressed bioproductivity, particularly at the more northerly latitudes, by extending and deepening periods of winter weather. Modern land use practices, such as those making use of chemical fertilizers, would also work towards boosting bioproductivity levels detected in the twenty-first century, particularly in northern waters where ecological eutrophication has become a point of real concern. As a result of these logical, albeit crude, deductions, the difference between Europe and the tropics over the period spanning the sixteenth to eighteenth centuries was most likely far greater than it is today. I examine this critical difference and its role in driving Europeans towards globalization more closely in the next chapter.

The general lack of long-distance shipping from Asia and the absence of a claim to the New World from any society considered "advanced" by fifteenth-century Europeans made the notion of *mare clausum* seem reasonable. The Iberians were – in their own minds – first to travel the deep seas and they had a right to protect their navigational discoveries – a sort of intellectual property. But the major maritime nations of Europe – England, France and the Dutch Republic – vehemently disagreed. How can two nations claim ownership of the world's oceans? In *Freedom of the Seas*, Hugo Grotius (1609) established the grounds for their opposition:

1   Trade with East Indies does not belong to the Portuguese by title of occupation.
2   Trade with East Indies does not belong to the Portuguese by virtue of title based on the Papal Donation.
3   Trade with the East Indies does not belong to the Portuguese by title of prescription or custom [prior use].
4   The Portuguese prohibition of trade has no foundation in equity [unfair competition].

5 The Dutch must maintain their right to trade with the East Indies by peace, by treaty or by war.

Grotius argued that the Iberians had no claim to the open seas and an exclusive trade in the East Indies based on these four natural laws and that if they were unwilling to rescind their claim, or pay damages *in lieu* of foregone trade, then the Dutch Republic would be legally and morally entitled to overturn their monopoly through force (or the taking of "prizes"). The reality of seventeenth-century Holland was that they were boxed in by their former master and the only way to break out was through piracy and war. When the Iberians naturally balked at the Dutch claim, *Freedom of the Seas* offered the legal justification for the use of force that the VOC had already began to employ. Of course, not everyone outside Spain and Portugal automatically supported Grotius' arguments. The Englishmen William Wellwood and William Beecher would publish counter-arguments to the notion of free seas for all, as it became apparent that the Dutch would make use of this precept to justify an expansion of their fisheries. But *mare liberum* offered more than a simple pretext for piracy, war and fishing. Grotius' work laid the foundational stone in a legal framework that would inevitably give way to the concept of public goods and services – of which international waters are one example – and the right to use these to transport and trade private goods. In this context, it fundamentally opened the world to globalization, although only a handful of nations at the time had the means to make profitable use of the seas at a global scale.

Dutch ascent

The VOC and their counterpart, the Dutch West India Company, or WIC, proved extremely successful at disrupting and displacing Portuguese interests in the tropics, particularly the East Indies. This litany of campaigns would become known as the Dutch–Portuguese War and last over six decades. In 1603 they attacked Goa (India), Amboina in the Spice Islands (Indonesia) was captured in 1605, then in 1610, they attacked the Philippines and, again, Goa, both unsuccessfully. Jakarta (Indonesia) was taken in 1619 and re-named Batavia after the Dutch ancestral line. Numerous failed attempts were made to take the Portuguese trade colony at Macau (China). They invaded Brazil in 1620 and took control of the sugar plantations, again attacked the Philippines unsuccessfully in 1624, formed a trade colony on the island of Formosa (Taiwan) in the same year and claimed the islands of Aruba, Bonaire and Curacao in 1634. They expelled the Portuguese from Guinea in 1637, captured the strategically important Malacca (Malaysia) in 1641 and Angola in the same year. A final failed attempt was made to capture Manila in the Philippines in 1642. By the end of the same year, the Dutch had conquered the entire island of Formosa, expelling the Spanish. After a bitter campaign that saw the Dutch form an alliance with and then betray the native Kingdom of Kandy, they took complete control of Ceylon in 1658. By 1663, the Malabar Coast along the southern tip of India had been won, including the important trade ports of Cranganore, Cochin and Nagappattinam. A half-century after the publication of Grotius' pamphlet and the usurpation of the Portuguese presence in the tropics was complete – the extent and depth of the Dutch takeover was staggering. This was particularly so in the East Indies. Only the important trading centres at Goa and Macau remained, despite numerous attempts by the Dutch to gain control of these too. The Spanish also successfully repelled several attacks on Manila. The Philippines, Goa and Macau would constitute the only Iberian presence in Asia by the end of the seventeenth century, remaining this way for another two to three centuries.

As early as 1526, French corsairs and English frigates had persistently squared up against the Iberian fleet in the Americas (Marley 1998). This was simply state-sponsored piracy designed to make the Iberian monopoly in the region a little less profitable. After the United Provinces declared their independence, the Dutch joined these forays. The privateers pillaged Iberian trade ships, even plundered their port towns, but never with the means or intent of occupying their territorial possessions permanently. The establishment of the joint-stock holding companies in England and Holland changed this strategy. Soon after a charter was issued, the WIC put into action a new plan to carve out a tropical empire of its own. The plan by-passed the heavily defended Spanish ports of the Caribbean and involved a direct assault on the Brazilian captaincies with the intent of permanent occupation. They knew that the growing sugar exports reaching Lisbon and Antwerp were the key to extracting value from the neotropics and they rapidly took control of the regional seat of government at Salvador over a two-day period in the winter of 1624. Initial Dutch success would prove fleeting. After the significant losses sustained in the East Indies to their rivals, the Iberians were determined to prevent the same from happening in Brazil and assembled a massive force to extricate the Dutch from their provincial capital. Despite a counter-force of thousands and a fleet consisting of more than two dozen ships, the Dutch stay in Salvador would not last more than a year in the face of overwhelming opposition. The capture of the city was the first successful attack on an Iberian territory in the New World and both the occupation and the subsequent loss proved a great revelation in Holland, where the Dutch had recently suffered a number of domestic setbacks in their prolonged fight against the Spaniards. The occupation sparked the publication of a steady stream of less-than-factual news, positing claims of great wealth amassed by the victors and the triumph of Protestantism over Catholic idolatry (van Groesen 2011). The subsequent loss shocked polite Dutch society, but would also engender a second attempt by the WIC to establish a Dutch presence in the sugar belt of the Americas. They would return to Brazil in the spring of 1630, registering several temporary successes before being definitively expelled by the Portuguese in 1654 (Boxer 1957). During the brief life of Dutch Brazil, or New Holland as it was called, the WIC made every effort to expand their control in the region and to begin reaping the anticipated profits from tropical commodities that drove them to undertake such a risky – and costly – overseas campaign. But tribal rivalries within the native population ate away at plans for agricultural expansion and failure to fully expel the Portuguese from the continent meant the possibility of an attack remained very high. The effort employed towards growing the sugar industry simply could not account for the costs of the region's defence and financial profit from the venture remained elusive (Schwartz 2004). In the end, it was this expenditure that ultimately saw the WIC collapse. By 1674, with the loss of Brazil still lingering, the WIC was dissolved and with it, any prospect of Dutch colonial dominance in the tropical Americas.

By the time of the WIC dissolution, the Portuguese had also emerged victorious after a 30-year effort to regain their national sovereignty, ending nearly a century of Spanish extra-territorialism. But while the Brazilian campaign and the WIC bankruptcy loomed as a disaster for the Dutch Republic's efforts to build a global empire capable of competing with the Iberians, they achieved some measure of success in the region north of the Amazon along the coast of present-day Guyana. The failure of the WIC to establish a stake in the New World inevitably presaged a decline. But this would not come at the hands of the Iberians – perilously in debt and with little desire to further expand their territories. It was the French, and, above all, the English that would supplant any notion of Dutch dominance and usher in a new age of colonial patrimony centred on sugar in the neo-tropics. The Dutch would realize the costs of their adventurism and instead opt for trade with those that chose to take on the financial and logistical nightmare of administrating a large number of overseas territories. What emerged was a nation well versed in global trade and the knowledge necessary to act as a profitable intermediary – a role that they learned to play through practice, but also necessity. Sir William Temple observed in 1669 that

> never any country traded so much and consumed so little. They buy infinitely, but 'tis to sell again, either upon improvement of the commodity, or at a better market. They are great masters of the Indian spices and of the Persian silks; but wear plain woolen, and feed upon their own fish and roots.
>
> (Temple 1720)

Despite losses in the Americas, the Dutch would consolidate their trade around their possessions in South-East Asia and linkages with China and Japan. Their foray into the tropics also played an important role in assuring the continued survival of the Dutch Republic in the face of an antagonistic, Catholic France and Spain by providing them with a commercial outlet that was free of the confining geography of Europe. In 1672, when forces of Louis XIV overran the Netherlands – an invasion that would come close to extinguishing the Republic's independence – and reached the very outskirts of Amsterdam, plans were made by the wealthiest Dutch citizens of that city to flee the country – not to another part of Europe, but to the Dutch East Indies (Board of Trade 1843). In the words of Voltaire: "Holland would have subsisted but a very short time, had she looked no further than the seizure of the Spanish plate [silver] – fleets, and had neglected to have laid the foundation of her power in India" (Smollet and Francklin 1761).

Banking on the tropics

Of course, as the Dutch and the Iberians were engaged in a bruising confrontation that lasted nearly a century, the English and French were rapidly acquiring a foothold in the tropics. While the Dutch laid waste to the Portuguese trade network, the Anglo-French assault would focus on territory claimed principally by the Spanish. Their successes would confirm, particularly alongside those of the Dutch in the East Indies, that any plan the Iberians may have had towards consolidating their control over the trade in tropical commodities was at an end. Moreover, by the close of the seventeenth century it was clear that the Anglo-French ambitions to carve out tropical empires in trade and agriculture would overshadow even those of the Iberians. With the monopoly on tropical trade broken, *mare clausum* and the treaties signed at Alcáçovas, Tordesillas and Zaragoza were nothing more than words on paper, destined for the historic archives at Seville.

The actions of the British and French would bring the timeline of colonial expansion in the American and Asian tropics to its apogee. Both would prove to be particularly adept at capturing and colonizing tropical islands, especially in the Caribbean. Although the Spanish claimed the entire archipelago, only half of Hispaniola (Santo Domingo), Puerto Rico and Cuba remained by the start of the Georgian Period in 1714. Tropical products were soon being shipped in British and French vessels sailing directly for northern European ports, not Lisbon or Cadiz. But these interests hinged overwhelmingly on a steady flow of sugar, and the economic impact of this commodity at home. Enjoying an insatiable growth in demand across Europe, the trans-Atlantic sugar trade was beginning to make clear to the merchants and brokers in European ports the financial opportunity that lay before them now that the Iberian grip had been weakened.

The Georgian Period in Britain was one of lavish economic expansion with all of the social and financial upheavals that tend to accompany such periods. The financing of capitalist enterprise, once the exclusive remit of renowned medieval European families, was beginning to "diversify". The Goldsmiths, Barings and Rothschilds of London and others, particularly in the financial establishments of the Dutch Republic, sought alternative methods of sourcing capital and earning a yield from it. The Goldsmiths began to innovate, offering bullion-backed exchangeable promissory notes, the precursor to bank-noted paper currency. Initially, their business was designed to simply ensure the secure retention of clients' holdings of precious metals. They would issue a record of these deposits in the form of a note that attested to the purity, weight and form of the metal held on their behalf. But the Goldsmiths realized that if they could lend money to third-party borrowers underwritten by the massive deposits of gold and silver that they had in their vaults, they could increase their own profits significantly. Depositors received interest on their deposit and the bank earned the margin through the difference in interest received by the borrower and issued to the depositor. If the borrower failed to pay the interest, the bank was on the hook. Goldsmiths only expected a small percentage of borrowers to default and they loaned money as a multiple of the bullion reserves, expanding their profits by magnitudes. Fractional-reserve banking – the same system of lending used by virtually every banking establishment today – was a revelation in the eighteenth century. The expansion of this form of finance fuelled agricultural, mining and industrial output at levels previously unknown in Europe.

At the same time, the success of joint-stock companies such as the Dutch VOC and the British East India Company (EIC) opened the door for the receipt of dividend income in return for the capital loaned to the companies. Companies dealing in regional trade with the Middle East (Levant), Russia and Africa joined those focused on the East and West Indies and a rapid rise in the exchange of marketable company shareholdings began to take place. Domestic ventures were also often shareholdings, particularly those investing in natural resources. Initially, shareholdings were created to document the resources of a small pool of investors in a venture, seeking a return on their excess capital that otherwise was exposed to price inflation. Shareholdings could be exchanged, but these were generally not marketable and any transaction was a private matter between two or more parties seeking to invest/divest their interest. As wealth began to spread in seventeenth century Holland and Britain, however, the demand for larger investments increased. Company shareholders began to consider offers for their stakes at points of exchange. Coffee houses in London and Amsterdam became well known as meeting points between sellers and buyers. Individuals seeking to service this exchange began to broker deals by authenticating holdings and ensuring a record of transfer. Fuelled by generous amounts of strong coffee, the transactions were raucous – even rude by libertine Georgian standards – as arguments broke out while attempting to secure potentially lucrative positions or offload failing ones. Speculation began to take hold as a way of quickly profiting from the trade in shares, rather than the profit of the underlying venture. The Royal Exchange, set up in Tudor London to facilitate the exchange of goods, banned these impolite share-traders initially and most transactions were being conducted at Garraway's, Jonathan's and Lloyd's (of London) coffee houses on or near (Ex)Change Alley, a small road wedged between the Royal Exchange at Cornhill and the Post Office on Lombard Street (Braudel 1982). Speculation and fraud prospered in these unregulated settings and the British and Dutch governments made several attempts to regulate the trade in marketable securities by restricting and licensing brokers and levying penalties against those found brokering without a licence. Eventually, the semi-formal exchanges in Amsterdam and London gave way to restricted settlement rooms and the beginnings of the modern, regulated stock exchanges, but not before they experienced their first share crashes.

One of the best documented incidents was the speculative bubble in shares of the South Sea Company (SSC) in 1720. When the formation of the South Sea Company was announced in 1711, it was met with jubilation. Britain was emerging from yet another war, this time with France over the successor to the Spanish king, Charles II. The British, under Queen Anne, and the Dutch Republic had much to lose if the French monarch, Louis XIV, succeeded in placing his son Philip on the throne as the first Bourbon monarch of Spain. This would consolidate vast power over much of Western Europe in the hands of Catholic France. It would also create a combined French and Spanish tropical empire of significantly greater size than that of their British or Dutch rivals, potentially suffocating British and Dutch interests in the commodity trade. In a bow to the complexity of the shifting political and military alliances pursued at the time, the British and Dutch chose to support the claim of the Hapsburg line of their former arch-enemy, Philip II, in order to repel the French ambitions. They plied this collective position after having been at war with one another across much of the latter half of the seventeenth century, mainly over their tropical territories in the East Indies. Queen Anne's uncle, Charles II, had in fact supported Louis in his war against the Dutch in 1672, despite his own nephew being the great grand-son of the Republic's founding father, William (the Silent). The end of the war with France came with a series of agreements that gave Britain sovereign control over Gibraltar, Minorca and St. Kitts in the Caribbean. In exchange for these concessions, among others granted to their allies, Britain agreed not to contest the ascent of Bourbon Philip to the Spanish throne. The peace also granted the British Crown a right of Spanish *asiento* – the permission to supply African slaves to the Spanish. Charles II and the Dutch had been trying to "capture" the prize of the *asiento* for decades, since the Spanish remained one of the largest importers of slave labour and the contract gave – on paper – exclusive rights to the trade across the

Atlantic. The administration of Queen Anne appears to have been seeking a much broader remit to trade with the Spanish colonies, but this was rebuffed on the grounds that such favourable terms would not be tolerated by other European nations (Anderson 1764).

Thus, it was the *asiento* concession – a 30-year contract to supply 8,000 slaves and 500 tons of dry goods per annum – that formed the basis of the South Sea Company. The terms of the *asiento* meant that the Spanish Crown was also paid a portion of the proceeds from each slave sold ($33^{1}/_{3}$ pieces of eight) as customs duty, as well as an upfront allowance of 200,000 pieces of eight (Anon 1713). The upfront allowance would be paid back in the final ten years of the *asiento* through an offset of the payable duty. The king of Spain would also hold a quarter of the shares entitling him to a quarter of any profit made each year on the sale of slaves and the annual allowance of goods. An additional 5 per cent of the remaining three-quarters of the profit from the sale of goods was also allocable to the Spanish. Moreover, the goods could only be sold at "Fairs" and not before any goods arriving from Spanish ships had already been brought to market. To ensure compliance, goods arriving in advance of the Spanish fleet would be held in a customs house under lock and key. If conflict broke out, Spain reserved the right to seize and retain all SSC property until cessation of hostilities.

The grant of *asiento* was handed over to the SSC by the Crown for a fee. It had previously been granted to the French Guinea Company in 1701 and before them, the Portuguese, under almost identical terms. Both the French and Portuguese efforts were beset with financial difficulties from the start. The terms for the SSC were unlikely to improve prospects, such that the Spanish explicitly acknowledged the failings of past ventures in the treaty itself:

> Besides the foregoing Articles Stipulated on behalf of the English Company, His Catholick Majesty, considering the Losses which former Assientists have sustained, ... to allow to the Company of this Assiento, a Ship of Five hundred Tons Yearly during the Thirty Years of its Continuance, to Trade therewith to the Indies.
>
> (Article XLII, *The Asiento*)

A local British diplomat wrote that the terms of the *asiento* were: "... the worst that I ever saw and the most calculated for captiousness and chicane" (Palmer 1981). Anderson (1764) writes:

> It was universally known, that the Portuguese Company first, and next the French one, were undone by their Asiento contracts for supplying ... the Spanish West-Indies ...: and this sugar plumb of an annual trading ship granted to our company was too much clogged with the above difficulties to prove of any certain advantage, more especially

considering how much the Court of Madrid had it in their Power to suspend the licence for any such annual ship, &c. as they often practised, and to seize on the company's effects in America at pleasure.

Yet, the chicanery perpetrated against the British was not limited to the terms of the *asiento,* nor to the Spanish. The lion's share of this would arrive via the British themselves. At the time, the British government found themselves, like the Dutch and most other monarchies, deeply in debt due to relentless war in Europe and the cost of securing their foothold in the tropics. For the beleaguered British public purse, this amount exceeded £50 million, an unfathomable sum for the time and the cause of deep concern for Parliament and the Crown.[3] The SSC was in part conceived as a means of financing this debt mountain domestically at a lower rate of interest, thereby affording the government more room to find a way to cover their liabilities, most of these being accrued through supplies to the military. Again Anderson:

If therefore a Fund could be established for the regular payment of the interest of the said large arrear(s), and at the same time plausible means could be devised to give the creditors the hope of farther advantages by a new and alluring commerce, he prudently thought he should obtain his principal end.

When Queen Anne's administration negotiated a peace with France and accepted Philip to the Spanish throne in exchange for the *asiento,* it was in part with the debt servicing potential of the SSC already in mind.

The British government, through faith – or desperation, had already began to offload public debt to the SSC a full three years before the final terms of peace had been ratified. A series of runs on the finances of their largest creditor, the Bank of England, dented investor confidence in public credit and government bonds began to trade at a discount. The Bank of England, established in 1694 under William III, was thought to be the solution to the chronic difficulties in funding the public purse. Originally a corporation, the Bank acted as a debt consolidation facility that married the excess capital of the wealthiest citizens with the on-going borrowing needs of the British government. It allowed the government to raise the funds necessary to rapidly expand the Navy, and through this investment, compete with the other European powers for territory in the tropics while maintaining a viable defence of the homeland.

Armed conflict was more common than not during the late seventeenth and eighteenth centuries and this was the primary driver of government borrowing across Europe. The wars against France (and Britain) over the past half-century had left the Dutch Republic's finances in a similarly terrible state. These conflicts tested the limits of their continued independence from the great powers of Europe, exposing the terminal frailty of the

Dutch plan to build a global empire. Although they met with considerable military success in the tropics, this could not compensate for the vulnerability manifest in a homeland that was too small, and with too few people to resist the might of their much larger neighbours, leaving the heart of their overseas project continuously at risk. Despite provisioning a "central" bank in Amsterdam that acted as a model to the Bank of England, the Dutch were simply unable to financially sustain a military capable of competing with larger neighbours that had also realized the power of naval might and invested accordingly.

Organizing the Bank of England proved a stroke of genius in this regard, since previous governments were obliged to spend an inordinate effort fund-raising bond by bond and could now tap the financial liquidity of the nation through the Bank alone. The Bank opened a "subscription" for the public to invest (as shareholders). It raised £1.2 million within a fortnight and loaned this to the government almost immediately at a rate of interest lower than those paid previously. The Bank was commissioned with a royal charter in return, giving it the right to borrow money from subscribers against future government revenue. The capital of the Bank grew accordingly, rising from the initial amount to around £6 million – a 500 per cent increase – in the 20 years between its founding and the signing of the *asiento* contract (Sinclair 1804). This consolidation opened the door to a massive investment in the military, a prospect that would see the Navy grown exponentially and gain the British a great advantage in trade and commerce overseas. Yet, after a series of duplicitous schemes and failed ventures tested the liquidity of the Bank, the British government found itself once again in need of further funding. Political risk rose to the forefront. A snippet here of a Stuart insurrection,[4] a rumour there of impending war with France, drove bond traders to lose faith that the government would make good on its debts. Under these multifarious pressures, the Bank simply could not fund the borrowing needs of the government. The SSC, for all of its failings, was adopted as the means of relief. Anderson's 600-ton "sugar plum" along with the slave trade and fishing rights were designed to entice further subscription to government-backed debt by offering the prospect of a dividend attached to the trade profit of the SSC. The apparent purpose of the SSC was to slave, trade and fish in the New World. The true purpose of the SSC was as a financier of public debt, a bond broker charged with converting public bonds into company shares on the grounds that it would be paid interest by the government on the converted loans. El Dorado had again appeared, this time in the waters of the South Seas, the plantations of the New World and the coffee houses of Exchange Street.

The *asiento* and trade with Spanish America were, of course, a ruse. Details of the charter had been published in 1712, officially promulgating the vast geographic range of the SSC's overseas remit and exclusive rights. Yet, four years after the treaty was signed (in 1714) and a ship of dry goods

– as prescribed under the *asiento* – had finally arrived in the Americas, the profitable sale of its cargo was immediately hindered by the Spanish bureaucracy. The trade in slaves, as expected, was also beset with logistical and financial difficulties and just as the prospects began to improve in 1718 and the SSC's second trade ship arrived in the Americas, war broke out between Spain and Britain. All of the SSC's assets in New Spain, including the ships and 600 tons of dry goods, were summarily seized. Certainly no profit had yet accrued to the company since chartering, let alone a dividend been paid.

In contrast to this miserable commercial performance, the volume of government debt converted by the SSC's finance arm grew rapidly. Several share subscriptions were successfully filled in the years following its initial absorption of £10 million, reaching a crescendo in 1720. The past successes of the SSC and Bank of England in converting public bonds into shareholdings convinced the government that this course of financial wizardry could deal with the remaining outstanding public debt of £30 million. Much of this was in the form of payable annuities from past lotteries – irredeemable debts that paid annual sums for long periods – but also more conventional long-term bonds (Anon 1811). The SSC offered to convert these at a much lower rate of interest, but found itself in competition with the Bank of England. A second round of proposals from each institution was submitted to Parliament that forced the SSC to revise its offer to a low interest rate (i.e. a larger payment to the government). The government agreed to sell the debt to the SSC in January, 1720. This sale would expand its capital base from £10 million to £43 million in less than a decade. The news was greeted euphorically by stock brokers. Public shares in the SSC had traded at £77½ shortly after its launch in 1710 (Anderson 1764). But in the wake of the news that the SSC had secured the massive government bond swap and the subsequent announcement of a dividend raise, prices of new subscriptions quickly rose to £130 in February of the same year and then to £300 in April, £400 in May. By the end of June, SSC stock was trading at a significant premium to its face value at £1,000 per share, then briefly to £1,050 by August during a two-month summer recess to administer the company's books (Dale *et al.* 2005). By the time official trading re-commenced in August, the price had declined to £820 and then to £520 in September, and £170 by mid-October (Carswell 1960). The South Sea scheme had become the South Sea Bubble.

The losses accumulated by many of the investors were undoubtedly devastating. A veritable who's-who of British society had invested, including Sir Isaac Newton, George I and members of the royal household, dukes and lords, the majority of Parliament, wealthy land-owners, merchants and private bankers (Carswell 1960). Of course not all investors lost money and some purported losses have probably been overstated (Hoppit 2002). But in the aftermath of the rise and fall of South Sea stock over the summer of 1720, it became clear that the scheme was a ruse designed to enrich the company directors and those flipping shares – disparagingly referred to as "stockjobbers" – through share price appreciation. Government officials were given stock promises at the issue price that could be sold in the future for a profit if the shares appreciated. The SSC recklessly loaned shares to encourage shareholders to retain or increase their positions. They bought back shares at higher prices. They issued dividends in the form of additional stock. The directors of the SSC aggressively lobbied for the passage of the Bubbles Act of 1720, designed to eliminate the issuance of stock by any company that did not have a royal charter, believing this would free up liquidity to invest in their own shares. The sole aim of these tactics was to increase the price of SSC shares. It is unclear precisely why the bubble burst. But the passage of the Bubble Act may have ironically played its part by dampening a much broader speculative exuberance that gripped Europe after governments began participating in the new art of financial engineering that had emerged during the final decades of the seventeenth century. Leading up to the SSC share price collapse, an abyssal complex of small, publicly traded shareholding companies, referred to as "bubbles", were being vigorously traded on Exchange Street. Shares of many of these companies, numbering in the hundreds and often absent any commercial activity whatsoever, were the subject of speculative price-rises, engineered solely to deliver a trading profit. The Bubble Act put an end to this den of duplicitous stock-gambling (at least until the dot. com bubble of 2000) and *inter alia*, may have facilitated the fall in the share price of the SSC by quelling the activity of those also involved in the bubble trade.

The losses in the scandal of 1720, however, were not unique. Other "schemes" promising inflated returns from the tropical trade in commodities paved the way for the South Sea event. In 1698, the government of Scotland embarked on a scheme to establish their own tropical empire to compete with the interests of England and other European powers. Scottish nobles and merchants had no access to the buoyant trade in sugar and other tropical commodities that had propelled others towards greater wealth and power and they believed that they would lose a grand opportunity to elevate Scotland if something was not done about it. After some consideration, it was decided to establish a colony off the coast of the Darién, in present-day Panama. The aim was to establish a trade colony that could control a shorter route for the movement of goods between the Atlantic and Pacific oceans – a prescient strategy in light of the Panama Canal, built a little over two centuries later. Funds were initially raised from investors in Scotland, England and the Dutch Republic, but the East India Company reminded the government of its exclusive rights to the tropical trade and the English and Dutch investors pulled out of their commitments to the Scots' project. The capital needed was ultimately raised from Scotland alone and accounted for nearly a quarter of all the liquid assets of the country at the time. After several attempts to establish a colony met with resistance from both Spanish and English colonists, bouts of disease, disagreement and a Spanish siege finally confounded the company's efforts (Ridpath 1700; Entick 1757). The surviving settlers – less than a fifth of those that had arrived – returned to Scotland or made their way to the small coastal settlement of New York.

To say that the wealth of most families in Scotland had been severely damaged is an understatement. Many had been coaxed into staking everything on the scheme and now found they were nearly bankrupt. The pressures on the Scottish government in the wake of the spectacular gamble were immense and the consequences much greater than the loss of financial solvency among the better-off clans. It has been suggested that the Darién Scheme acted as a primary driver in the final political union of Scotland and England shortly after the disaster in 1706/7. The Act of Union, voluntarily entered into by both Scottish and English Parliaments, was the basis of the modern United Kingdom. As a result of the Act, many of the Scottish families that had suffered losses from the Darién debacle were able to seek compensation from Parliament. This took the form of a fund, referred to as Equivalent Money that was set against national debt. Rather than seek to sell on or redeem the debt position, the Scots instead incorporated their holding and accepted an annuity of £10,000 from the government. This liquidity would be used to obtain a charter in 1727 for the establishment of the Scottish Banking Company, later to be re-named the Royal Bank of Scotland.

The South Sea Company enticed naïve investors on the promise of immeasurable profit from trade in the tropics. The premise of this scheme, as well as the Scottish Darién Scheme and dozens of much smaller London "bubbles", worked its ill effect in part due to the reputation the tropics had gained as the means to achieve significant wealth quickly. Most investors did not have the means to assess the veracity of commercial proposals – certainly not in the distant tropics – opening up a window to misinformation and lies. Ultimately, the South Sea Bubble was formed as part of a much broader irrationality that set in during the period leading up to its share price collapse. Shares in the Bank and the East India Company also experienced a rapid rise and fall in their share prices consonant with those of the SSC. The tropics had played a huge role in precipitating the financial calamity. By applying a veneer of commercial authenticity to a stock market that was little more than a pyramidal process of price appreciation, the promoters of these schemes simply fed the prospect of future share price gains, not meaningful commercial profits.

The consequences of these events did not terminally impact investment in the tropics or the development of capital markets in Europe. On the contrary, they led to a more sober appraisal of the role that financial houses should play in managing public debt. In Britain, the conclusion was that the Bank of England should be the primary steward of government borrowing and, more importantly, should retain a "reserve" to offset any future financial crises. In the aftermath, the Bank of England would see its capital doubled and paying a dividend yielding 5 to 7 per cent (Sinclair 1804). The Bank would act as a backstop and lender of last resort, much as it does today. This new prudency was an offshoot of the ruinous debacle that arose initially from the South Sea Bubble and other schemes in the early 1700s. Public debt formed the bulk of legitimate trade in bonds at the time and the government began to understand the paramount need to ensure that this trade was not compromised by complex chimeral schemes that badly married public finance with private enterprise. After the stock run of 1720, the capital of the SSC was separated into company stock and government annuities in 1733 (Anderson 1764), ending any scope for further "stockjobbing" based on specious claims of profit arriving from the tropics. Going forward, public debt would remain public debt underwritten by the assurances of tax revenue alone, eventually transforming into the modern gilt by the late nineteenth century. The experience of 1720 also highlighted the need for government intervention and regulation of the growing industry of finance. Funding government, infrastructure and the military was no longer an isolated affair between monarchs and family-owned banks. George I could not simply renege on a debt to a private banker – as many other European monarchs had done previously – and expect a modest economic response.

This transformation paved the way for fiscal expansion and the broader availability of credit, particularly to the government. It also saw the advent of fiat (paper) currency – backed by the reserve of the Bank of England – as a substitute for the system of bullion payments that dominated trade previously. Other nations soon followed by implementing similar systems in dealing with credit and public debt. Credit, paper currency, a reserve bank – all of these enhanced Britain's capacity to fund both military and private ventures by increasing investor confidence. Both would weigh heavily in the pace and extent of colonization in the tropics and its first transformative economy – cane sugar.

The sugar islands

Thirty years after the crisis of 1720, the trade in sugar and other tropical commodities had grown significantly. The *asiento* contract had terminated and after suffering through several wars and the confiscation of property, the final accounts showed an accrued loss of £100,000 to the SSC (Anderson 1764). The company had been regularly ridiculed in the press since the stock crash and became a byword for failure. But the British and French no longer viewed access to Iberian trade in the tropics as paramount to their success and the revocation of the *asiento* registered little more than a footnote in the periodicals of the day. What was once seen as a British and French insurrection by the Iberians had gradually, if not quietly, transformed into domination and the two growing powers were now in full-blown competition with one another. The sugar trade was no exception. The Iberians had managed to establish a measure of constancy in production of cane sugar from their New World possessions since the first shipments left Brazil in the early 1600s, but the industry could not keep up with the rise in demand. Nearly 17,000 mt of sugar was exported from the region in 1600, but this amount had only grown to 30,000 mt by the time the SSC was established and it hovered around this number for decades thereafter (Simonsen 1944) (Table 6.1). Sugar exports from New Spain followed a similar trend of declining growth. In part, a renewed focus on precious metal mining in the Americas during the 1700s distracted investment in sugar after a series of rich deposits were discovered in Brazil. But competition was also starting to suppress the high prices that the Iberians had enjoyed under a near monopoly. High prices had offset many of the inefficiencies in the Iberian management of the trade and as these declined so too did profitability. Many countries that once sought Iberian sugar had developed their own plantations and began to levy high tariffs on foreign produce while exporting their own surpluses. Much of this competition arose from the gradual rise of the Caribbean plantations over the latter half of the seventeenth century – now managed by French and British growers under the protection of their respective naval forces.

Every sober person seriously involved in the tropical trade knew that success or failure in the colonies hinged overwhelmingly on a steady flow of sugar. For example, Barbados had already transitioned towards a "pure" sugar economy by the late 1600s with over 90 per cent of its export value between 1665 and 1700 attributed to sugar (Eltis 1995). This small island on the edge of the Caribbean arc was exporting more economic value in relation to land area and population than any other colony or country at the time. David Eltis, Professor Emeritus at Emory University, refers to seventeenth century Barbados as the "Hong Kong of the pre-industrial era" – but based almost exclusively on sugar (Eltis 1995). Enjoying an almost insatiable growth in demand across Europe, the trans-Atlantic sugar trade had made it clear to European merchants and brokers the financial opportunity that lay before them now that the Iberians had lost their grip.

Brazilian sugar continued to account for the bulk of the trade up until the 1730s, although British exports from Jamaica and Barbados were by this time accounting for at least half this amount (Eltis 1995) (Table 6.1). British society was also clearly hooked on sugar and this was stimulating investment. The public was consuming on average nearly 5 kg of sugar per person per year, up from only 1.5 kg a mere three decades earlier (Figure 6.2).[5] While still below the 9 kg pp pa recommended by the World Health Organization, and well below the average consumption in most OECD (Organisation for Economic Cooperation and Development) countries today, this was a remarkable rise, coinciding with the spread of cane plantations in the British and French Caribbean. Combined with the

Table 6.1 Cane sugar production in the major European export colonies over four centuries (in average tonnes per annum)

Years	Brazil	Cuba	Barbados	Jamaica	Guyana	Mauritius	Haiti	Martinique	Guadeloupe
1600–625	***13,846***		–						
1625–1650	***23,223***								
1650–1675	**24,491**		9,724	809				–	2,037
1675–1700	**28,805**		8,095	4,229				–	1,753
1700–1725	***13,982***		7,954	7,700			10,182	–	–
1725–1750	40,625	300	7,000	21,685			***42,468***	20,000	2,003
1750–1775	36,864	2,037	7,445	32,737	1,150		***56,099***	11,615	3,443
1775–1800	22,413	18,610	5,521	53,160	1,150		***57,824***	10,300	9,000
1800–1825	33,972	43,241	11,464	**79,198**	10,595	5,726	8,030	20,596	21,091
1825–1850	80,158	157,636	18,055	45,894	36,303	37,272	7	25,322	30,699
1850–1875	123,725	**549,280**	36,337	25,287	57,826	109,172	–	31,840	28,740
1875–1900	219,773	**606,280**	50,440	23,760	104,640	127,240	–	36,760	47,880
1900–1925	408,172	***2,660,400***	46,920	29,320	109,720	211,920	7,625	32,960	33,000
1925–1950	1,371,025	**3,692,846**	93,769	122,692	159,269	278,423	31,577	39,231	41,846
1950–1975	4,116,372	**5,569,525**	161,536	397,071	303,246	585,602	60,671	53,762	131,487
1975–2000	***10,789,372***	6,350,649	79,495	230,041	255,711	611,960	32,957	7,366	63,079
2000–2010	***29,830,000***	1,925,270	36,233	153,072	269,620	510,837	9,000	4,899	63,661

British Navigation Acts

'*Colbertism' & l'Exclusif*

Notes
Leading producer for each period in bold and italicized, abolition of slavery is in grey and national independence underlined. Period of "mercantilistic" policies in England and France outlined. Periods of sugar production with no data indicated by dash; periods before sugar production left blank.

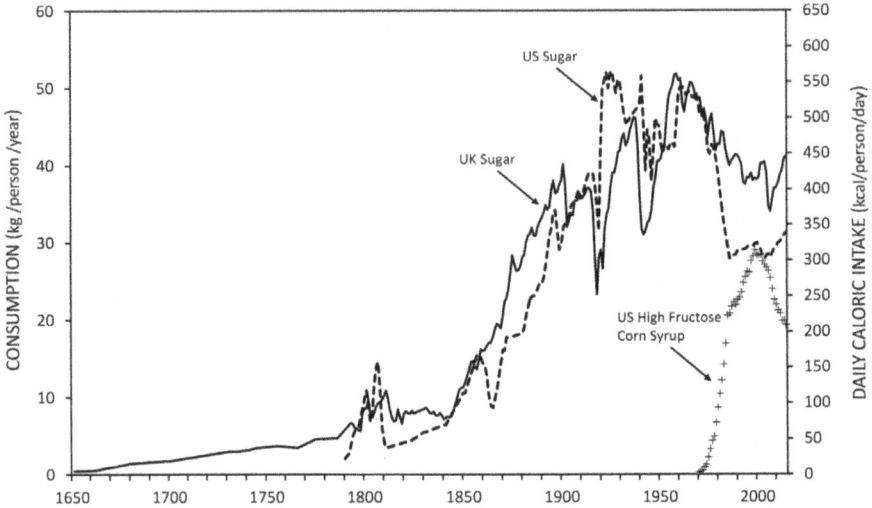

Figure 6.2 The history of rising per capita sugar consumption in the United Kingdom and United States and the equivalent change in average daily caloric intake.

continued supply from Spanish and Portuguese territories, as well as from traditional sources in India, Africa and the Middle East, sugar prices continued their multi-decadal decline. Between 1650 and 1710, the price of sugar outside London had dropped more than 40 per cent (Simonsen 1944).

Laissez-faire, no

But not everyone involved in the sugar economy was pleased with the boom in production and consumption. Rumblings began to emerge from the British Caribbean over the government's management of the trade. Plantation owners and their advocates based in London launched a public campaign aimed at convincing members of the government that the economies of the sugar islands were in a perilous state – or worse, on the edge of imminent collapse – due to the dis-advantages placed upon them by the government restrictions on trade. They argued that the French and Dutch producers were not burdened by similar restrictions and were able to offer their sugar – mainly from Guadeloupe, Martinique and the Guianas – at much lower prices. To make matters worse, the North American colonies were by-passing the pricier British sources and purchasing sugar and molasses – the latter used in the important rum distilleries in the colonies – directly from the French and Dutch at considerably reduced prices. Rumblings about the way that trade was being conducted between Dutch producers and the North American colonies were already in publication just as the British Crown was sardonically celebrating the *asiento* award as a breach of the monopoly on trade to the Spanish Main. William Cleland – commissioner of customs in Scotland at the time – explains (Cleland 1719):

> Whereas the Inhabitants of the Continent of America, though on some other Account they may be Advantageous, yet they are in the above Manufactures Rivals to this Kingdom, will in a little Time want nothing from hence, and are already become the very Carriers of Merchandize, and can lay out their Improvements and Riches in their own Country, and so extend their Dominions, whereas the Inhabitants of the Sugar Colonies must have almost everything they Eat, Drink, and Wear, from this Kingdom; and all the necessaries for carrying on the Sugar Works; and what Money may be over and above such Charge and Expence, is laid up in Britain; therefore its indispensibly necessary for the Government, to cherish and support this valuable Trade, and to discourage all Attempts to wrest it out of our Hands.

Cleland's description tells us that the British sugar islands were operating within a closed economy. Overseas territory was initially seized through naval force, secured through the army, colonized by planters or their agents, and capitalized with slave labour supplied by British agents. Sugar was produced and transported exclusively aboard British ships to the homeland or other colonies. Any manufactured goods needed on the islands arrived exclusively from Britain or its other colonies, even if these were produced in another European country. And the islanders were legally bound to this system through a series of parliamentary Acts specifying the restrictions placed upon their trade. The most significant of these, the Navigation Acts, were passed in the 1660s as a brake on the competitive advantage the Dutch Republic had accrued in the tropical commodity trade after independence from Spain. These Acts forbade any direct trade between the American colonies and other European nations. It had to first pass through British ports. They also censured the use of foreign (mainly Dutch) ships and crews to transport product between Britannia and her colonies and placed a financial bond on merchants to ensure that the law was followed. Although the notion of laissez-faire – or trade without government interference – was beginning to swirl in French economic circles by the early eighteenth century and had been the operational basis of Dutch trade for decades – largely due to their incapacity to militarily defend their monopolies – there would be none for the foreseeable future in Britain or any other parts of the rapidly expanding area where it held dominion.

Cleland was alluding to Britain's trade with her sugar colonies being bound by the protectionism embodied in the Navigation Acts; a policy the North American colonies were now circumventing by trading with the

French and Dutch. The sugar colonies wanted this vulnerability patched through stronger enforcement of the barriers called for by law but being largely ignored. The plantation owners argued that the North American colonies must be forced to buy sugar – and especially molasses – exclusively from the British West Indies. Molasses or black treacle is the main by-product of the crystalline sugar extraction process. It is also the primary ingredient in the distillation of rum. Rum rapidly became an everyday drink of choice among sailors and fishermen, and its popularity grew to rival ale in the taverns of eighteenth-century colonial America. As a result, rum distillation emerged as an important industry and the demand for molasses from the sugar estates of the Caribbean grew rapidly with the rising population in the American colonies. The colonists enjoyed rum to such an extent that they were consuming some 340,000 hectolitres, or nearly 7.5 million gallons, of it by 1770. This amounted to nearly 14 litres (3.1 gallons) per person each year (Butel 2002) – a figure nearly six times greater than the per capita consumption in Britain and Ireland at the time and an even larger multiple of the amount consumed in France, where distillation had been banned to prevent competition with the significant wine and brandy industries. Molasses had also become a favoured cheap substitute for crystalline sugar in the American colonies where it was used widely in food. Consequently, imports of sugar and molasses were burgeoning by the time that concerns over trade leakages began to flood parliamentary letterboxes.

This problem of trade within the British colonial system and its various leakages was driven almost entirely by price distortions induced by government taxation. The British colonies were one of the largest producers and exporters of cane products in the eighteenth century, but the French, Dutch and Spanish sugar colonies were offering molasses at considerably lower prices. The French, unable to export rum to metropolitan France due to the government's prohibition, were simply discarding the molasses as waste. On the one hand, the British Crown had made it clear to all concerned within their sugar industry that trade would be controlled in line with rules established under the Navigation Acts and that this was necessary to pay the cost of protecting and administering overseas territories. On the other hand, it was also clear that the consequence of this policy was to advantage the Crown and British shipping merchants and brokers at the expense of West Indian sugar producers, New England distillers and consumers in Britain and the American colonies. In addition to restricting where sugar could be shipped and who could ship it, Parliament created a complex customs system to ensure government revenues – in the form of duties, excises and licences – were collected at every entrepôt.

Income tax did not exist at the time – it was introduced by William Pitt the Younger in 1799 – leaving three primary avenues to financially sustain public expenditure in Britain. The first consisted of land rents – the equivalent of property taxes – that were levied on the value of the real estate held by its owner, whether domiciled or rented out. The second means was through additional borrowing. The British government chronically overspent from the early 1600s onwards, mainly due to increasing military costs. The need for borrowing was often so large that ministers and their advisers created a wide range of schemes aimed at coaxing private capital into government coffers. In exchange, some of these offered annuities, others were run as lotteries. Bonds were issued at fixed rates (the coupon). Finally, funds were borrowed secondarily through the Bank of England, joint-stock companies and other schemes whereby the principal loaned to the government was converted into equity stakes in the companies holding the debt who would then receive a dividend as a shareholder. Some of these were very poorly constructed – perhaps due to a lack of financial acumen – and at the very least did not serve the public interest well. Some bonds, annuities and lottery prizes lacked a specified term and were set to make annual payments in perpetuity, while the South Sea Company illustrated how badly organized some of the other capital raises proved. Of course all of the borrowings were ultimately secured against future tax revenues and where schemes proved flawed or interest rates too high, it was tax rates that were reviewed towards alleviating debt or underwriting further borrowings. Land rents were raised, but doing so too frequently invited significant ire from the gentry and noble classes that would invariably be hit the hardest, but also wielded the greatest influence over the political establishment. In this context, it was perhaps seen to be more expedient and less risky to raise and expand the tax base through the third route available to meet the financial needs of the government – taxes on the trade and consumption of goods. Expanding taxation on commodities produced in the colonies was an obvious outlet and had the effect of raising capital while inducing a change in trade behaviour – a practice still regularly employed today in the implementation of public policy. In this case, the requisite behaviour was to uphold a monopoly on the trade of goods within the empire with the ultimate goal of benefiting the government and capitalists in Great Britain. This two-edged approach to taxation might have appeared a sensible means to pay for the increased military expenses while ensuring rivals did not profit from the economic expansion that followed in the wake of colonial growth. But it also smacked of arrogance and a common self-belief among the upper classes that they alone knew how best to organize trade, invariably to their own benefit. This attitude was not unique to Georgian Britain, as events in France and her colonies at the end of the century would confirm. But taxation would unleash the first major intra-imperial dispute since Pizarro led a revolt against the Spanish reform to *encomienda* in the 1550s with sugar, molasses and rum resting squarely in the eye of the storm.

By the 1730s, the competing demands of the sugar islands, government and the American colonies were forcing Parliament into action. London-based agents from both sides attempted to sway the argument before a critical vote in the spring of 1733. Representatives for the sugar estates argued a comprehensive set of trade barriers needed to be put in place to avoid economic catastrophe:

1   That all Sugar, Rum and Molasses, from America, be prohibited being imported into Ireland
2   That the Duty on all Foreign Sugars, imported into the British Colonies on the Continent be 4 s. per cent [£0.25], on Rum, 9 d. [£0.0375] per Gallon, and 6 d. [£ 0.025] per Gallon on Molasses
3   That the remaining 9 d. per Cent on Sugar from the British Plantations, be drawn back on Exportation from England

(Robertson 1733)

Vested interests in colonial America urged Parliament to consider otherwise, insisting that the plantation owners were largely embellishing their economic plight while in reality enjoying a standard of living far greater than themselves, the French islanders, and most importantly, the average person living in Britain (Anon 1732).

The state of the British sugar islands as they neared the close of the 1830s was invariably more complex than either side cared to admit. The New England rum distillers were admittedly obtaining nearly 90 per cent of their molasses from French, Dutch and Spanish suppliers, brazenly ignoring the restrictions imposed by the Navigation Acts (McCusker 1989). Raynal (1776), in his great history of the East and West Indies, notes in the case of the French islands that "... as early as 1719, they [Americans] carried off 20,000 hogsheads [5.74 million litres of molasses]; and that by the year 1733, this navigation employed 300 ships and near 3000 sailors". The rum produced in the American distilleries was not of particularly high quality and by all measures considered inferior to the West Indian product, but it could be made in vastly greater quantities and distributed more easily within the colonies (Butel 2002). More importantly, it was also around 20 per cent cheaper than West Indian rum (US Bureau of the Census 1975). By the 1730s, the New Englander product was meeting most of the North American demand, deflecting an important "vent", or export market, for the superior but significantly more expensive rum produced in the British West Indies (McCusker 1989).

American colonial distillers and merchants defended their trade with the French and Dutch on the basis that the British sugar islands did not have the capacity to meet the demand for rum in the north and that any subsequent prohibition would only result in damage to their economies and a reduction in their ability to purchase British manufactured goods. They also did not trust the islanders' motive:

Upon the whole, the secret and real view of the Sugar Islands, is, to gain the absolute Monopoly of Sugar and Rum (with respect to the

Subjects of Great-Britain) to themselves; that so they may have it in their Power to exact what Prices they shall please from the Buyers.

(Anderson 1764)

The Americans knew that they had far less influence over matters in London than their counterparts in the Caribbean. Many of the early planters in the islands – Jamaica in particular – had met with significant financial wealth. Returns on capital invested of 15 to 20 per cent were common and those that had managed to hold on to their estates began to ascend the economic ladder (Butel 2002). This rise in wealth was particularly rapid during the boom years from 1730 to 1760. Over this period, the British sugar islands had collectively overtaken Brazil as the region's major producer. With their wealth in hand, a large proportion of these British plantation owners decided for various reasons to move back to Britain. Added to the ranks of these new absentee owners were those that simply had invested in plantation sugar but never planned to be in involved in its production nor had any intention of moving to the Antilles. They lived off the "passive" annual income generated by their estates in the West Indies and those inherited or acquired in the British Isles and this allowed them to pursue other activities, including parliamentary politics. Through their shared commercial interests, these two groups formed a new "planter class" in British society. They also quickly became a formidable lobby in Parliament advocating for the interests of the Caribbean planters and slave-traders. The sugar islands were settled, developed and managed with little more than an extension of the traditional English estate in mind, but with the added advantage of being able to utilize slave labour. They were remote from British society and no other constituency existed on the islands except that of the planters, merchants and investors profiting from the trade in cash crops. This meant that the primary nexus of society in the Caribbean was aligned with crop production and export to Great Britain and its colonies. Any plans for the future of the islands were drawn up almost entirely with a view towards securing and growing a steady flow of agricultural commodities.

Different latitudes, different attitudes

The American colonies had evolved differently. Virtually all of the planters, tradesmen and merchants engaged in the commodity trades were permanently domiciled in North America. A large fraction of the seventeenth-century immigrants had left Britain and other European countries on their own accord for religious and economic reasons. They considered the colonies as their home and few – until the American Revolution at least – gave any thought to returning to Great Britain. Many had wagered the future of their families on an ambition – land ownership – that could not be fulfilled in Britain due to the suffocating class structure. Although many of the earliest immigrants died from starvation, disease and violence, the generous grants of land offered to those that would agree to immigrate to the colonies meant that an entirely new class of land-owning traders and businessmen developed over the latter half of the seventeenth and early eighteenth centuries. Many families became land-owners on a scale only matched by British aristocracy, as initial land grants were extended based on family size and the number of servants supported and were combined with initial tax-free periods to allow people time to develop their holdings (Rabushka 2008). The belief was that the Crown would ultimately benefit from this development. Every colony was bound by law to provide a fraction of any precious metals discovered, akin to the Spanish *quinto real*, and to pay *quitrents* – the colonial equivalent of British land rents, or property taxes – to the king or his sanctioned proprietors, among other tokens of fealty unique to each colony.

Most residents of the colonies were free men and women. In contrast, the majority of West Indians by the end of the 1730s were enslaved or bound by indentured servitude. Although Georgia – the youngest of the American colonies – was initially chartered as an alternative penal destination for criminals living under inhumane conditions in Great Britain[6] – a role transferred to Australia after the revolution – these numbers were small in comparison to the flow of people freely immigrating to North America each year on religious and economic grounds (Raynal 1776). The slave-driven model of plantation agriculture was also being adopted in the southern colonies, but the proportion of residents that were enslaved remained relatively small compared to the numbers interned in the West Indies. By 1700, slaves constituted at least 80 per cent of the resident population of most sugar islands and remained so for most of the eighteenth century (Figure 6.3).[7] By comparison, the fraction of the population in any of the American colonies would struggle to exceed 70 per cent and these levels would never be sustained for more than a decade or so. The long-term contribution of slavery was in most southern states less than 50 per cent and less than 5 per cent in the northern ones (Galloway 1989). The striking difference between the levels of enslavement in the West Indian and American colonies was certainly not driven by any collective social consciousness limiting slavery in the latter. The number of plantations and slaves in fact grew at a much faster rate in the American colonies over the latter half of the eighteenth century, but so too did the number of free-born immigrants. People arrived in the West Indies for one purpose – to grow cash crops – and the modest level of governance was primarily in place to repel foreign invasion, ensure crown taxes were collected, protect merchant fleets from piracy and back-stop the planters' management of their enslaved workforces. Most American colonists were only fractionally involved in cash-crop production and the exploitation of slave labour, if at all, and certainly not dedicated to this pursuit, as was the case in the Caribbean. South Carolina's large cotton, indigo and rice plantations and high ratio of slaves to free persons (Figure 6.3) perhaps came

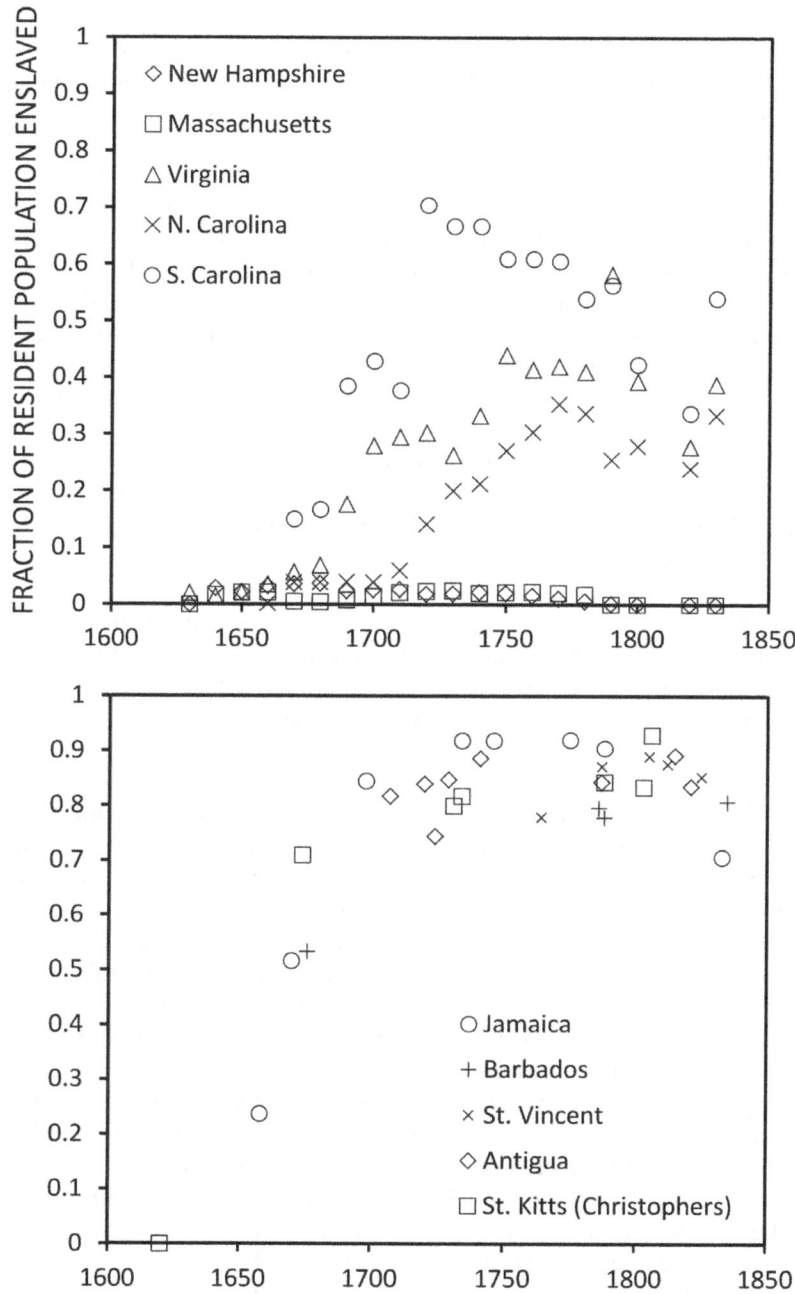

Figure 6.3 Relative proportion of resident population enslaved in West Indian sugar islands and North American colonies/states between 1600 and 1850.

closest to replicating the West Indian model of a plantocracy in the eighteenth century. But even this cash-crop colony lacked the firm social and political connections to Britain that were evident in the sugar islands, although these were certainly greater than those in New England. The colonial and revolutionary-era historian Jack Greene suggests:

> neither in their materialistic orientation, their disease environments, their number of African inhabitants, their concern to cultivate British values and institutions, nor perhaps even their commitment to the colony was there a sharp break between island and mainland societies. Rather, there was a social continuum that ran from the Caribbean through Georgia and South Carolina to the Chesapeake through Pennsylvania and New York to the urban and then rural New England.
>
> (Greene 1996)

While available data on African enslavement support Greene's observations, the suggestion that southerners were more likely than northerners to return to Britain or had greater vested interests there is not supported. Adam Smith commented that

> I have never even heard of any tobacco plantation that was improved and cultivated by the capital of merchants who resided in Great Britain, and our tobacco colonies send us home no such wealthy planters as we see frequently arrive from our sugar islands.
>
> (Smith 1776)

Certainly the southern colonies, due to their plantation economies of cotton, indigo and tobacco, maintained a continuous trade surplus with Britain compared to chronic deficits of the northern colonies. But the notion of the "planter class" of the West Indies "[cultivating] British values and institutions" disregards the bulk of work by the likes of Hobbes, Locke, Smith, Burke and other notable British reformers of the Enlightenment that did not whole-heartedly share the views of the plantocrats and self-serving capitalists that invested in cash-crop production. These were more in line with the views of the northern colonies.

Most importantly, the southern colonies could not cultivate sugar cane in any meaningful amount due to the vicissitudes of winter weather, while the Caribbean was almost entirely dedicated to its production, but still capable of growing the main crops of the southern colonies. The southern colonies lacked the climatic monopoly that sugar brought to the American tropics at the time. The bioproductivity advantage that most of the West Indian islands, the Guiana Coast and the Brazilian sugar-growing regions held over the North American colonies, particularly those in Canada and New England, was enormous. Virginia, the Carolinas and Georgia represented the best growing conditions in the American colonies, but still offered only half of the productivity potential found in the very best cane-growing regions of the neotropics. Akin to the situation in Europe, a significant proportion of the bioproductivity in Canada and New England rested with the surrounding oceans and this marine NPP diminishes southward while terrestrial productivity takes off at the lower latitudes within the tropical belt. Canadians and New Englanders in particular had little choice but to turn to the sea to survive and it was the flourishing fisheries industry that spawned a large part of the demand for rum.

The inability of the North American colonies to cash-crop north of the Mason–Dixon line left many of them with a chronic trade deficit with Britain – a negative balance that would be sustained throughout most of the eighteenth century and the early decades of the nineteenth century (Figure 6.4). The small surpluses generated by the southern plantation colonies were insufficient to mask the much larger deficits of the import-dependent and more populous New England region. By the 1770s, exports to the American colonies were accounting for roughly 20 per cent of Britain's total exports (McCusker 1971) and only the demands of a growing British military presence in the colonies and the subsequent loss of markets during the ensuing wars witnessed a reversal of this deficit for any significant length of time. These deficits were, of course, also a consequence of the Navigation Acts and the moratorium placed on international trade of North American manufactured goods, but find their roots in the bioproductivity constraints to cash-cropping in the region and, importantly, Europe. In contrast, the British West Indies consistently produced sizeable trade surpluses that grew (in nominal terms) over the same period, either matching or exceeding the total trade value of their North American siblings (Figure 6.4). Smaller populations of free citizens in the sugar islands invariably put a cap on demand for British manufactures while the production of sugar, coffee and other tropical commodities benefited from increased demand and investment prior to emancipation (Table 6.1). But the trade surplus that the West Indies sustained over the eighteenth century needs to be seen in a different light than the deficits accruing in the north since only a modicum of financial benefit from these accounts actually translated into improve infrastructure and capital expenditure beyond the plantations. Much of the value was taken off by the various shipping agents and importers, customs and excise officers in Britain and the plantation proprietors – often domiciled in Britain – preferred to spend profits there and in Europe than in the sugar islands.

The nature of governance in the American colonies was also very different from that prevailing in the West Indies. Individual colonies were linked to the Crown in more administratively complex and diverse ways, some of these approaches providing substantively greater opportunity for home-rule than was evident in the sugar islands. By the mid-1700s, most had witnessed modifications to their underlying charters from the British

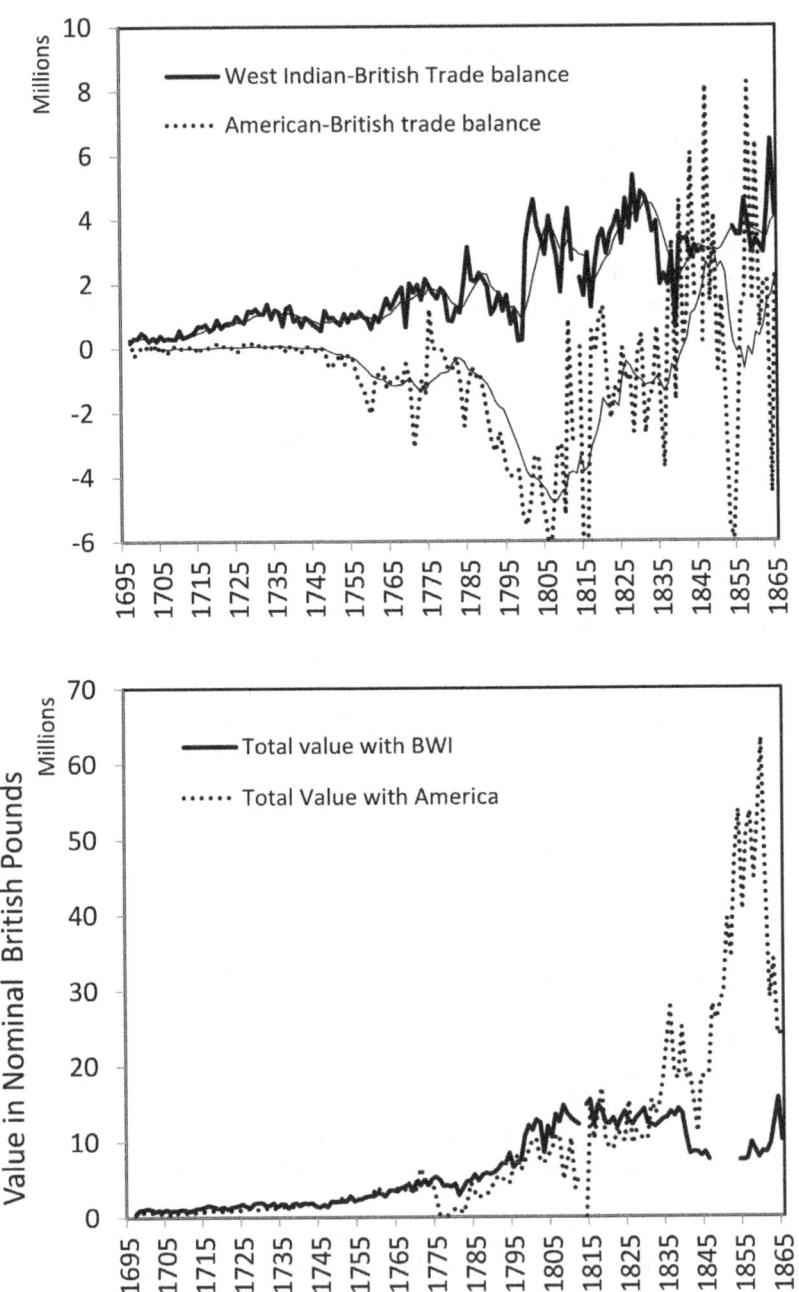

Figure 6.4 A comparison of the balance of trade and total trade value between Great Britain, British West Indies and British North America/United States.

government. This was in part as a response to changes in the way that over-seas possessions were administered in the tumultuous aftermath of the English Civil War, the subsequent restoration of the monarchy and the Glorious Revolution that dominated British politics from the initial estab-lishment of colonies in North America up to the Hanoverian takeover in 1714. During this period a sea-change in the way power was wielded in Great Britain was underway. The Crown was stripped of its royal preroga-tive and the functions of the Privy Council were transferred to Parliament, effectively ending any vestige of monarchical control over the administra-tive affairs of the nation. These developments did not, however, lead to greater conformity in the colonies nor a liberalization of authority. Some measure of supervised self-governance was politically expedient given the distances between Britain and her colonies. It may in some way also have encouraged immigration that was, at the time, of primary concern to Britain in her efforts to secure North America against the French and Spanish. But it may also have encouraged more independent-minded souls rather than Crown loyalists to take up the challenge of forging a new life. Unlike the West Indies, these underlying political tremors became ensconced in the American colonies and none more so than in Massachusetts Bay.

Most colonies were established after a promoter had received a royal patent or charter. Charters detailed the rights and privileges of the propri-etor, the Crown and the colonists and always ensured that the European aristocracy received a portion of future economic activity and preference in land transactions. Colonies that were won through war customarily con-formed to policies constructed by the Crown and Parliament in anticipa-tion of a future grant being issued (Edwards 1743). But the royal charter assenting to the establishment of a new colony in Massachusetts was not completed before the Pilgrims set sail and this void – although hardly unusual for the time – was exploited by the Puritans as an opportunity to develop their own. Their final document – the Mayflower Compact – was completed during the first crossing and was distinct in that it conformed more closely to their own vision of a new society and less so to the eco-nomic interests of the British Crown. The Pilgrims – perhaps contrary to popular representations – were an intolerant religious sect that had yet to land in the New World and were already challenging the established order. They had sailed to America seeking not economic advancement but greater social independence, although the former would invariably prove fundamental in achieving the latter.

The Massachusetts experience was not replicated throughout the entire colonial assembly. Others, such as Maryland and Pennsylvania, were originally Swedish and Dutch settlements that under subsequent English grants made to the Calvert and Penn families became havens for Catholics, Quakers and other minor denominational groups. The Calverts and Penns held proprietary charters, effectively making them gentrified land-holders at a scale unseen in Britain beyond the Crown itself. Cecil Calvert, the second Baron of Baltimore, managed his lands without leaving Britain while relying on land rents to provide a return on his investment. The Penn family moved to Pennsylvania upon receiving a charter, where they actively participated and managed their affairs while seeking to grow the social and economic fortunes of the colony. They made payments to indigenous residents in lieu of lands granted under the charter and facilitated small leaseholds for less well-off families for a quitrent of a penny an acre alongside more substantive grants of 1,000 acres costing 14 and a half pounds or more (Raynal 1776). The land-grant incentives had created a large group of real property holders. More importantly, very few of the colonists that became land-owners in America viewed their residency as merely a conduit to financial wealth that would buy access to privileged British society at a later date.

It is clear that the social transformations taking place in many of the American colonies were not duplicated in the West Indies – a region that did not care to make changes to its lifeblood of cash-crop production or the methods employed to ensure that it was sustained. Fewer and fewer Europeans immigrated to the Caribbean as the plantations matured and the prime growing lands became locked into a landlord estate model. The bulk of population growth continued to accrue through African lines to such an extent that some islands, such as St. Kitts, were increasingly home to Africans that numbered many multiples of the resident Europeans, a disparate demography that began to instil a deep sense of vulnerability among planters who increasingly feared a revolt not only from slaves, but from the increasing number of escaped Africans, or *maroons*, that sporadically battled with colonial forces and planters. These fears of course would be realized in the 1759–1760 Jamaican slave uprisings and in those that subsequently broke out in the islands during the American Revolution and led to far greater consequences on the neighbouring French colony of St. Domingue (Haiti) in later years. But these violent remonstrations against enslavement were less commonly encountered in the tobacco and cotton plantations of Virginia, Carolina and Georgia (O'Shaugnessy 2015).[8]

These distinctions between plantations in the West Indies and America – demographics, trade relations and governance shaped by differences in bioproductivity – loomed large in the way the colonial regions evolved. The consequence of these growing differences between the two regions was that American interests were rarely represented in Parliament while West Indian interests were fiercely defended. Richard Sheridan, the eminent economic historian of the West Indies, calculated that before the American Revolution, only five North Americans had sat in Parliament while more than 70 West Indians, or those with financial interests in the plantations, had been members (Sheridan 1974). The superior sugar constituency won over the law-makers, resulting in the passage of the Molasses Act of 1733. Virtually everything that they were seeking was enshrined.

A public statement from a parliamentary office in Whitehall explained the outcome of the vote:

> The warm Dispute between the British American Sugar Colonies, and the British Northern Continent American Colonies, which had lasted from the Year 1731; concerning the Trade of the latter, with the French, Dutch, and Danish Sugar Colonies, of whom they took off (in Exchange for their Provisions, Horses, Lumber, &c.) considerable Quantities of their Sugar, Rum, and Molosses [Molasses]; and of which, under the said Year 1731, we have fully treated; was, in this Year 1733, finally terminated, by a prudent Temperament of an Act of Parliament of this said 6th Year of King George the Second, For the better securing and encouraging the Trade of his Majesty's Sugar Colonies in America.
>
> The Preamble to which sets forth, That whereas the Welfare and Prosperity of your Majesty's Sugar Colonies in America are of the greatest Consequence and Importance to the Trade, Navigation, and Strength of this Kingdom. And whereas the Planters of the said Sugar-Colonies have of late Years fallen under such great Discouragements, that they are unable to improve or carry on the Sugar Trade upon an equal Footing with the foreign Sugar Colonies, without some Advantage and Relief be given to them from Great Britain. Be it therefore enacted,
>
> I. That the several after-mentioned Rates and Duties be granted, viz. upon all Rum of the foreign Sugar Colonies which shall be imported into any of the British Plantations in America...
>
> II. That no Sugars, Paneles, Syrups, or Molosses, nor any Rum or Spirits of America, except of the Growth of his Majesty's Sugar Colonies, shall be imported into Ireland, but such only as shall be laden and shipped in Great Britain in Ships according to the Navigation Laws; under Forfeiture of Ship and Cargo...
>
> The Sugar Colonies, viz. Jamaica, Leeward-Islands, and Barbadoes. By the last Returns which we have had from those Islands to our circular Queries, we do not find that they have any other Manufactures established, beside those of Sugar, Molosses, Rum, and Indico [Indigo], of their own Produce...
>
> From the foregoing State it is observable, That there are more Trades carried on and Manufactures set up in the Provinces on the Continent of America to the Northward of Virginia, prejudicial to the Trade and Manufactures of Great Britain (particularly in New-England) than in any other of the British Colonies; which is not to be wondered at: For their Soil, Climate, and Produce, being pretty near the same with ours, they have no Staple Commodities of their own

Growth to exchange for our Manufactures; which puts them under greater Necessity, as well as under greater Temptation of providing for themselves at Home: 4 &c. [and] Whether it might not be expedient to give those Colonies proper Encouragements for turning their Industry to such Manufactures and Products as might be of Service to Great Britain, and more particularly to the Production of all Kinds of Naval-Stores. – Whitehall, Feb. 15, 1731–2. Paul Dockminique &c.

The message from Parliament to the American colonies could not have been any clearer – abide by the rules established by Parliament under the Navigation Acts, work towards the over-arching benefit of the British Empire, re-consider your economy towards supplying raw materials needed by Britain and refrain from trading with other European nations, except through the explicit consent of the mother country. It also hinted at the impact that an underlying low level of bioproductivity across New England was having on relations – the climate could not sustain cash-crop production and was inevitably driving the region towards a manufacturing economy that would "[be] prejudicial to the trade and manufactures of Great Britain …". The Molasses Act also extinguished any lingering doubt in the colonies as to the most-favoured region in the Americas.

During the early to mid-eighteenth century any direct trade with colonies was largely conducted within the confines of empire. This was the primary reason that the award of the *asiento* during the reign of Queen Anne was celebrated in Britain – it was thought they had broken the imperial confine of their main competitors in the New World. The Molasses Act was Britain's stop-gap measure to prevent this from happening to their imperial confine. But the wheel of free commerce was spinning faster and the burgeoning demand for rum in the American colonies was turning towards a supply – the French islands – that had no imperial price support for molasses. The Americans – more precisely, the New Englanders, New Yorkers and Pennsylvanians – were exploiting the interstices between British and French imperial trade preferences and protectionism to cultivate growth in the first great American manufacture – rum.

The apparent favouritism expressed in the Molasses Act underscored the economic model chosen by the British – one focused on cash-crop and raw material production in the colonies, one intent on exporting British manufactures exclusively to those colonies, and one designed to deny their European rivals the *specie* that was considered the bedrock of a nation's financial wealth. In effect, mercantilism. The West Indies conformed to this model, but the American colonies were increasingly disrupting it. It is important to understand that the severe attitude that many British appeared to take towards North American industry had been entrenched well before passage of the Molasses Act in 1733. In 1719, a Bill entitled *For rendering the laws concerning the importation of naval stores from the British American plantations more extensive* was tabled at Parliament to allow a wider range of North American timber species to be exported to Great Britain (Anderson 1764). In part this was aiming to relieve ships arriving in the ports of New England laden with manufactured goods from having to return to Britain unladen – an uneconomic situation for the British shippers – but it was also a symptom of the chronic trade deficit between the two regions. Further, certain clauses in the Bill went as far as to destroy a nascent metalwork industry by ensuring:

> That none in the Plantations should manufacture Iron Wares of any kind whatever, out of any Sows, Pigs, or Bars whatsoever; under certain penalties.

To which was added by the House of Peers (Lords):

> That no Forge, going by Water or other Work whatsoever, should be erected in any of the said Plantations, for the making, working, or converting of any Sows, Pigs, or cast Iron into Bar or Rod Iron, upon Pain, &c.

A modest ironworks had been developed in Virginia in 1714 – the first of its kind in North America – and it was clear that British interests wished to destroy any form of metalworking industry before it prospered. To the relief of the colonists, the Bill failed to pass, in part out of fear that it would further push New England towards their French rivals, but the over-arching intent of their British patrons must have begun to make itself clear, leading up to the later assault on their growing rum industry. Thirty years later, a modified Act "to encourage the importation of pig and bar iron from his Majesty's colonies in America" would successfully pass through Parliament (Anderson 1764). This Act acknowledged the need for some latitude to be given to the colonists, but it also aimed to put a cap on the extent of any ironworks industry by "prevent[ing] the erection of any mill or other engine for flitting or rolling of iron; or any plating forge to work with a tilt hammer; or any furnace for making of steel, in any of the said colonies". If the New Englanders could not grow cash crops due to natural bioproductivity constraints, but could not produce rum, iron or other industrially manufactured goods, what were they to build their economy around? Apparently, the raw materials needed to feed British artisanal, industrial and naval works for the production of refined goods that would in part be exported by return to New England.

The American colonies supplied vast quantities of masts, planks, pitch, tar and turpentine to the Royal Navy, both in Britain and the West Indies. In 1711, the British government under Queen Anne had made the unli-censed felling of any pine tree in New England a crown offence, punish-able by a fine of £100 (£12,700 in 2015) (Anderson 1764). The industry was vigorously promoted in the colonies by Britain as a means to end their dependence on Scandinavian supplies. This source was seen as risky, since the supply of materials from Sweden had been abruptly curtailed in the past during times of war (Raynal 1776). These materials – referred to as "naval stores" due to their critical importance to ship maintenance – were mainly supplied from the pine and hardwood forests of New England and, later, the southern-pine forests of the Carolinas. Queen Anne's Act of 1711 *For the preserving of white and other pine-trees, growing in her Majesty's colonies … for the Masting of her Majesty's Navy* proved extremely useful to the Navy while saving funds that otherwise were being sent to the king-doms of Norway and Sweden. But it only placed further restrictions on New Englanders that were unlicensed or had other purposes in mind for the coniferous timber resources of the region.

The French and Dutch colonies deep inside the tropics were ideal loca-tions for the production of sugar cane, but the surrounding forests were not a good source for naval stores or materials needed by coopers or mill operators and the American colonies were able to supply both in abun-dance. They also suffered from shortages of livestock – often needed to power sugar mills – that the North Americans were able to supply within a voyage of four to six days compared to a trip of nearly three weeks from France or the Netherlands. The Molasses Act was designed to ensure that the French and Dutch colonies were isolated from the growing American economy, while furthering the British West Indies, their planters and British merchants through forced trade loyalty to London.

The Molasses Act did not have the intended effect. Instead, it drove the American molasses trade deeper into a shadow economy powered by smuggling. Partly due to the Act, this practice not only became the default approach to feeding America's rum industry but also sparked an increase in the volume of manufactured trade goods sourced outside British entrepôts, further undermining efforts to rein in the rapidly expanding trade between the American colonies and other European nations. Cor-ruption on the part of merchants and customs officials – in both the sugar islands and the American colonies – ran rampant. Fines as a portion of the value of the cargo, when imposed, often amounted to less than the cost of meeting the prescribed import duties. Rum prices continued to reflect this duty-free cost of molasses and became an even more lucrative business for many colonial merchants. Many simply did not believe the hefty increase to the import duty on foreign molasses and sugar to be socially just and were dismissive. Adam Smith wrote about the smuggler that he

> would have been, in every respect, an excellent citizen had not the laws of his country made that a crime which nature never meant to be so. In those corrupted governments where there is at least a general misapplication of the public revenue, the laws which guard it are little respected.
>
> (Smith 1776)

Abbé Bernay wrote "It is thus tyranny that has given birth to contraband trade. Transgression is the first effect produced by unreasonable laws" (Raynal 1776). Colonial governments, merely seeking to appease the sugar island constituency, were not eager to enforce the Act and very little effort was expended by the Crown itself to curtail the illicit trade.

The primary intent of the Molasses Act was to alter the economic behaviour of the American colonies while allaying concerns of the West Indian planter class. It did just so, but in the perverse form of widespread and systematic smuggling rather than expunging the trade in illicit foreign goods. But by the time the Act was due to expire in 1764, circumstances had changed and the British government, already carrying an enormous public debt of more than £75 million [£1.1 trillion in terms of UK GDP in 2015] found itself in yet another expensive war with one of its European rivals. The Seven Years' War would galvanize the British Empire and curtail French ambitions in the Asian and American tropics, but at considerable cost. It was a war fought almost simultaneously on four continents, beginning with India. Competition for control of the important trade ports of the Carnatic region in the southern provinces had already seen the French and English forces clash sporadically since the 1740s and the Seven Years' War was in this context more of an escalation and conclusion to a multi-decadal struggle between the two powers. This conflict ran in parallel with a full outbreak of hostilities in North America over competing claims to the large expanse of the Ohio River Valley while simultaneously seeking to reduce the French influence over trade in sugar, coffee and cotton from the Caribbean. The consequences of the war are often attached to shifting fortunes in Europe, but the wider implications of French losses in India, the Caribbean and North America made it clear that the conflict promoted the British to the apex of a new order. Global trade supremacy, which was at first Portuguese, then Spanish, Dutch and soon to be French had instead now become British.

Sugar-fuelled revolution

The financial burden of achieving this position – nearly doubling the outstanding public debt to more than £134 million by 1764 [a staggering £1.6 trillion today] (Mitchell and Deane 1962) – re-directed Parliament's legislative objectives from trade protection towards revenue collection. When the Molasses Act expired, London promptly replaced it with the Sugar Act of 1764 in an effort to counteract the outflows. At the very least, the Whig Prime Minister Grenville hoped to stem the flood of deficit spending, if not claw back some of the expense already incurred to consolidate control over the colonies. It appeared to have at least some effect on reducing the debt, which declined from the peak of £134 million in 1764 to £127 million in 1775 (Mitchell and Deane 1962). The Sugar Act halved the customs duty on sugar from six pence to three pence per gallon, but also made clear that collection of duty would be vigorously pursued and punished. The Act was complex and detailed in comparison to its predecessor, more than three times as long and containing 37 different enactments (sections), compared to only 13 in the Molasses Act. It also initiated a complex system of checks and verifications – many that are standard procedures today – by introducing bills of lading, customs certificates and registrations. No stone was left unturned. Surety bonds were required for each ship registration. This bond, amounting to £1,000 – a princely sum in the eighteenth century equivalent to just over £123,000 in 2015 – would be forfeited upon any infringement of the Sugar Act. Any counterfeit, forged or modified document would incur a fine of £500 (£61,500 in today's money) and attempts, successful or otherwise, to bribe customs officials would be subject to a £50 (£6,150) penalty. Moreover, any unsuccessful effort to plead against unfair seizure by customs officials in court would demand that the plaintiff pay three times the court costs to the defendant. As one can imagine, such dramatic change in crown policy towards the molasses, rum and sugar trade in America only further alienated the colonials, particularly those in the northern colonies that had finally found an industry that they could grow while continuing to meet with the tight controls on manufacturing the Crown had mandated. The insidious nature of the new Act's demands was interpreted as punishment, particularly in New England.

The Sugar Act, perhaps better referred to as the American Revenue Act, established new duties on a long list of commodities and manufactures commonly entering and leaving the American colonies, of which sugar and its by-products were selected for special consideration. These remained among a select group of products, along with cotton, silk, tobacco, tea and coffee, that was subject to a constantly growing demand in Europe and North America and made for an obvious target when revenue-raising was necessary to defray the exorbitant costs incurred during wars. But like the Molasses Act, it was clear which colonies were being favoured and which were – if not being punished – then being conveniently exploited:

> An act for granting certain duties in the British colonies and plantations in America, for continuing, amending, and making perpetual, an act passed in the sixth year of the reign of his late majesty King George the Second, (intituled, *An act for the better securing and encouraging the trade of his Majesty's sugar colonies in America,*) for applying the produce of such duties, and of the duties to arise by virtue of the said act, *towards defraying the expences of defending, protecting, and securing the said colonies and plantations;* for … more effectually *preventing the clandestine conveyance of goods to and from the said colonies and plantation,* and improving and securing the trade between the same and Great Britain.
>
> (Whitehall, 5 April 1764)

The Act was designed to secure the sugar colonies and cover the cost of British troops in the Americas by making the northern colonies pay for it, mainly through a curtailment of the illicit molasses trade, but also through new levies placed on most other trade goods. The colonies that had been most engaged in the fight against the French – New England, New York and Pennsylvania – were also the main home to the rum industry and were most heavily hit by the restrictions placed on the importation of French and Dutch molasses. Tax collection under the new Act sky-rocketed. Revenues from the enforcement of the existing statutes under the Navigation Acts typically yielded between £1,500 and £3,000 per year. The introduction of measures under the Sugar Act of 1764 and its replacement of 1766 brought in an additional £14,000 to £40,000 per annum over the period leading up to the American Revolution (US Bureau of the Census 1975). The inevitable result was increased dissatisfaction in the colonies with the way that Parliament viewed them, particularly from Massachusetts. The famed Bostonian lawyer and libertarian James Otis Jr. wrote:

> I think I have heard it said, that when the Dutch are asked why they enslave their colonies, their answer is, that the liberty of Dutchmen is confined to Holland; and that it was never intended for Provincials in America, or anywhere else. A sentiment this, very worthy of modern Dutchmen; but if their brave and worthy ancestors had entertained such narrow ideas of liberty, seven poor and distressed provinces would never have asserted their rights against the whole Spanish monarchy, of which the present is but a shadow. It is to be hoped, none of our fellow subjects of Britain, great or small, have borrowed this Dutch maxim of plantation politics.
>
> (Otis 1764)

By default or design, the British government had followed precisely the Dutch example in building their empire, and the American colonies would continue to be subject to the same centralized authority as the other regions. Additional Revenue Acts followed the Sugar Acts of 1764 and 1766. Seeing the consequences of these policies being somewhat blindly promoted in the House of Commons, Thomas Pownall, a former crown-appointed Governor of Massachusetts turned MP, wrote of his deep misgivings on the course being furrowed by government:

> It makes no difference in the matter of the truth, whether the government of England should be averse to the extending of this privilege to the Colonies; Or whether the Colonies should be averse to the receiving of it: –
> That the British isles, with our possessions in the Atlantic and in America are in fact united into a one grand marine political community: And ought therefore, by policy, to be united into a one

Imperium, in a one center, where the seat of government is: And ought to be governed from thence, by an administration founded on the basis of the whole; and adequate and efficient to the whole.

(Pownall 1764)

Pownall was speaking of a more complete political union wherein Britain's colonies across the Atlantic would share the same rights and privileges – including political representation – as subjects living in Britain. He had lived and administrated in the colonies for many years and understood the daily issues that were most important to those living there – an experience that could rarely be ascribed to other members of government, including the king himself. Instead, the bulk of parliamentarians and politicians chose to expand their knowledge of overseas life via a Grand Tour of Europe, not the expanding overseas territories.

In the wake of the Stamp Act, New Englanders promptly organized a meeting with the other colonies, where they agreed to boycott the purchase of British paper goods, forcing repeal the next year. By this time it should have become clear to those in government that the American colonists were not going to simply relent and conform to the demands of Parliament. But in response, Parliament instead passed the American Colonies, or Declaratory, Act of 1766, stating that Parliament's power to tax the American colonies was the same as its rights to do so in Britain. Of course, any right to representation in Parliament was not extended to the American colonials – these would continue to be heard by the British government through their appointed governors. The British West Indies – seemingly impervious to the growing disturbances in New England – had been engaging this plea approach for decades in their attempts to redress concerns over the American purchase of molasses, increases in the export sugar duty and the re-export sugar trade from Ireland (O'Shaugnessy 2015). But the Americans were far less inclined to view this beggared approach to political discourse as an acceptable substitute for a debate by equals that was customary in their colonial assemblies. James Otis Jr. famously described the rapid change in British views on duties as "taxation without representation is tyranny". The Declaratory Act was quickly followed by a series of bills submitted by the Chancellor of the Exchequer, Charles Townshend. Eponymously referred to as the Townshend Acts, these five measures put paid to any prospect of reconciliation between the American colonists and the British Crown by applying further duties on various imported manufactures and, importantly, tea. Tea did not figure as importantly as coffee and rum in the trophogeography of pre-revolutionary America, due to the fact that it was produced in Asia and coffee and molasses were produced in the Caribbean and subject to lower duties. Nonetheless, it remained a staple of many colonial families and up to half a million tons were imported annually from Britain. This was particularly true in the north colonies. New England and New York alone accounted for three-quarters of the tea imported from Britain during the 1760s and the Townshend Acts made this staple more expensive (US Bureau of the Census 1975). More importantly, this increased tax on tea in the colonies was designed to partially offset a reduction in duty on tea imported to Britain by the East India Company – the other offspring that was showered with unrivalled favouritism by a Parliament with strong connections to shareholdings in the company.

The Townshend Act duties were also being levied on British goods, not Dutch or French, and did not have any other narrow purpose than to increase revenue flowing to the British government and its overseas agents. The broader purpose – embodied in the wording of the Revenue Act – was to remove the salary-payment facility of the colonial assemblies that had been in place since founding and to repatriate this important function to central government in London by applying the new revenue towards paying crown administrators and judges overseeing matters in the colonies. Further, the Acts established a series of new military courts that would – absent any jury – prosecute trade-related crimes. But these Acts, instead of establishing control and curtailing insubordination, galvanized rebellion. By the mid-1770s, Massachusetts – the colony most clearly intent on suffering no overlord since the signing of the Mayflower Compact and most aggrieved by the policies of Parliament – was on the cusp of open rebellion. Sir William Pulteney, a Scottish MP and Baronet said of Massachusetts in his assessment of the factors that led to the revolution:

> The constitution of the Massachusetts Bay is, by far, the most republican in America, Rhode Island perhaps excepted. It is that Colony which has always been considered as the most averse to the English government, and to have been chiefly instrumental in lighting up the present flame in America.
>
> (Pulteney 1776)

The dissatisfaction of leaders in Massachusetts with Parliament's clear intent to project its unchallengeable authority over the colonies would eventually precipitate a series of violent acts in Boston, Lexington and Concord, transmuting social and political disquiet into the American Revolution.

Breaking king cane

Sugar cane – the only practicable source of sweetener able to meet the burgeoning demand in Western Europe and North America up to the close of the eighteenth century – would lose its dominant grip on the tropical trade by the middle of the nineteenth century. A series of events would transpire to irrevocably alter the economic landscape of Europe's mainstay tropical cash crop in the wake of the American Revolution, increasing the costs of production while placing downward pressures on prices. Combined, these factors cast a dark shadow on anyone seeking to achieve the sort of profits accrued by their predecessors and, as a result, cane sugar production declined in many of its former strongholds, particularly across the West Indies.

The first of these factors was the increasing uncertainty of slave-driven plantation systems propagated by the Haitian Revolution. The French-controlled portion of the island of Hispaniola hosted nearly 8,000 plantations spread across more than 9,250 square kilometres; dedicated to the production of sugar cane, coffee, indigo, cotton and cocoa, these dominated the coastal lowlands (Rainsford 1805). Within the context of the global trade in tropical commodities during the late eighteenth century, Haiti was a titan. Various estimates suggest that it produced almost half of Europe's total imports of sugar and coffee. Nearly 67,000 tonnes of sugar were produced on average each year during the 1780s, amounting to around half of the total sugar production across all of the British colonies in the Americas – a staggering amount relative to the area of land under cultivation (Edwards 1798). Rainsford (1805) estimated that the network of agricultural fields supporting these exports was worked by approximately half a million Africans with another 40,000 French residents on the island and over 24,000 free persons of mixed race. The combination of growing republican sentiment, over-sized economic influence, a large population of slaves and mixed-race freemen at the geographic heart of the Caribbean primed the island for violent change. A timeline that would begin with the French *Declaration of the Rights of Man and Citizen* published in 1789 would end 14 years later with the expulsion of numerous European forces sent to quell rebellion and re-establish the plantocracy. The emergence of the first tropical possession to achieve independence would prove to be an inflection point in colonial control of the Americas. Without the hard hand of slavery to run plantations, sugar production in Haiti collapsed, never to recover the pre-eminent position that it had enjoyed during the latter half of the eighteenth century. In the same year that Haiti's new leaders declared independence, Jamaica was exporting the largest amount of sugar ever, more than 100,000 tonnes (Deerr 1950) (Table 6.1). In part this came through the interdiction of trade with Europe, although Americans continued to trade with Haiti for sugar and molasses both during and after the revolution. Various efforts by the British to re-take the island would fail, as well as those of the Spanish, the restored French monarchy and, later, the forces of Napoleon. Haiti's resistance would become a rallying point and shaping influence for many of those remaining enslaved in the Americas or those that sought independence from European control, including Simon Bolivar, the leader of the independence movement across northern South America.[9]

A series of slave revolts in Barbados, Demerara (Guyana) and Jamaica in the 1820–1830s would ask further, serious questions of the British sugar economy in the New World. The Baptist War in Jamaica proved particularly brutal as plantation owners and managers sought to prevent future insurrections through a series of judicial executions (Rodriguez 2007). But various efforts back in Britain had also been periodically stirring a change in public sentiment towards slavery since the 1780s. A number of well-regarded Protestant reverends in Britain had been publishing their moral arguments against enslavement throughout much of the eighteenth century and petitions were sent to Parliament from time to time urging abolition of the practice (Thomas 1997). But action in Parliament by members seeking to legally abolish slavery did not begin in earnest until after the American Revolution. Led by William Wilberforce, William Pitt and Sir Charles Middleton, a move to abolish the slave trade, but not slavery, gathered momentum when Pitt submitted to Parliament in 1788, "that this house will, early in the next session of Parliament, proceed to take into consideration the circumstances of the slave trade, complained of in the said petitions, and what may be fit to be done therein" (Martin 1834). The slave trade was debated for nearly two decades thereafter, but still with considerable opposition from those representing their own interests in the trade of slaves, sugar, coffee and cotton or those of their constituents, particularly in Liverpool and Bristol. Motions to vote on abolition were persistently put to Parliament by Wilberforce between 1792 and 1799, each time failing to pass in the House of Lords. Efforts were renewed in 1804 and with initial support for interdiction of slave imports to British colonies arriving via the Privy Council, an Act of Parliament was passed in 1806, with subsequent laws making the trade in slaves a criminal act and imposing severe penalties on British subjects defying the new laws. A similar Act was passed by the Congress of the United States in 1808. Over the subsequent decades, the British would sign treaties with most of the European, African, Middle Eastern and American nations to mutually abolish the trade in slaves and permit confiscation of any ship carrying human cargos (Anon 1844). A majority of parliamentary members were convinced to debate and pass the Emancipation Act in 1833, particularly after it became clear that slave-owners – being well represented in Parliament – would receive generous compensation from a government manumission fund amounting to £20 million. The gradual manumission of slaves prescribed in the Act offered freedom to those that had been forcibly enslaved, but little else of material use.

This Act brought to an end the only economic model employed on the sugar islands. A transition period of "apprenticeship" extended slavery in all but name for a further six years, although political pressure would see it discontinued by 1835. Moreover, the Act did not end British involvement with slave labour since it wholly excluded the East Indies and colonies owned and operated by the East India Company. Universal emancipation across the British Empire would not take place until the signing of the Indian Slavery Act (V) in 1843. Nor did these measures curtail

British and other European importers from making use of the sugar, cotton, tobacco and coffee that continued to be produced using slave labour in Brazil, Cuba and the southern United States. The Spanish and Portuguese (Brazil after 1822) – originators of the Atlantic slave trade – would continue to use slave labour in their plantations into the 1880s. Emancipation was the second factor leading to the breaking of king cane as it began to deprive the sugar islands of their primary labour force. As the impact of emancipation set in, the Iberians re-asserted their control over cane sugar production via their slave-worked plantations in Brazil and Cuba. By the 1860s, they once again had become the largest global producers of cane sugar (Table 6.1). The centre of the British sugar trade – a title held by the sugar islands of the Caribbean for nearly 200 years – shifted to British Guiana [Guyana] and Mauritius, buoyed by large influxes of indentured workers, mainly from India (Knight 2014).

The continued use of slave labour in Brazil and Cuba undoubtedly provided an economic advantage in the production of cane sugar relative to the emancipated West Indies. Gentrified Cuban families formed an oligarchy, controlling land distribution and most of the capital and investment that arrived, predominantly through trade. Sugar production increased rapidly from the 1780s to the 1820s, just as the amount being produced across British and French West Indies was beginning to decline (Knight 1977). Advancements in agricultural engineering – particularly the use of steam engines and pumps to drive extractors in refining – simply increased production in Cuba, Mauritius and other colonies with economies that continued to be monopolized by sugar cane. But technological advancements also led to a much more profound change that few involved in the highly profitable cane sugar trade of the eighteenth century would have seen arriving, particularly since it emerged most improbably from central Europe.

It began when Andreas Marggraf, a German chemist, successfully extracted sugar from the root of *Beta vulgaris* in the mid-1700s under patronage from the king of Prussia. The sweetening property of beet root extract had been known for millennia, but Marggraf was the first to confirm that this had the same chemical composition as cane sugar and that it could be systematically extracted. Subsequent efforts by Marggraf and colleagues to leapfrog the barrier between scientific demonstration and industrial production proved unyielding, principally due to the very low sucrose content of the variety he used for testing. Little changed until his student, Franz Achard, expanded Marggraf's line of enquiry in 1789 through a series of experiments incorporating a much larger number of candidate varieties and different fertilizer treatments. His results would identify *Beta vulgaris ssp. vulgaris* – the white Silesian sugar beet – as the plant with greatest industrial potential, a variety that that would become the progenitor line to the global beet sugar industry. He published his findings (Achard 1809), but sugar beet production lacked the investment needed to scale-up capacity towards a national industry. Investment needs an impetus and this arrived in the form of the Napoleonic wars. The loss of Haiti, the runaway control of the sugar market by British producers and re-sellers in its aftermath and a series of trade blockades run throughout the conflicts between France and Great Britain dominated European affairs at the beginning of the nineteenth century. It was this runaway competition that provided the spark for the initial expansion of beet sugar production. The central and eastern European nations did not have tropical colonies and relied entirely on the flow of cane sugar via their westerly neighbours. The embargo of British cane sugar dried up this flow, prompting the efforts by the Prussians to further develop their own sugar beet industry. In 1811, Napoleon decreed the financial support of the French government for the development of a beet sugar industry to the amount of one million francs and demanded that thousands of hectares of farmland be consigned to the project. Factories were built and crops planted, but the end of Napoleon's reign several years later coincided with the release of sugar from the French embargo and the enormous, relatively inexpensive volumes emanating from the cane industry quickly re-exerted its dominance over the sugar economy of Europe. Faced with the renewed imports from the tropics, the sugar beet industry failed before it had started.

Many proprietors simply abandoned the industry and resigned their efforts to defeat from the overwhelming capacity and bioproductivity of the sugar cane plantations. One French manufacturer, Crespel-Delisse, defied this trend and continued to explore the prospects for beet sugar production by focusing on the introduction of cost-saving mechanization during the critical extraction process (Francis 2006). By the 1830s, the advances in sugar processing they achieved would provide a firm economic foundation for growth in the industry. Innovation proved crucial, but the consequent implementation of new tariffs on foreign-sourced cane sugar added another, more potent catalyst to investment in sugar beet production across continental Europe. Moreover, many tariffs on cane sugar were being increased just as those levied in the United Kingdom were being phased out in the latter half of the 1800s. Hundreds of refineries had been established in most northern European countries by the 1870s and beet root was accounting for nearly half of the sugar produced annually. France produced nearly four million quintals (195,800 tonnes) of raw sugar and nearly two and a quarter million quintals of molasses (110,137 tonnes) across 528 refineries in 1872 and Belgium was producing nearly four times as much sugar from its own domestic industry as it was importing from elsewhere (Levi 1876). Trade flows in sugar also promptly changed course. Prussia, once a net importer of sugar, became a net exporter by the 1860s, even running a trade surplus of six to eight thousand hundredweight with Great Britain (Board of Trade 1869). The German provinces equally demanded less and less cane sugar as their domestic industries began to produce beet sugar well beyond their annual consumptive demand. Germany's annual consumption in the 1870s was estimated at 277,000 tons while its beet sugar production hovered around 400,000 tons. At the same time, production in France, Holland, Belgium and Poland was running at or near double their domestic consumption (Anon 1876).

The emergence of the sugar beet industry in Europe – a region that continues to account for three-quarters of global beet sugar production and continues to levy significant duties on foreign-sourced cane sugar – was a watershed moment in the history of the tropical trade. Through technological advancements, it fundamentally released the extra-tropical nations from an exclusive reliance on the tropics. Combined with the epic overturning of the plantocratic system in Haiti and the subsequent manu-mission of plantation slaves across most of the sugar islands a quarter of a century later, the re-alignment of production towards central and eastern European nations in the middle of the nineteenth century destroyed the conventional rationale that argued towards monopolization. The British government, recognizing the consequences of these developments to its plantation system, opted to push for tariff-free trade after three centuries of castigating foreign-sourced goods. The relinquishment of tariffs and trade restrictions combined with the onset of industrial-scale beet sugar production sparked a massive rise in sugar consumption in both the United Kingdom and the USA (Figure 6.2).

The end of king cane though did not coincide with any collapse in demand or a re-allocation of capital away from sugar production. Produc-tion continued to grow at a steady rate up until the 1960s (Table 6.1), par-ticularly where it continued to be run through indentureship, as was the case in Mauritius and Guyana. Demand for cane and beet sugar only failed to keep up with the growth in population when technological advance-ment in agricultural chemistry brought yet another natural source of industrial sugar to market in the early 1970s – high fructose corn syrup (Figure 6.2). Instead, the end of king cane arrived through its own success. Production expanded exponentially and the gradual reduction in tariffs prodded consumer demand, particularly across Western Europe and North America where sugar had become a mainstay of national diets. Revenue for both the producers and governments declined consequently and sugar fell by the wayside of economic and political consideration. Moreover, the Industrial Revolution of the early 1800s had created new opportunities in the realm of textile manufacturing. Queen cotton became the new ruler of the tropical plant trade and tea – almost entirely from India, China and Japan – and tobacco – almost entirely from the Americas – slowly began to contribute greater customs and excise revenue in Britain than sugar over the latter half of the nineteenth century. But there can be little doubt as to the Herculean role that sweet salt – as a foremost component of the tropical trade – had played in re-aligning societies, economies and cultures over the 400 years leading up to the dawn of the twentieth century.

Consequences of the tropical trade

Economically, it acted as the lever that catapulted trade from isolated, regional markets to global networks of exchange. Early trading between northern and southern Europe and the Middle East was utterly transformed by the Iberian discoveries and the developments that followed. Vast resources were sunk into navigation, naval technologies and armaments in order to improve the efficiency of long-haul voyages and better compete against European rivals. Naval and merchant fleets became significant sources of employment. Fishing villages and small port towns were transformed into powerful port cities through the expansive trade in tropical commodities, slaves and maritime supplies. Liverpool, Bristol and, later, much of the East End of London, expanded in the wake of the increasing trade generated with the East and West Indies. On the continent, the French ports of Nantes, Bordeaux, Le Havre and Cadiz and Seville in Spain, Portugal's capital Lisbon and the twin behemoths of Rotterdam and Amsterdam in the Dutch Republic flourished largely as a consequence of the tropical trade.

Tropical commerce was also the primary intent behind an entire class of new joint-stock companies established throughout the seventeenth and eighteenth centuries and these contributed significantly to the formation of early stock-markets and development of prototypic "multi-national" corporations. Along with these institutions came many of the earliest forms of financial innovation and malfeasance. The South Sea Company and "bubbles" of the late eighteenth and early nineteenth century utilized the profitability of the tropical trade to create a chimera of public borrowing ensconced within a shell of embellished financial prospects. Ultimately, in the wake of the South Sea fiasco, the Bank of England emerged as the first "lender of last resort" to the government and, along with the Dutch and French central banks, evolved into the ultimate arbiter of monetary policy, as we know it today.

Perhaps most importantly, governments quickly discovered that they could extend their domestic taxation policies to the trade in tropical commodities, increasing their revenues by orders of magnitude. The birth of custom and excise duties was a master-stroke of public finance. In England, the levying of duties on the production of beer and ale in 1643 and those on tobacco, silks and soaps soon thereafter brought the realization that accounting for product and consumption could create a stable, reliable, yet entirely fungible basis for financing the military and civil expenditures of government (Ashworth 2003). It also made clear the effects that duties could have on promoting certain products over others. Within a year, additional excise duties had been placed on "foreign goods", mainly wine, raisins, currants and, significantly, sugar. Fifteen years forward and duties were extended to tea and drinking chocolate at 1s 4d per gallon and coffee at 8d. By 1660, all of the major beverages being consumed in Britain would be subject to customs and excise and sugar was an integral ingredient or accompaniment to most. With these expanding revenue streams, excise duty collection in particular became a serious undertaking with opportunities for considerable financial gain. These attracted both significant financial innovations – that would play a formative role in defining modern finance – and corruptions – that would remind those exposed to innovation that these were rarely devoid of risk. By expanding the revenue potential of their governments through customs and excise, the European nations primarily engaged in the tropical trade were also able to extend their "creditworthiness", allowing them to borrow increasing amounts of capital. For example in 1690, sugar and tobacco – the principal tropical imports at the time – accounted for no more than 14 per cent of customs and excise in England (Whitworth 1731). By 1795, tropical commodities amounted to 20 per cent of all customs and excise duties collected by the British government and sugar alone accounted for 26 per cent of all customs duties, making it the most lucrative revenue-generator of any imported product. Only excise duties on domestic beer and ale remained greater (Dodsley 1800). By the mid-1800s, sugar and molasses continued to account for more than a quarter of the revenue collected from customs duties. Tropical products combined – mainly sugar, tea and coffee – were making up more than half of customs revenue and nearly a fifth of the government's total annual revenue (Fonblanque 1870). The capacity to borrow in Western European nations against their future revenues played a crucial role in their successful establishment of the first global trading empires. Monopolistic mercantilism grew with the expansion of naval power and this, in turn, was the most obvious product of the government's expanding capacity to borrow on credit. Funding the military and civil list were the primary functions of government expenditure up to the middle of the nineteenth century in Great Britain (Mitchell 1971) and sugar, tea, cotton and tobacco, subject to ever-growing demand, became the jewel in the crown of taxation during the 150 years leading up to the take-off of the Industrial Revolution. Consensus regarding the precise role of the plantation economy in the lead up to the Industrial Revolution is difficult to find among modern historians, but many regard the tropical commodity trade as paramount in the development of the markets and business structures that would underpin later industrial expansion. As D'Avenant writes at the time:

> Whoever looks strictly and nicely into our affairs, will find, that the wealth England had once, did arise chiefly from two articles: 1st, Our plantation trade. 2dly, Our East-India traffic. The plantation trade gives employment to many thousand artificers [skilled craftsmen]

here at home, and takes off a great quantity of our inferior manufactures. The returns of all which are made in tobacco, cotton, ginger, sugars, indigo, &c. by which we were not only supplied for our own consumption, but we had formerly wherewithal to send to France, Flanders, Hamburgh, the East Country and Holland, for 500,000l. per annum.

(Whitworth 1731)

Through the tropical trade, Europeans had for the first time in human history globalized the connection between centres of raw material supply and those of secondary manufacture and had developed – largely through trial-and-error – the economic, business and military instruments needed to sustain such far-flung linkages. No other had achieved as much as the Western Europeans in this regard, although as we have seen, this also exacted a significant toll on those societies and ecosystems supplying the materials – a toll that in many tropical nations has proved difficult to overcome.

Politically too, the expansion of commerce into the tropics changed the landscape in Europe. A litany of wars was fought in an attempt to parley territorial concessions in the tropics or to deter rapid successes of one nation in monopolizing trade over another. Less than a half-century after Columbus' arrival in the Caribbean, the Spanish and Portuguese had already engaged in direct conflict with each other in Brazil during the Iguape War. This was the first of many conflicts that would see the age-old rivalries between European nations exported beyond their geographic borders. But the growing recognition of the importance of the tropics to European national ambitions also played a fundamental role in altering the political geography of Western Europe. The new territory and trade linkages acted to embolden European leadership while providing the economic means to implement their ambitions through military conflict. In this context, the tropics acted as the fuel driving Western European ambitions as it simultaneously vented the exhaust of its myriad conflicts. The most transformative among these conflicts was that sustained from 1568 to 1648 between Spain and the breakaway Dutch Republic. The break-up of the Spanish Lowlands would lay the framework for the modern nations of the Netherlands and Belgium. The ability of the Dutch to establish a vibrant tropical trade across South-East Asia and the Caribbean – largely at the expense of the Portuguese and underpinned by *mare liberum* – provided them with the financial and political clout necessary to outlast Catholic Spain's persistent attempts to regain control of their lost territory at a time when the future of an independent Dutch state was far from assured. Conversely, the failed Scottish attempt to gain further autonomy by establishing a commercial colony of their own in Panama only led to them forming a closer union with England in an effort to cover financial losses.

Europe's tropical possessions extended and diversified the natural resources that were available for development and this in turn was reflected in their markets and the amount of economic capital. By the early 1800s, Great Britain was importing upwards of 75 different classes of goods and at least a third of these arrived from foreign and colonial lands in the tropics (Powell 1825). Even wool and linen – the stalwarts of Europe's textile industry since the Middle Ages, would give up their dominance to cotton (calicoes). By the middle of the nineteenth century, it would surpass wool and linen as the textile material of choice, transforming Europe's cloth trade and the garment industries supplied by it. As with cane sugar, the productive potential of cotton would first be recognized in India, only to later be industrialized by Europe and North America through plantation agriculture and factory-based manufacturing. By 1838, estimates suggest more than 1,800 cotton mills were operational in the United Kingdom, employing more than a quarter of a million people, particularly in the coal-rich and poverty-laden Lancashire region (Crafts and Wolf 2013).

It is impossible to overstate the catalytic effect these new materials had in the expansion of European economies, and consequently, the foundations of our modern, global economy. It is clear that Europe had reached a "terminus" as late as the sixteenth century in terms of resource exploitation and the tropics offered a vast new expanse of possibilities. In the Americas, much of this expanse was turn-key, due to the indigenous effort that had been put into identifying, selecting and improving native plants, such as chilli, cocoa, vanilla, cassava, maize and tobacco, towards agricultural production. In Asia, the process of adding value was also highly advanced, but unlike the American experience, many Asian societies were able to establish a reasonable basis for trade, primarily due to the size and skill of their indigenous labour force. Their appreciation and use of precious metals as economic instruments – particularly the Chinese – provided some sort of basis for exchange that was not as easily initiated with the indigenous inhabitants of the New World. Trade between nations rather than outright usurpation of resource rights became the de facto model of interaction with China, Japan and other kingdoms, such as Siam (Thailand). The colonization of Africa did not really jump into overdrive until the mid-1800s, a period that had already given way to successful independence movements in many colonial regions of the Americas. This is surprising given its close proximity to Europe. But for reasons explained in the next chapter, Europe's unchallenged expansion into the tropics – along with the sweeping social, economic and environmental change that entailed – was, at its most basic level, driven by the vicarious nature of bioproductivity and how it is distributed across our planet.

Notes

1 Between £0.10 and £0.18 per lb on the Sugar, Free Market, Coffee Sugar and Cocoa Exchange (CSCE). Contract no. 11 nearest future position.
2 This is based on Simonsen's export volume from Brazil for the year 1600 (1944, p. 171) and the London price of sugar for the same year costing 30 troy ounces of gold per tonne at £639 per troy ounce in 2016 GBP. This equates to a tonne of sugar in London in 1600 costing £20,831 in today's terms.
3 Adjusting for the change in size of the British economy (using GDP) from 1711 to 2015, this amount is equal to nearly £1.1 trillion in today's terms. Unlike today, British public debt was not considered "undoubted" and was open to the real prospect of default and re-negotiation, as was the case during the reign of Anne's uncle, Charles II where crown debt was only made good by half. See Dale (2004).
4 The Stuart kings were notorious for their poor track-record of making good on the funds they borrowed.
5 Based on compilation of consumption data from Mintz (1985), Deerr (1950), Mitchell and Deane (1962), Mitchell (1975).
6 Many were "purchased" by planters and deported to work on the plantations under a 7–14 year term of servitude. During this time, they were without rights to marry or travel under penalty of an incremental extension of service.
7 Population data for free and enslaved residents in North American colonies derived from US Bureau of the Census (1975) and Raynal (1776). Data for the British West Indies derived from and cross-checked against Anderson (1764), Edwards (1743), Raynal (1776), Powell (1825) and Martin (1834).
8 The Stono Rebellion of 1739 in South Carolina was the largest to occur prior to the American Revolution.
9 Toussaint L'Ouverture became an icon of slave resistance. After leaving his role as the head of Haitian forces, he and his family were kidnapped by agents of Napoleon and transported to France. He died several years later, incarcerated in the Fort de Joux.

7   Productivity's promise

Let me propose this: sixteenth-century Europeans colonized the tropics to break out of the bioproductive isolation of their continent. They had known of the great agricultural diversity of the Asian tropics since the time of Alexander the Great and the spices, medicines and fibres it later yielded from the trade trickling through the markets of Basra, Aleppo, Alexandria and Constantinople. But this trickle of a trade was just that: a thin stream of tropical products that made their way overland to Mediterranean ports and were then transported to Europe using galleys ill-suited for bulk shipping. The linkages established by medieval Venetian and Genoese merchants allowed these two city-states to control most of this small market and cultivate a nascent demand in Europe. But the voyages of da Gama and Columbus were without doubt the twin events that led to an incipient break-out. These two voyages did not occur spontaneously, having built upon a half-century of Iberian maritime exploration along the West African coast, but their success, more than any other, opened up the prospect of direct marine trade with source-producers in Asia and an entirely new tropical realm in the Americas. In the previous chapters, I attempted to show the magnitude and extent of impact these events had over the ensuing centuries not only within the tropics but reflexively on people, institutions and affairs in Europe and North America. We know that the tropics was transformed from a region dominated by local commerce connected by small threads of inter-regional trade to one driven by cash-crop production, natural product extraction and mining based on a universal reserve currency of gold and silver coinage. European institutions were built, often incorporating or building upon existing political and social hierarchies. Slavery became an institutionalized factor of production and an inextricable link in the 300-year process of exchange between Europe, Africa, Asia and the Americas. Conflicts across the tropics transformed from local and regional rivalries to ones between these indigenous societies and Europeans and finally, between European imperial powers themselves.

Yet, historical analysis on its own has not really provided an answer to perhaps the most important question: why Western Europe? Why did this region emerge as the singular force on which our modern global system for carrying out trade and conducting business is based? Why did this region and its complex chimera of nations and national identities ultimately delineate the path towards globalization? As is the case with so much historical enquiry, the primary precepts that govern empire-building and dominance have been examined and re-examined continuously over the past half-century. What has emerged is a series of enabling factors that have been meticulously probed and prodded to show how they influenced outcomes. Technological advantages, religious zealotry, superior military tactics, social stability and inclusiveness, geography, nationalism and cultural self-belief, an Asian legacy, epidemic disease, and concepts of value, property and ownership are some of the more prominent factors that have been put forward to explain, directly or indirectly, the rise of Western Europe as the first bastion of global superpower (Diamond 1997; Landes 1998; Clark 2009; Morris 2010; Acemoglu and Robinson 2012; Hoffman 2015 and many others). It is worth briefly discussing – without being too polemical – these myriad explanations, particularly to show how their connections ultimately lead back to Europe's bioproductivity constraints. I believe they can be pooled into three categories that I refer to as devices, destinies and diseases.

Devices, destinies and diseases

Devices. A device is a piece of equipment made or adapted to a specific purpose. But it is also a plan or method engaged towards a particular aim. Not only are functional objects such as weapons, wheels and windows the devices used to achieve an objective, but so are currency, title deeds, patents and licences, architectural plans and Newtonian mechanics. Societies create equipment, plans and methods with the intent to build, manage and destroy elements of their own and those of other societies over the course of their respective life histories. Dominance, perpetuation, even survival, hang in the balance. Differences in the breadth and complexity of devices available to European and indigenous societies in the sixteenth century have been frequently highlighted as the primary factor leading to European success in colonizing the world. There are good reasons behind travelling down this path. Some of these devices, such as those championed by Professors Jared Diamond (1997) and Phillip Hoffman (2015), were directly involved in the initial process of subjugation. Gunpowder and steel weapons allowed relatively small European forces to gain a military advantage over much larger indigenous populations still relying predominantly on wood and stone implements for their defence. The historic record of military technology tells us that the advantage in a singular fight would be overwhelming, if opposing forces were of comparable size. But it is difficult to consider how a small force of one or two hundred conquistadores or marines armed with matchlock

muskets, swords and pikes, unsupplied and without fortifications or knowledge of the landscape could survive an opposing force of thousands with
arrows, darts, clubs and knives made of wood, stone and bone. Borrowing
an axiom from ecosystem management, we can say that at a critical difference in force size (density of extraction), any effort to survive (to sustain
the resource) will fail, all else being equal. Gunpowder and steel, particularly used within the crude armaments manufactured in the sixteenth
century, would not have provided a long-lasting survivorship advantage in
open battle with indigenous societies if numbers were drastically unequal,
particularly since muskets and cannon at the time would have taken at
least 20 seconds to re-load, were unwieldy, inaccurate at distances beyond
10 metres and subject to misfiring due to moisture. Having shot a flint-
lock rifle – the direct descendant of the matchlock – I can attest to the
unwieldiness and inaccuracy of the gun. Moreover, across South Asia and
the Middle East, guns and steel were equally employed by the powerful
Ottoman, Safavid Persian and Mughal empires – in some cases to better
effect than early European forces. Europeans seeking control over production and trade in the Indian subcontinent would not have been able to
rely solely on any superior technological capability to achieve their ends.

It is far more likely that conquest of the dominant civilizations in the
tropics by relatively small European forces was due in large part to
cooperation with other subjugated communities. Hernan Cortés, conqueror of the Azteca, explains to his king, Charles V:

> Although they were subjects of Moctezuma, yet according to the
> information I received, they had been reduced to that condition by
> force, within a short period; and when they had obtained through me
> some knowledge of your Highness, and of your great regal power, they
> declared their desire to become vassals of your Majesty, and to form
> an alliance with me. They also begged me to protect them against that
> mighty Lord, who used violent and tyrannical measures to keep them
> in subjection, and took from them their sons to be slain and offered as
> sacrifices to his idols.
>
> (Cortés 1485–1547: *38–39, Second Letter*)

While Cortés underplays the vital role of local tribes, they clearly provided
his small force of several hundred with the critical support and knowledge
needed to take on a superior force that was clearly not pacifistic. Similarly,
when Pizarro arrived on his third and final attempt at subjugating the
Incan empire, he discovered that a recent civil war had left many subordinate tribes resentful of Atahualpa's rule. Many historians now surmise that
these tribute-paying groups may have altered the balance in favour of the
Spaniards, despite their inadequate numbers and isolation. Likewise, the
European play on deep historic animosities between a vast complex of alliances that dominated the Indian subcontinent may have been the primary

driver of success in the region as opposed to any advantage accrued through superior devices. By the time the British became potentates of India in their own right, the previous ruling dynasties had exhausted themselves through constant rivalry and conflict (James 1997). Equally, this play on local rivalries secured many of the resource concessions in Africa. For example, when Jameson arrived in the region of East Africa that now forms part of Zimbabwe, he quickly surmised the inflammatory relationship between the conquered Ma-shona and conquering Ndbele peoples. The Ndbele were related to the fearsome Zulu and had migrated northward to the Limpopo river region after their chief, Mzilikazi, broke ranks with King Shaka. They quickly carved out a new kingdom from the traditional Shona lands and coalesced a kingdom built upon tribute payment from the subsumed villages. By the time that the British began to take an interest in the region after discovery of gold in the north, Mzilikazi's son, Lobengula had ascended to the Ndbele throne. Shona tribal leaders may have seen the presence of the British South Africa Company as an opportunity to dissent against the Ndbele, claiming they no longer needed to pay tribute. The result was a cataclysmic destruction of a Shona village and a pretext for British annexation of Ndbele lands on the basis of "protecting" the Shona. Simmering resentment between the two tribal ethnicities as a consequence of this history continues to this day in modern Zimbabwe.[1] Likewise, the alliance of the Crow tribe with the US cavalry in the 1880s out of their fear of Lakota Sioux attacks, along with many other earlier examples of North American tribes forming alliances with British, French and US invading forces against other indigenous tribes, attests to both the complexity of inter-tribal relations and the critical role these played in facilitating European successes.

It is also worth remembering that many of the other devices that allowed Europeans to arrive and survive in the tropics were not indigenous to Europe. The lateen sail and rigging that played a pivotal role in early navigation of the Mediterranean Sea and coast of West Africa had been built upon technologies originating from the Middle East. As had the astrolabe, a device for measuring latitude. Yet there is no credible record of any voyages across the Pacific or Atlantic by Arabs, Persians or Ottomans despite their frequent trading forays along the coasts of South Asia to acquire cinnamon, black pepper, indigo, sugar and silks. This may have been due to the Arab *dhow* lacking the square sail rigging that was common to Viking *knarr* or the Frisian *cog*. Square rigging had been integrated with the triangular, lateen sail into Portuguese ships, such as the *caravel*, by the late Middle Ages. A square rig device afforded less manoeuvrability than the lateen sail configuration and did not perform as well when making short-distances along rivers and coastlines that required greater agility. But it was far more efficient in open seas where a rapid change in direction was less important and propulsion was the prime objective, particularly as vessels increased in size. The installation of a hinged rudder to replace steering oars improved the manoeuvrability that larger, square-rigged ships needed in open seas. It seems unlikely given the long-standing trade ties between Europe and the Middle East that square sail rigging or stern rudders remained an innovation secreted away from the prying eyes of West Asian societies by the Iberians. In fact, many historians remain divided in their views of the evolution of sail geometry and rigging, suggesting the square-rigged sail was displaced by the lateen sail in the Mediterranean or that the latter was in fact transferred to the Arab *dhow* from Iberian vessels and not the converse.

Dynastic China had already developed its own form of square rigging and stern rudder long before the Portuguese navigators began exploring the west coast of Africa and a route to Asia. Chinese civilization was also, in many other ways, more advanced technologically than fifteenth- to sixteenth-century Europe. They had developed institutional structures and record-keeping that was equal, if not superior, to the best known from medieval Europe. A full two centuries before Columbus' voyage, the Song Chinese invented gunpowder, began using paper money (fiat currency), and developed the magnetic compass – three integral elements in the subsequent expansion of European trade and colonization. By the time Yunglo ascended to the throne of the Ming Dynasty, the Chinese had built massive, square-rigged junks several times larger than those used by the early Portuguese navigators (Peterson 1994). Historic evidence indicates that a great Chinese naval fleet of these ships under the admiralty of the legendary eunuch Cheng He successfully crossed the Indian Ocean in the early 1400s, reaching the Persian Gulf, Red Sea and east coast of Africa before returning to China (Duyvendak 1939). The grand Chinese fleet made seven voyages between 1405 and 1430 and probably left crew members behind at their various ports-of-call (Wilmott 1960). On the seventh and final voyage, Cheng He died while visiting the great spice emporium of Calicut (on the Malabar Coast of India). Yet, despite having the devices necessary to undertake long-distance navigation and colonize remote regions, there is no credible evidence to suggest that this fleet or any other had an interest in or successfully undertook a trans-oceanic voyage across the vastly larger Atlantic and Pacific oceans before Europeans began arriving en masse in the Americas during the 1500s, although some studies have suggested alternative histories (e.g. Levathes 1996). Available historical evidence suggests either that plans for longer-distance voyages died with Cheng He or that the objective of those undertaken was simply to establish a regional sphere of trade and tribute across South Asia (Peterson 1994). It may have been a political decision by the emperor to focus on traditional Chinese values and the past, a fear of liberalizing influences among the eunuch class that ran most of his empire, or simply a lack of knowledge of the extent of the two great oceans.

It would be equally fashionable to dismiss Europe as the staging ground to a globalized economy and argue that it was in fact China and its hyper-consumption of Spanish American silver in exchange for (mainly) silks, porcelain and lacquer-ware that drove incipient globalization in the sixteenth century (Gordon and Morales 2017). But this theory too ignores evidence, this time born from trade records. Evidence such as the value of Chinese silk and porcelain imports to Europe and their importance relative to sugar, coffee, tea, tobacco, cotton, cocoa, indigo and spices. For example, Abbé Raynal (1776) notes:

> The annual purchases made by the Europeans in China, if we compute them by those of the year 1766, amount to 26,754,494 livres [£1,170,500]; this sum, *above four-fifths of which is laid out on the single article of tea,* has been paid in piasters [Spanish pieces of eight], or in goods carried by twenty-three ships.
>
> (A. Raynal, Vol. II, Book V)

Finished goods from both China and Japan were immensely profitable to British, French and Dutch trading companies, contributed to retail consumption and re-exportation in European markets and certainly influenced social and cultural trends, but by the seventeenth century they were being copied or substituted. Domestic artisanal manufacturers successfully created a demand for products made with a European interpretation of Chinese art and decoration – *chinoiserie* – and many of these industries eventually supplanted further growth in demand for Asian manufactured products. The blue and white porcelain produced in Delft, Holland and the *blanc de chine* production in Meissen, Germany shamelessly aimed to mimic Ming styles and innovation. The former was even subject to its own imitator in the form of English Delftware out of London. The tremendous popularity of these European knock-offs attests to the true demand for Chinese styles in Europe at the time and the displacement effect they were having on imports. Still, Chinese and Japanese craftsmen did maintain a trade by copying the copiers and producing wares designed specifically for the European market or, in some cases, creating fakes designed to appear as the originals. The typical *famille rose* styles, decorated with European figures and classical objects such as Greek urns and Roman columns became part of an "export-class" porcelain made in China beginning in the eighteenth century. Yet a great European demand for porcelain and silk does not attest to any centrality of silver in jump-starting globalization. The main problem with the silver theory rests in the occluded nature of the metal's relationship with economic expansion and diversification.

Silver, like other currency-based assets, is not a consumable good or service. Yes, there are various manufactures made of sterling silver and gold and these became prominent industries in their own right, but the "family silver" and jewellery has been traditionally seen in the same context as bullion – as a repository of wealth – to be bought and sold and transferred in relation to need. Silver is supremely storable and in tandem with gold was the primary means of financial saving prior to the development of government-backed credit and loan instruments. But the attributes that make precious metals ideally suited as a means of payment – rare and resistant to decay – also predispose them to hoarding and in medieval England at least, these bullion hoards were frequently accumulated as "dry powder" to be used during future periods of agricultural dearth or conflict rather than deployed as security in lieu of credit to yield an interest payment (Briggs 2008). This, in turn, meant that the value of non-circulating bullion was not meaningfully contributing to economic production. In economic parlance, precious metals were the most liquid assets available prior to the use of paper currencies. When these were not readily used to transact purchases for less liquid assets, economic activity would slow down due to an increase in time it would take to transact an "in-kind" exchange. Liquidity was the reason for adopting silver both as means of payment and as a means of saving. Credit thus emerged when savings or future revenues were used to underwrite payments through loans. The amount of gold and silver in circulation played a fundamental role in determining the levels of credit available in medieval economies, but hoarding had a counter-effect on expansion as excess bullion was soaked up by risk-averse landlords and their agents. In some instances, it was literally buried in the back garden.

The true power of bullion resided in its convertibility and as the first universally accepted means of payment for goods and services. Silver and gold were acceptable forms of payment in the Middle East, India, South Asia, China, Japan and of course Europe. Only indigenous societies in the Americas, Austral-Pacific and Africa saw little intrinsic value in the lustrous metals preferring to use them for ceremonial purposes or adornment. Silver adeptly fulfilled a transactional role, but demand in China up to the mid-eighteenth century was largely internal and the Spanish bullion shipped to the country in exchange for its products rarely left its shores. At a global scale, rather than lubricating the wheels of commerce, China's national demand for silver equated to hoarding. As we saw in previous chapters, the silver-vortex of China remained a great concern to European trading powers well into the nineteenth century. It eventually pushed the British to steal away living samples of the tea plant to India and begin mass production of the opium poppy in an attempt to drive demand in China as an "in-kind" means of exchange, ostensibly to close the growing trade gap created by their demand for Chinese goods. Two brief conflicts – the Opium Wars – led to the inimical "opening up" of Chinese markets to their manufactures, the re-balancing of payments and an end to the bullion imbalance. Interestingly, these events still resonate profoundly with many in China aware of their history and wary of Western commercial interests. This was made crystal clear to me in amiable discussions with my Chinese flat-mate while studying at university in the late 1980s.

Why, then, were tropical agricultural commodities providing a greater impetus to European expansion? While they were of course more perishable than precious metals, they were also consumable and their influence in driving European expansion, both geographically and economically, was far more profound. In contrast to the transactional role of bullion, tropical agricultural products not only began to shape the daily habits of Europeans, but lacked viable substitutes from the extra-tropics. By contrast, gold and silver supply is a function of a specific set of geological processes that are independent of bioclimatic conditions.[2] Mines across Europe, Africa and the Middle East had been supplying markets for millennia before Columbus and the search for new deposits was taking place continuously. By comparison, maritime access to the tropics in the sixteenth century offered a vast range of new commodities never seen in Europe. The great trade companies profited principally through the trade in foods, spices and dyes. The domestic industries that grew around the secondary processing and refinement of tropical products strongly affected national economic expansion. They also sparked technological innovation, such as the cotton gin, that made processing the raw agricultural material more efficient and – as we have seen in the case of cane sugar – contributed more to government finance through excise and customs duties. The impact of Asian textiles and technologies was inspirational, that of silver was transactional, but the impact of the trade in tropical commodities was transformational.

It is an unshakeable fact that technological introductions and innovations, both native and acquired, played a sequential role in advancing European economies and colonization across the tropics. They were the means to an end and there are many other factors that have been proposed but I have not discussed. What we see in the case of Cortez's improbable defeat of the Azteca, Ming China's highly developed maritime technologies, Pizarro's success on his third attempt against the Inca, and the development of the Raj in India, is that superior devices – both material and intellectual – were in some regions important arbiters of European success in the early stages of globalization while in others, less so. Devices, technologies and innovation alone cannot explain how Europe became the birthplace of globalization because most of them were not exclusive to Europeans.

Destinies. Were fifteenth-century Europeans simply more motivated? Reading through the journals of many early explorers, mendicants, governors and soldiers, it becomes immediately apparent that they shared a common motivation underlying their actions – a sense of predestination. Explorers and soldiers had a belief in their manifest entitlement to lands that they saw as unoccupied, either for themselves or their monarchs. Monks and governors had a belief in their manifest responsibility to bring spiritual and administrative order to those already inhabiting the lands they colonized. This belief in destiny, unlike devices, provided an intangible advantage to believers. It allowed early Europeans to continuously seek advantage in tropical lands where numerically they had none. Pizarro irreverently accepted an invitation from the emperor Atahualpa into the heart of the Incan lands with a force of only a couple of hundred soldiers and a few harquebuses and small cannon. He did so in the belief of Spanish cultural superiority and his destiny to rule his own empire – a goal he would achieve through Queen Isabella's royal patent and the relationship he formed with Atahualpa's young wife. Cortez' *cartas de relaciones* convey a similar narrative in the case of Mexico and Central America.

Indigenous perceptions of European insurrectionists are far more difficult to understand. We lack independent evidence attributed directly to their voices, free of European interpretation and instruction, and what accounts there are were often written many years after conquest was completed. Most of what we do know is conveyed through these filters. Bearing this in mind, early accounts suggest a converse pre-ordination. Many tribes are thought to have bestowed demi-god status on the early entrants, invoking elements of their own belief systems to explain the presence and purpose of the alien arrivals. A predisposition to self-defeatism or fatalism among many indigenous societies in relation to the way Europeans viewed their own destinies has equally been invoked as one of the stark differences that proved advantageous (Hennessy 1993).

Diseases. Did a superior European immune system support their takeover of the tropics? Certainly the extent of human loss in the wake of disease carried by the Iberians to the Americas in the early decades of colonization is frequently cited, although numbers remain difficult to properly estimate. Very little evidence beyond Iberian tribute records, journals and *cartas* is left to assist in the calculation. Many of these sources were only published several hundred years afterwards and were subject to various revisions. They often require significant interpretation since many words that we used today to identify understood diseases, such as "bubonic plague", "smallpox" or "measles" were not used in the sixteenth century. Forensic archaeology has also made significant advancements in compiling physical evidence of pathogenic infection and origin. But this too is restricted by the availability of suitable material for analysis. Consequently, the population size of New World inhabitants prior to the arrival of Columbus and the impact of new, infectious diseases the Iberians brought with them remains the subject of intense debate. It is very probable, however, that losses ran into the millions across the neotropics given our modern understanding of communicability and morbidity of epidemic disease combined with minimum estimates of population size. Small pox, measles, plague, and influenza are most likely to have accounted for the bulk of deaths during the initial wave of epidemics. Yellow fever and *falciparum* malaria parasites followed (Packard 2007). But was indigenous vulnerability to disease the primary reason Europeans flooded into the Americas? Did it, as some suggest, leave continents largely devoid of their populations and open to European appropriation with minimal conflict? Did it provide the invaders with a genetic advantage?

There is plenty of evidence to suggest that smallpox and measles were chronic in sixteenth-century Europe but typically carried relatively low morbidity among those infected. This low-grade infectivity and low morbidity among children would have perhaps made Europeans less concerned when outbreaks occurred (Riley 2010). The virulence of the strain running through Europe at this time may also have been less deadly than those that would infect the region later and may not have carried the same reputation as a lethal disease, perhaps being more akin to modern chicken pox (Carmichael and Silverstein 1987). Equally, the standard notion that indigenous residents of the Americas were genetically maladapted to cope with exotic viral infections does not pan out (e.g. Lindenau *et al.* 2016).

So why then did Native Americans suffer such devastating mortality once in contact with these diseases? Our ability to combat infection by pathogens is related to our two immune systems – one innate to us as a species and another that is acquired through our interactions with viruses, bacteria and protists. One of the most logical theories suggests that Europeans were better primed to combat viral infections as a consequence of the relatively heavy reliance on livestock within their agricultural systems (Diamond 1997). Most infectious diseases, including smallpox, chickenpox, measles, and influenza, are known to have "jumped" from domestic livestock to humans. Many others, such as malaria, yellow fever and yaws, contain known variants or strains that are endemic to primates, birds and reptiles. According to Diamond's theory, the absence of livestock within the agricultural system of Native Americans did not afford them the same opportunities to acquire immunity to these diseases through repeated exposure to these other variants. Others suggest that Native Americans were predominately exposed to bacterial and protozoal rather than viral diseases prior to European arrival (e.g. Guerra 1993). The most devastating losses are thought to have occurred through epidemic viral infection – smallpox, measles and influenza.

Both of these theories are substantive and inter-related. There is no evidence of widespread domestication of animals by pre-Columbian societies in a manner akin to that undertaken by virtually every other society outside the Americas. The few examples of domesticated animals, such as the guinea pig, alpaca, llama and dogs, are restricted to the temperate climates of the highland Andes or Altiplano of Mexico. Lowland forest societies, including the Maya, did not appear to rely on livestock. I know from my own experience that many lowland indigenous communities remain unattached to livestock, although I have frequently seen a menagerie of animals, particularly birds and small monkeys, such as tamarins or trumpeter birds, kept as pets or in a semi-domesticated condition. There are several good reasons why lowland societies may not have domesticated livestock despite the many species available to them for food production, ranging from tapirs and deer, capybaras and pacas to curassows and Muscovy ducks. First, many lowland communities were capable of obtaining adequate protein from alternative sources, primarily fish and the New World black bean, *Phaseolus*. Second, maintaining domesticated livestock within a milieu of complex, evergreen tropical forest would have proven immensely difficult compared to the high-latitude regions of Europe. Clearing large swathes of forest in the neotropics would have rapidly depleted available nutrients and pasture quality would have declined within a year or two, as many European ranchers discovered centuries after Columbus' arrival. In addition, forest recovery in the wet tropics is rapid compared to high-latitude regions, such as Europe, where tree recruitment and growth is much slower due to seasonal dormancy. Predators – largely exterminated over Western Europe – would have also proven immensely difficult to control in lowland forest environments. This is in addition to the significant place many of these animals – such as the jaguar – occupied within their traditional belief systems. Third, indigenous farmers in the Americas were, in my opinion, more botanists than zoologists. They had a vastly larger pool of plants to work on and the number of domesticated and semi-domesticated plant species is far greater than the number produced within European agricultural systems.

In summary, indigenous peoples living in the American tropics did not need to domesticate livestock because they had access to a much larger, reliable and diverse source of nutrition – plants and fish. Among all of the potential reservoirs of disease likely to jump to humans, those attached to plants and fish are arguably the most remote. As a result, they were less likely to come in contact with the variant strains that conferred a better antigen response in Europeans. Why did Europeans have such an expansive coterie of domesticated livestock compared to indigenous Americans? This has more to do with the bioproductivity of their homeland and the surrounding regions than any advanced state of agricultural development, but more about this later.

The American tropics of course was not entirely devoid of disease and indigenous contact with potent reservoirs of zoonotic viruses, such as the arbovirus group responsible for haemorrhagic fevers, would have occurred through hunting and co-habitation with tamed pets. The frequency of exposure, however, was considerably lower than in agrarian Europe where contact with livestock occurred on a daily basis. As a result, some geneticists have suggested that the acquired immune system of Amerindians produced fewer varieties of our main anti-viral agent – the histocompatibility antigen – and this in turn did not bestow on native populations the capacity to combat rapidly mutating viral strains that typify smallpox, measles and influenza (Black 1992). The histocompatibility antigens are agents that identify and bind with foreign pathogens, preparing them for elimination by antibodies, such as T-cells. Acquired immunity occurs through exposure – often as a child – to various foreign antigens, building up a "library" of histocompatibility bodies primed to detect these pathogens. If there was any superior adaptability to disease conferred to Europeans, then it would have been strictly limited to their acquired immune response to the viral strains prevalent in Europe at the time. This adaptation would eventually be conferred upon the surviving inhabitants of the Americas, but like Europeans, Africans and Asians, they too would also remain vulnerable to new viral strains, often referred to as emerging infectious diseases, albeit without the same velocity of infection that occurred at first contact.

The vulnerability to Afro-Eurasian diseases that is believed to have been devastating to indigenous Americans of course did not provide any unintended advantage to Europeans in colonizing other continents. In fact, it is very easy to view Europeans as victims of a devastating demographic loss in the century and a half leading up to Columbus' arrival in the Caribbean. Europe at that time was on the receiving end of a pandemic of plague that had originated and already run its course across a densely populated Asia. Giovanna Morelli and colleagues recently undertook a DNA typological study of the plague bacterium, *Yersinia pestis*, in a fascinating study that shows that the Black Death event devastating Europe in 1348–1351 originated in Asia (Morelli *et al.* 2010). Phylogeographic evidence in the study indicates that the ancestral line of the plague bacterium evolved in China more than 2,600 years ago and spread to various other parts of Asia, India and eventually, Europe. Even more surprising is the highly plausible thesis that links the spread of the "2.MED" Black Death strain to transmission along the Silk Road and the subsequent pandemic that hit Europe in the 1430s to admiral Zheng He's voyages across the Indian Ocean.

Similar to the plague bacterium, most recent variant strains of seasonal influenza have been traced back to Asia (Wen *et al.* 2016). The devastating flu pandemic of 1917 that killed around 50 million people is also believed to have travelled with Chinese workers to Europe and North America based on historical records (Humphries 2014), although it is important to note that, unlike the plague, analyses of the genetic provenance of this non-avian type of flu have not proven to be definitively linked to any single region. Our knowledge of influenza prior to this time is limited to anecdotal evidence drawn from journal records describing symptoms and outbreaks that are consonant with those of influenza, but records describing a pandemic spreading from Asia to Africa and then the Americas via Europe in 1580 appear to support the notion that many of the diseases that ultimately decimated the native inhabitants of the Americas had already led to similar losses in Europe, as they spread westward across the planet (Beveridge 1991). A similar tale can be told in the case of the malaria parasites *Plasmodium vivax* and *P. falciparum* that appear to have originated from Australasia and Africa before travelling to southern Europe, and then to the Americas, although at different times (Rodrigues *et al.* 2018).

This tells us a great deal about global processes. Particularly in relation to how the long-standing narrative that indigenous populations were decimated by "European" sicknesses needs to be seen in the context of global transmission patterns and the behaviour of these and other diseases during the sixteenth and seventeenth centuries. Plague and influenza should have been the most lethal of those carried to the New World. Morbidity rates were high and surviving the onslaught did not confer any long-term immunity to survivors. In some instances, such as in Santo Domingo, these are thought to have been the main killers. Without immunity, plague and influenza could continue to spread across surviving populations, mitigating the reduction in host availability that is the cornerstone of modern vaccination programmes. Certainly Europeans continued to suffer repeated epidemics for centuries after the Black Death arrived in 1348, often carrying these to the Americas. Yet, despite diseases such as smallpox and measles conferring immunity upon survivors, the initial strains proved particularly deadly to Native Americans. As was discussed earlier, this may have been due to a smaller "library" of available antigens capable of tackling these previously unencountered variants – a condition caused by immunological isolation, possibly from non-human reservoirs, but not any genetic predisposition towards an "inferior" immune system. Over time, this library grew and morbidity rates declined to levels in Europe and Asia (Ramenofsky *et al.* 2003).

What we can conclude from the history of these diseases is that Europeans did not have superior immune systems to those of indigenous Americans – they were simply exposed to these diseases first, with similarly shattering impacts on their own populations. It was not first contact between people, but first contact with an exotic disease that mattered. The human species consists of a multitude of different attributes, of which immunological response is one. We are engaged in a continual host–parasite arms race at the cellular level, but not all of us have the same defensive capabilities at the same time. This difference occurs both between individuals and within individuals over their lifetimes. If this was not the case – as in the case of the clonal propagation of bananas and the demise of the *Cavendish* variety or the potato blight that rocked Ireland in the 1800s – we would be exposed to a catastrophic extinction-type event once a strain unlocked our shared immunological safeguards. If vulnerability to disease had been a precursor to successful colonization, one might also logically argue that the Black Death of 1348–1352 would have equally advantaged other societies to exploit Europe in the aftermath of a four-year onslaught that led to a halving of the population in some countries and the loss of at least a third in many others. The reality of the plague, influenza and smallpox was one of near global transmission that left most potential conquerors suffering the same impacts in their own lands. While Europeans did acquire selective immunity by surviving other diseases, any adaptability to most, but not all, diseases was probably the same across the planet. Instead, diseases transported by Europeans to the Americas and brought back to Europe from these and other regions during the sixteenth and seventeenth centuries were simply being globalized, like trade, culture and commodities.

Questioning accepted pedagogy is part of forming new narratives. They challenge us to examine the linkages between past events and how these affect the present. But they often lack scale and being new does not make a narrative more compelling. Historic narratives can only be evidence-based and where evidence is unavailable, so too should be the narrative. Compared to the interplay of institutions and actors, forces driven by environmental processes are foundational. Remove secure access to clean air, water, food and energy and many of the complex social or economic factors that provide some sort of Ricardian advantage of one society or place over another rapidly disappear. The extent to which one region produces more of these environmental goods than another is primarily governed by the same processes, described in Chapter 2, that situate the tropics as the planet's powerhouse and drive the productivity and diversification of life on our planet. Of course we still do not fully understand everything about these processes but they are undoubted. More importantly, they possess the needed scale because the importance of labour, innovation and capital formation at a given point becomes subsidiary to nutrition and hydration. Buildings do not build themselves, for now. Historically, the fundamental role of nutrition in driving worker productivity and vulnerability to disease has been emphasized repeatedly by economic and medical historians (e.g. Scrimshaw 2003; Rawcliffe 2013; Alfani and Ó Gráda 2017). If we consider the factors that provided the root impetus to Europe's march across the planet and the establishment of the first truly global trade network, we invariably arrive at the worldwide environmental context of pre-colonization and how a bioproductivity problem for demographic and economic expansion paradoxically set the scene for European ascendancy.

Europe's bioproductivity problem

Andalucia, the southernmost region of Spain, is dry. At least, the landscape conveys to anyone arriving from northern Europe that it is dry. It actually registers a similar amount of rainfall each year to that measured in southern England, but, unlike most of Britain, it is highly seasonal with most precipitation occurring during a few winter months. Travelling from Malaga north-westward to the grand city of Sevilla takes you through countless hillsides of *matorral* – a landscape predominately made up of naked rock and soil, sporadically covered with low-lying sticky thickets of the rock-rose, *Cistus*, inter-mixed with a simplified mixture of other low-lying shrubs and herbs adapted to arid environments. Deathly dry flatlands are farmed mechanically with irrigation from the Guadalquivir river, but large areas along the way remain under *dehesa* – the region's traditional practice of grazing livestock amongst small groves of holm and cork oak (Cabo Alonso 1998). Holm oak has been highly prized for its acorn crop and the role this played as forage for pigs since medieval times. The other agricultural mainstay – olives – can be seen covering most of the available farmland while polyethylene-covered greenhouses growing smaller volumes of fruits and vegetables cluster here and there in-between. Olives, cork, citrus and dried fruit have formed the bulk of Spain's modern agricultural growth (Harrison 1989). But despite clear indications that agriculture in Andalucia has experienced something of a boom of late, I cannot escape the hard, hot reality that this region's bio-productivity is teetering, more typical of Morocco and Algeria than north-ward towards France, the British Isles or Holland. Examining the geographic distributions of the natural vegetation of southern Spain sup-ports these suspicions. Most of the principal components are species that reach their northernmost limit in southern Europe. To find the natural distribution of the flora in Andalucia we need to focus our gaze to the east and the south (Grove and Rackham 2001). *Pino carrasco*, also known as Aleppo pine (*Pinus halepensis*) is one of many dominant species that are equally as well known from these regions as they are from Andalucia. *Quercus suber*, the cork oak locally known as Alcornocales, also finds its natural range extended southward. The matorral shrub *Anthyllis cytisoides*, a local plant found also on the Balearics but not much further north than Andalucia, heavily inter-mixes with clusters of *esparto, Stipa tenacissima*, a tall drought-tolerant bunch grass that is found in equal measure across western North Africa. Most of the native plants display characteristics that are typical of hot, dry environments. Many have thick bark and extremely coriaceous leaves. Pines, oaks and some shrubs such as *Anthyllis* produce copious amounts of resin – a trait believed to reduce tran-spiration and water loss. These attributes reflect on the region-wide chal-lenges that a deep seasonal dearth in rainfall combined with sky-rocketing temperatures presents to agriculture. Over 25 per cent of the current agricultural area in the region is under irrigation, compared to 20 per cent worldwide (FAO 2016a). Olive groves likewise account for a quarter of farmland, with *dehesa* oak and livestock systems occupying a fifth. The importance of irrigation, olives and cork to the region's agriculture shadows the predominant attributes of the native flora – both are attempt-ing to find a way to maximize bioproductivity under otherwise relatively constrained environmental conditions through a process of selective adaptation.

Arriving in Inverness, Scotland – nearly 2,200 kilometres north of Andalucia – you quickly sense a distinctly different set of environmental conditions. The default capital of the Scottish Highlands, Inverness grew originally on its fishing and later on the proliferation of "official" whisky distillation that arrived in the wake of the 1823 Excise Act and its legalization of spirit manufacture. Today, with over 120 distilleries in operation, the industry consumes near half of the barley produced in the country and brings in much of the rest from England. Most of the whisky production clusters near Inverness, particularly in the Speyside area along the coast where more than half of the distillers have chosen to base their production. For good reason – virtually all of Scotland's arable land supporting barley and wheat production is found huddled along its east coast between Edinburgh and Inverness (Macaulay Land Use Research Institute 2010). Most of central and west Scotland is agriculturally unproductive and only capable of supporting – at its very best – some form of livestock grazing or low-yielding timber production. At its worst, peatland bog and heather moorland dominate – both habitats characterized by some of the lowest NPP levels on the planet. Beyond the amenity and conservation value that these unproductive landscapes might deliver, the challenges in meeting the basic dietary needs of people are reflected in the historically low population densities, exacerbated by a century-long policy of consolidating land ownership that began in the early eighteenth century (the Highland Clearances) and remains in place today. Like Iceland, past attempts at agricultural diversification have met with limited success and a default system has remained largely the same since medieval times – one based on grazing unique varieties of sheep and cattle on marginal uplands while growing barley, wheat and oats in the limited area of coastal lowland amenable to cropping. Unlike Andalucia, moisture is rarely limiting to plant growth – most of Scotland experiences nearly a 30 per cent chance of rain every day of the year compared to 0 per cent to 20 per cent in southern Spain. But synchronous seasonal declines in day length and temperature that typify land at these high latitudes put to rest any notion of an extended growing season. The growing season in the lowland area between Edinburgh and Scotland is narrow; in fact so narrow that it registers roughly half the Growing Degree Days (GDD) of south-east England and only two-thirds of southern Norway and Sweden, largely due the region's weak rise in summer temperature.[3] GDD in south-western Europe further emphasizes the lacklustre growing conditions – Scotland's growing season is less than a third of that in the wine-growing region of Bordeaux and a fifth of that found in Andalucia. The great colonial powerhouse of sugar production, Jamaica, registers a GDD value 1,200 per cent higher than eastern Scotland and 600 per cent over the best in Great Britain, located somewhere near Hawkhurst, Kent.

The decline in GDD in the vicinity of Inverness, Scotland defines a northern, temperature-driven limit to the agricultural productivity of Europe. Similarly, the rapid drop in seasonal rainfall that affects Andalucia defines a southern, moisture-driven limit. Combined, north and south represent a set of environmental boundaries to growing the principal European dietary staples: wheat, oats and barley. Barley is the most heat and drought sensitive of the three, while wheat has the highest GDD

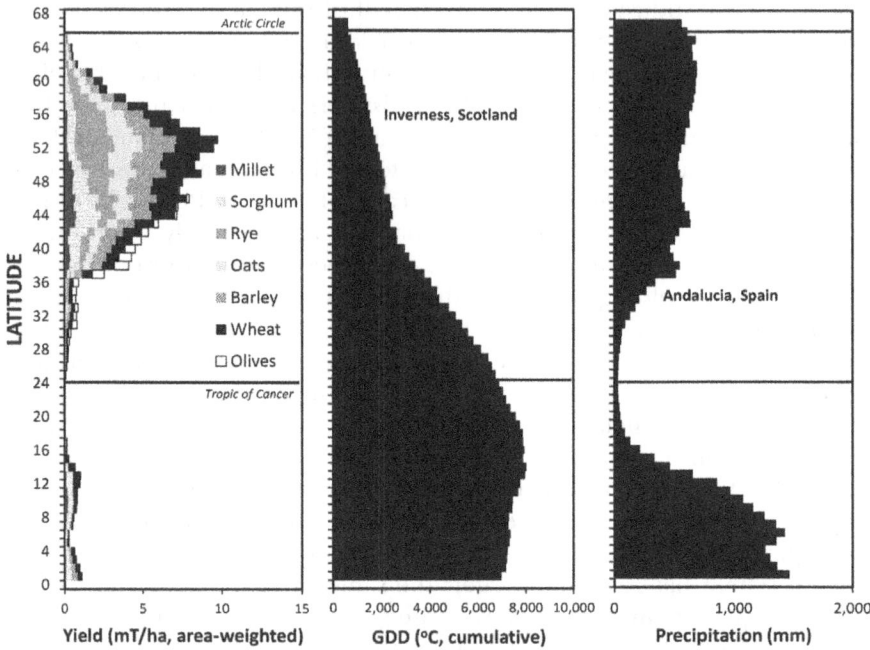

Figure 7.1 Changes to Net Primary Productivity (NPP), Growing Degree Days (GDD) and annual precipitation across Europe and Africa between the Arctic Circle and Equator.

requirement and is less sensitive to rainfall. In Figure 7.1,[4] we can see how these limits pan out in relation to the broader, latitudinal changes in average GDD and rainfall across Afro-Europe. Olives complement the decline to these staples along the southern rim of Europe; still further south, palm dates replace olives and then virtually nothing as the Tropic of Cancer is crossed until production of the drought-resistant, tropical cereals millet and sorghum begins to appear along the southern margin of the Sahara around 15 degrees north latitude. In the case of the north, crop species with lower GDD requirements, such as rye (550–800 GDD, Brennan and Boyd 2012) and the pseudo-cereal buckwheat (500–700 GDD) are sometimes planted in areas where the other three main staples begin to perform too poorly, particularly in the far north and east of Europe where the growing season becomes inherently constrained by decidedly colder and longer winters.

Figure 7.1 illustrates the average latitudinal change across a swathe of longitude spanning the Azores to Moscow. In many regards, these are crude depictions of changes in crop yields and the main environmental factors influencing their cultivation across Europe and North Africa since crop performance can also reflect the play of non-environmental factors, such as irrigation, harvesting efficiency and government financial support, among others. Yet, field-based agriculture is rarely supported by investment or continuously subsidized by government in areas that fail to meet the most basic GDD or moisture requirements because it simply does not make economic sense to do so. There are also areas that fare far better than the averages show – wheat and barley grown in the Egyptian Nile floodplain or in the cool, wet highlands of Ethiopia. But these pockets are restricted to anomalous geographic features. At latitudinal scales, they prove inconsequential in terms of the total land area and production. Bearing these caveats in mind, the depiction of crop yields by latitude does a good job at identifying the modern peak average yield of the three main European staples – hovering around 52 degrees north latitude. It also reveals that north and south of this optimum there is a steep, but monotonic decline in yield. Annual production and yield statistics for European and North African countries corroborate the position of these optimal latitudes (FAO 2016b), as do records of cereal production for the region dating back to the beginning of the nineteenth century – in Finland and Sweden, annual production of rye respective to wheat hovers around 100 to 1, in France it is 1 to 10, while in Italy this inverts further to 1 to 26 (Mitchell 1975). We can see that rye finds its optimum slightly north of the peak for wheat, barley and oats, while those of sorghum and millet – both now grown in various quantities across southern Europe – are further south. The pattern shown in the graph tells us that these yields are optimized as a trade-off between a slow northward decline in GDD and the countervailing southward trend towards much lower precipitation. More importantly, they also show the precipitous loss of cereal productive capacity to the north and south of this optimum. If we were to extend the analysis further eastward to capture the vast region of Siberia and central Eurasia, we would encounter the same dramatic decline in agriculture, but at the same latitudes where yields peak across Europe. This is due to a combined effect of latitude, elevation and continental climate on temperature and GDD in the vast landscape east of Europe (Moscow) and the stultifying effect these have on bioproductivity. The geography of cereal performance across Europe and its relationship with rainfall and GDD help us to understand two critical features of Europe's own bioproductivity problem leading up to and through the early centuries of its global expansion. Environmental, historic and economic evidence leads me to conclude that two critical features of the region's bioproductivity – isolation and hypodiversity – played an instrumental part in driving Europe towards the tropics.

Europe's isolation

This might seem an odd course to take since few would consider Europe as isolated. But this is a view distorted by centuries of post-Columbian expansion into the tropics and the enormous impact this in turn has had on the modern dietary habits and agricultural systems of Europe. The region's isolation also wasn't reflecting any lack of natural productive capacity. Much of Europe registers levels of bioproductivity similar to those found in other extra-tropical regions of the planet. The problem Europeans faced was attached to the isolated nature of its anomalous distribution – an isolate centred on the region west of the Alps. In the prehistory of Europe, this island of bioproductive capacity would have supported immigration and settlement from Africa and the Middle East in a way that the vastly larger regions north (Scandinavia), east (Siberia) and south (the Sahara) could not. It would have supported higher rates of population growth than these circumscribing regions, particularly as agrarian societies took hold through the expansive Roman Empire and its role in fostering the beginnings of regional trade during the medieval period. This early trade of course was almost entirely based on agricultural and forest products – grain, wool, linen, leather, beeswax candles, tallowed soaps, dried fruit and wine. Minerals, especially tin, lead and iron, were important as well but the capacity to transport these in meaningful quantities was limited by distances to the nearest port. Medieval Europe – but particularly the north – was regionally isolated and this meant the rate of agricultural expansion, improvement in crop yields or both had to stay ahead of population growth rates if population size was to be sustained.

There is much evidence to suggest that medieval agriculture and population growth in Europe were delicately balanced. The average yields depicted in Figure 7.1 reflect modern European farming conditions and technologies and therefore a pinnacle in crop performance. Medieval crop yields were considerably lower and prone to wild fluctuations from season to season, often driven by extended winters and subdued summer temperatures that pushed the GDD count close to or below that necessary for full crop ripening. The consequence was an increase in grain prices during these periods (Scott *et al.* 1998). Grain shortages were exacerbated by the peasant farmer's need to retain a fraction of each harvest for next year's seed stock, to pay tithes to the Church (10 per cent) and rents to the landowner. Consolidation of land ownership under baronial estates in Britain improved agricultural productivity and grain yields, but by 1300 this consolidation of agriculture under the seigniorage model had reached a peak. Crop yields would continue to peak for brief periods throughout the next two centuries, but within an overall declining trend in yields and production leading up to the sixteenth century (Broadberry *et al.* 2008). A body of high-quality, scholarly work by Bruce Campbell, Mike Overton, Gregory Clark and other agricultural historians fosters the conclusion that the tumultuous two centuries leading up to the Iberian navigational discoveries played a formative role in England's social and economic development. The French historian Fernand Braudel echoes this view for Europe in general, but he also emphasizes the developing role of urban-based commerce and the growing connections between ports in the north and south (Braudel 1984). Most economies across Europe were dominated by agriculture, forestry and fisheries in terms of employment, wages and trade and consolidation of arable farming in particular had supported a boom in medieval populations. The Italian city-states of Venice, Genoa and Florence began convalescing the imperial Roman trade links between the Levant, themselves and northern Europe during the late thirteenth and early fourteenth centuries. Fairs across France and Belgium were run as a means of bringing together supply and demand, growing trade links.

But then something very surprising happened. The gains made in population and agricultural production during the early Middle Ages were swiftly lost in the period leading up to the fifteenth century. A series of famines and epidemic sickness – led by the Great European Famine from 1315 to 1322 – would begin to shake the rural labour pool across most of the continent north of the Alps and Pyrenees. Prices of staples – wheat, barley, peas and beans – sky-rocketed. The monk and historian Thomas Walsingham notes in his chronicle, *Historia Anglicana* that the price of wheat sold prior to the Feast of the Assumption had doubled during the first year of the Great Famine (Walsingham 1574). It is estimated that the famine impacted more than 30 million people in an area extending at least 400,000 km^2 (Jordan 1996). Epidemics of famine-driven sicknesses followed.

Evidence suggests that these famines were driven by bouts of extreme weather that occurred as part of a longer climatological transition in Europe. Reconstructions using various proxy indicators suggest that temperatures increased from 800 AD to a peak around 1100 AD, often referred to as the Medieval Warming Period. Temperatures in Europe then began a more pronounced decline towards a minimum average temperature around the year 1600, sometimes called the Little Ice Age. This decline in temperature provides a cogent explanation for the declining grain yields over the fourteenth and fifteenth centuries in Britain, as well as the spectacular decline in Europe's population that occurred during the period (Lima 2014). Declining temperatures across the region would have slowly suppressed the latitudinal distribution of GDD that we see in Figure 7.1. Particularly in the north, the seasonal window available to bring the main cereal crops to full maturity would have contracted. Persistent rainfall and extended cloud cover during the crucial summer season and extended winter temperatures would have increased the incidence of crop failure. Europe's natural growing conditions are constrained, even at their best, and the impact of a climatological transition of the magnitude indicated for the period would have placed the balance of food on an increasingly thin edge. The records of famine in England and France and their dissipation around the early eighteenth century at least suggest that declining temperatures from 1100 to 1600 AD were working against food self-sufficiency across the north (Dupâquier 1988; Overton 1990). Famine driven by inconsistent agricultural performance was endemic to medieval Europe and outbreaks of sickness and disease continuously followed in its wake (Rawcliffe 2013). The Great Famine was perhaps the most extensive to hit the region, but was followed 25 years later by the Black Death from 1348 to 1352. The emergence of the plague along the shores of the Black Sea, its sweep across Europe and the series of recurrent, convulsive post-epidemic outbreaks that would continue to periodically terrorize its citizens for the next three and half centuries, witnessed the largest contraction of urban populations in Europe's history.

It also manifestly changed the rural landscape. In England, the economic effect was such that the size of the national economy contracted almost immediately, reaching a nadir around 1475 (Broadberry *et al.* 2011). At the same time, gross domestic product per person actually increased, reflecting the severe contraction in the labour pool. Wages and living standards of survivors received a significant boost from this shortage but the land that had been the bedrock of feudal seigniorial wealth in England could no longer be fully utilized towards crop production using cheap labour. Consequently, the value of the land declined and with it, the wealth of land-owners. The transformation from a seigniorial system to a more free-flowing capitalist variety as a result of these events was slow and catalyzed by a series of peasant rebellions (Hilton 1985). But in many ways the dramatic decline in populations during the Black Death laid the foundation for many Europeans – previously not privy to the economic privileges that arrived through a system of bestowed capital – to, at the very least, make an attempt to accrue capital themselves, if they would be so inclined. These changes were most pronounced in England, Holland, Scandinavia and parts of northern Germany. Changes to the old system across much of eastern Europe and Russia would remain bound by feudalism for another two to three centuries, but in the West they gained traction, untethering many from the yoke of debt peonage or worse, serfdom (Campbell 2006).

Ironically, the immense suffering of the Black Death made the surviving labour pool more dynamic. The sudden dearth of farm workers increased individual choice, allowing a more rapid re-distribution of labour towards more productive pursuits, both rural and urban. This stimulated trade within and between countries by allowing room for a more expansive merchant class and, in many ways, was an early primer to the tropical trade boom in the 1500s. The principal downside to these developments was a chronic loss in agricultural productivity. Tithing records from the late thirteenth century suggest that cheap labour working on feudal demesnes achieved some of the highest crop yields of the Middle Ages. In the wake of the Black Death, estimated yields for all of the major cereals in England declined over the next two centuries (Broadberry *et al.* 2011) and the famines that regularly impacted the much larger populations of a pre-plague Europe continued as food production declined in lock-step with the contraction of the population. When crops failed regionally and food shortages formed locally, much of the bulk trade in cheap grain between neighbouring nations also ceased. Prices of the principal grains – wheat, oats, barley and rye – typically rose in tandem, preventing the labouring poor from switching between cereals to avoid starvation (Scott *et al.* 1998). Only those able to afford the more expensive stock or in receipt of tithes were able to maintain a healthy dietary intake during these periods. Riding on the back of a climatological dip in temperature, crop failure, high grain prices, food shortages and famine remained an intrinsic component of life in Western Europe through the seventeenth century (Healey 2011), and in some cases – such as Finland – much later (Peltonen-Sainio 2012).

Europe's bioproductive isolation meant that trade in agricultural products remained intra-regional for centuries before and after the Black Death. Why did this matter? The "short-trade" in agricultural products that dominated early commerce between Western European countries acted to synchronize shortages internationally since bad weather that wrecked one country's harvest was also impacting those across the region, registering similar impacts on grain production and availability. Drought regionalization (Briffa *et al.* 1994) meant prices often rose together across the closely aligned Flemish, Dutch, English and French markets, squeezing affordability. Dearth in one country was met with draconian action by another suffering a famine of its own. Royal decrees were issued to ban the export of grain, supply contracts were cancelled or worse, grain was impounded and removed from ships in port. Moreover, much of the trade between regions was controlled by the Hanseatic League in the Baltic and Venice and Genoa in the Mediterranean (at least up to the Ottoman takeover) and these merchants could play the markets towards higher prices. Not surprisingly, these higher prices have been shown to be strongly correlated with increased mortality across Europe (Alfani and Ó Gráda 2017). With little bioproductive capacity to turn to north of the Baltic Sea and south of the Mediterranean during periods of hardship, pre-Columbian Europe was encapsulated in a geographic bubble that did not afford easy access to additional and non-synchronous natural productivity. Establishing a maritime connection with Asia and discovering the Americas acted to burst this bubble of isolation.

The geography of isolation

Why is Europe's bioproductivity isolated? It is a function of its unique geography; more specifically, its geographic position. There are four major north–south coastlines on our planet. The first of these coastlines

– starting at the prime meridian and moving eastward – is centred on Western Europe and runs southward along the West African coast towards the Cape. The second is the East Asian coastline that begins on the opposite side of the planet to Western Europe and runs south-westward from the west bank of the Bering Strait down through Russian Siberia and along the Kamchatka Peninsula to Japan, Korea, China, along Vietnam, around the Indonesian-Philippines island complex towards Papua New Guinea. The third runs south-eastward from the east bank of the Bering Strait through the United States, Canada, Mexico, Central America and then along the Pacific coast of South America. The fourth, runs erratically south-westward from Greenland and Newfoundland in Canada through the United States across the Greater Antilles and then south-eastward along the east coast of South America towards Patagonia. Combined, these four coastlines have been home to the bulk of humanity since Neolithic times due to the moderating influence of the adjacent oceans on climate, access to coastal and marine resources, and the relative ease of moving along coastlines to migrate and to trade.

The geography of these north–south coastal features on our planet is fairly self-apparent and in and of itself of little interest. That is, until we examine the pattern of bioproductivity in a swathe of land centred on each of these north–south coastlines as they run from the Arctic towards the equator, when an important distinction is revealed. Of the four, only the Euro-African axis shows a lower terrestrial NPP at the Tropic of Cancer than at the Arctic Circle (Figure 7.2). All three of the other major coastlines show a pattern consistent with what we would expect if moving from the highly seasonal and cold north towards the thermally aseasonal tropics. The boost in NPP that arrives when crossing over the Tropic of Cancer and into the tropical belt is apparent along these three. The clear outlier is Euro-Africa and the pattern we see is due to the blocking effect of the Sahara desert. This blocking effect, as we saw earlier in relation to crop-growing, is a function of rapidly declining precipitation, not a shortening of GDD. In the case of the other three coastlines, it is the rise in GDD moving southward that is largely shaping NPP levels. Of course this fact alone is, again, hardly a revelation. But when compared to the north–south axes centred on the other major continental coastlines of the northern hemisphere, the uniqueness of Europe's isolation becomes much more apparent. This latitudinal inversion in bioproductivity along the Euro-Africa axis is counter-intuitive. We expect growing conditions to improve as we move towards the equator, and for most of the world's coast-centred regions this rings true. But for a medieval Europe first venturing southward by sea, there was little prospect of finding another agricultural Shangri-la at their doorstep.

The inverted bioproductivity of Europe is a symptom of its geographic isolation. If we return to Figure 7.2, however, another attribute of Europe's bioproductivity also becomes apparent. Compared to the other three main

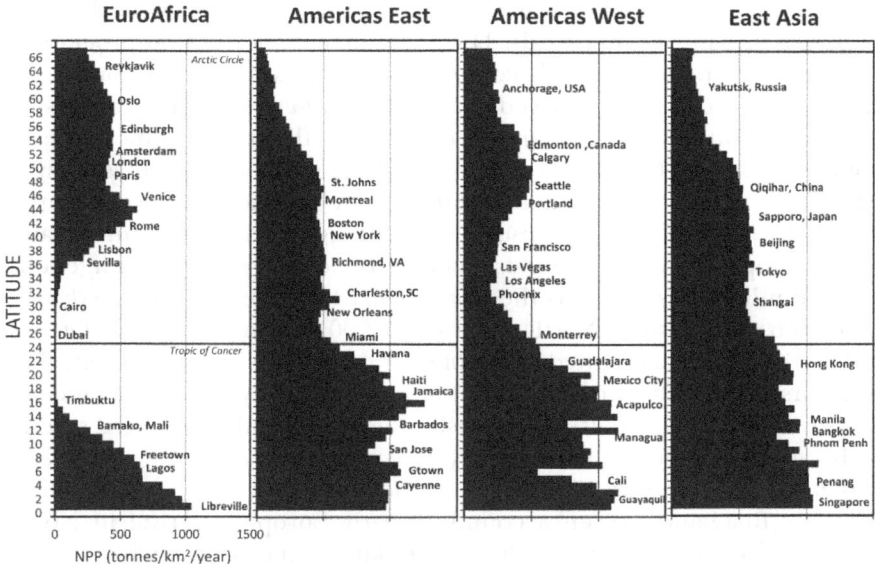

Figure 7.2 Changes to bioproductivity between the Arctic Circle and Equator centred on the planet's four major coastlines.

coastlines, average NPP levels are sustained above one ton of carbon produced per square kilometre per year at much higher latitudes. This level is breached around 52 degrees north latitude on the east coasts of the Americas and Asia and around 56 degrees along the American west coast. In Europe, average levels above one ton are sustained up to 63 degrees north – an extension that is found nowhere else on the planet. If you recall that Europe experienced a significant decrease in average temperatures between 1100 and 1600 AD, then we can reasonably expect the modern distribution of average NPP exceeding one ton to have contracted further south during this period, squeezing the region's agricultural capability at these high latitudes even further southward by the late 1500s (Lima 2014). As a result, the history of chronic food shortages in Europe is long – extending back through Greek and Roman times – punctuated by extreme famine and the need to carefully manage food supplies (Mattingly and Aldrete 2000).

By contrast, societies established along the east coast of Asia and the Americas were never constrained by a precipitous extinction of bioproductivity as they moved along a north–south axis. The transition in NPP was more gradual. Communities along the eastern coastal regions had the flexibility of transitioning north or south depending on prevailing conditions without any rapid decline in NPP. More importantly, many of the staple food crops were able to follow them. The impact of this less punctuated transition in NPP is reflected in the latitudinal plasticity of the regions' two main staples: maize and rice. Both grasses, they are broadly related to the European cereals. The ancestral complex of wild progenitor types to our modern maize is commonly known as *teosinte*. Although still debated, it is currently believed that ancestral *teosinte* was first domesticated in region between 14 and 18 degrees north latitude somewhere in southern Mexico, Guatemala and north Honduras about 9,000 years ago. Most archaeological evidence points to the Oaxaca–Tabasco–Chiapas region as the centre of domestication, although this remains uncertain. Radioisotopic evidence suggests maize was being grown by indigenous communities as far north as the lower Great Lakes through Pennsylvania to the coast of eastern North America by 900 AD (Smith 1989). Maize is believed to have augmented a complex of cereal, nut and fruit crops that were first cultivated or collected in the region as early as 4,000 BP, including sunflower (*Helianthus*) and squash (*Cucurbita*), and supplanted several indigenous varieties of knotweed (*Polygonum erectum*), barley (*Hordeum pusillum*) and maygrass (*Phalaris caroliniana*) (Smith and Yarnell 2009). The few first-hand written accounts of early European arrival in North America corroborate the view that maize, squash and probably beans were being cultivated in eastern North America. One such account by Edward Winslow, a leader among the Pilgrims, explicitly details the use of maize in their first meals with the indigenous Wampanoag tribe in 1621. These meals – the origins of the Thanksgiving tradition in the United States – took place at nearly 42 degrees north latitude.

The evolution and spread of domesticated rice varieties in Asia shows a similar pattern to maize, although the general debate surrounding the location and timing of its origins has been hotly contested (Liu *et al.* 2007; Fuller *et al.* 2009). Current views suggest that rice was first cultivated – but not in its domesticated form – around 9,000 years ago somewhere in eastern China between 29 and 31 degrees north. The modern distribution of two wild-type rice species (*Oryza nivara* and *O. rufipogon*), however, ranges considerably further south, between 8 degrees south and 27 degrees north across south Asia (Fuller 2011) and archaeological evidence indicates that domesticated rice was being cultivated as far north as 39 degrees north, near Tianjin and the Bohai Sea (D'Alpoim Guedes *et al.* 2015). The bulk of China's modern rice production occurs in Yunnan, Sichuan and Guizhou provinces between 23 and 31 degrees north.

What we see with maize and rice are parallel developments, albeit within quite different cultural contexts. Both maize and rice are believed to have jumped the boundary from being a purely wild plant to one of cultivation interest around 9,000 years ago. Wild-type progenitors of both share a tropical/semi-tropical distribution with incipient domestication taking place gradually along a movement towards higher latitudes. Finally, domesticated varieties migrated to the mid-latitudes while retaining a broad tropical distribution. Nearly 700 varieties of cultivated maize allowed this staple to be grown along nearly the entire length of the Americas, from Chile and the southern cone region of South America through to the modern border between Canada and the United States. Rice too split into various regional varieties that allowed it to bridge the disparate environmental conditions present across northern East Asia to Melanesia. By comparison, the evolution and prehistoric spread of wheat and barley along its own Euro-African axis is highly compressed. These cereals originated from wild varieties in west-central Asia and were first cultivated between 31 and 39 degrees north. They too are believed to have been domesticated around 9,000 years ago. But unlike maize and rice, by the sixteenth century they were being cultivated as far north as the Baltic Sea region but were not found in anything more than isolated patches south of the Mediterranean and were absent along the west African coast. Wheat did spread eastward into the valleys of Afghanistan, Pakistan and northern India from their geographic origin, but lacked the physiological plasticity or varietal diversification to make the transition to tropical climates further southward, unlike maize and rice, two inherently tropical crops that made the jump to temperate latitudes.

While all three staple crops – wheat, rice and maize – are currently believed to have undergone some form of early domestication during the middle to late Neolithic period, the latitudinal distributions of rice and maize before Columbus' voyage across the Atlantic were considerably more extended than those of wheat or barley. Archaeological evidence suggests that maize was being cultivated near the equator (e.g. in the Orinoco Basin of Venezuela, Roosevelt 1980) and, as we have seen, as far north as the Great Lakes and New England in the United States. This is a latitudinal span in the northern hemisphere of nearly 45 degrees. Similarly, rice is believed to have been cultivated as far south as Indonesia and extending northward to the southern edge of the Bohai Sea in north China, also a span of nearly 45 degrees. The latitudinal span of wheat at the end of the sixteenth century was only 25 degrees, ranging between its very northernmost limits around the Baltic Sea and across the southern rim of the Mediterranean into Persia, Turkey and north India and not much further south than 28 degrees north. Ranges for barley, oats and rye were even narrower. The variation in the distribution of these regional cereals – all members of the colossal family of grasses – reflects the differences in NPP patterns along the north–south axis of these three regions. Once again, in Asia and the Americas, a more gradual gradient reflecting mainly changes in GDD is evident, but the Euro-African axis is more sharply influenced by changes in precipitation and this had a pronounced effect on the ability to adapt staple food plants to rapid changes in moisture availability. For maize and rice to make the move north and south required a slow walk along a continuous gradient. Wheat and barley could not be adapted to account for the rapid loss of rainfall southward – the walk required a jump, and this jump southward was too large.

Instead, another grass better at handling unexpected drought conditions arose south of the Sahara – sorghum.

Thus, the distribution of the main staples (there were other regionals too, such as millet, amaranth, and manioc) reflects on the isolated situation of Europe. Europeans moving north were confronted with a decline in productivity due to the dwindling GDD, but this drop-off was not as rapid as the changes occurring in the Americas or East Asia at these latitudes. South of the Mediterranean, the dramatic loss of NPP across North Africa and the Arabian Peninsula was substantively different from circumstances prevailing in the eastern part of the Americas and East Asia and would have critically curtailed migration during periods of transitioning climate. Early Iberians would have had to travel a distance of nearly 3,000 kilometres across the Sahara before they encountered growing conditions at a scale comparable to central Europe (Figure 7.2). For people living around the North Sea and in the Baltic regions, this distance was considerably greater. Trade through Turkey, the Levant and Egypt connected southern Europe to the productivity of the south Asian tropics, but this trickle was never capable of sustaining further population growth, particularly in the north. People in eastern North America and East Asia were never far from wet tropical conditions, but the high-latitude isolation of the Europeans meant that for thousands of years leading up to the Iberian navigational discoveries during the fifteenth century, they were never near them. Mastery of the open seas and the colonization of the tropics changed this dynamic.

Europe's hypodiverse agriculture

As an island of relatively high bioproductivity, prehistoric Europe should have supported a relatively diverse, endemic flora offering a wide variety of progenitor crop plants compared to the adjoining regions. After all, the most productive biological communities on the planet – tropical rainforests and coral reefs – also register the highest number of species, even when comparing to other communities or habitats that are frequently found nearby, such as savanna and shrublands, or seagrass beds and mangroves. Yet, evidence suggests that Europe's relatively high NPP did not confer any greater propensity towards food plants. In fact, most of what we know about prehistoric European crops tells us that the overwhelming majority started as exotic cultivars in the low NPP regions south of the Mediterranean. Fewer than 60 plants currently contributing to the modern European diet can genuinely be traced back to a progenitor wild ancestor growing in Europe before Greco-Roman times. Of these, nearly half are secondary crops, mainly smallish herbs used as condiments or seasonings combined with some leafy vegetables, such as dandelion or rocket that were more suited to gathering from the wild or kitchen gardens than systematic cultivation (Table 7.1[5]).

Table 7.1 A collation of known food plants, their energetic content, consumed plant part and the timing of their introduction to Europe over the past 12,000 years based on archaeobotanical, genomic and documentary evidence

Plant family	Common name	Scientific name	Energy (kcal)	Plant part
NATIVE WILD-TYPE PROGENITOR PLANTS (> 10,000 years ago)				
Across Eurasia				
Asteraceae	Calendula	*Calendula officinalis*	11	Fl
	Dandelion	*Taraxacum officinale*	45	L
	Tarragon	*Artemisia dracunculus*	287	L
Lamiaceae	Oregano	*Origanum vulgare*	306	L
	Rosemary	*Rosmarinus officinalis*	332	L
Rosaceae	Woodland Strawberry	*Fragaria vesca ssp. vesca*	31	Fr
Across Europe				
Apiaceae	Caraway	*Carum carvi*	335	S
	Sea Holly	*Eryngium spp*	0	L
Asteraceae	Chamomile	*Matricaria chamomilla*	10	Fl
	Wild Chicory	*Cichorium intybus*	24	L
Betulaceae	Hazelnuts	*Corylus avellana numerous var.*	645	S
Brassicaceae	Black mustard seeds	*Brassica nigra*	454	S
	Wild arugula	*Diplotaxis tenuifolia*	25	L
	Arugula (rocket)	*Eruca vesicaria*	25	L
	Field pepperweed	*Lepidium campestre*	–	L
Fagaceae	Acorns	*Quercus robur*	510	S
Lamiaceae	Mint	*Mentha spp*	43	L
Rosaceae	Crab-apple	*Malus sylvestris*	72	Fr
	Raspberry	*Rubus idaeus*	53	Fr
	Sloe	*Prunus spinosa*	91	Fr
	Wild Pear?	*Pyrus pyraster*	48	Fr

continued

Table 7.1 Continued

Plant family	Common name	Scientific name	Energy (kcal)	Plant part
Mainly Southern Europe				
Amaranthaceae	Sea beet	Beta vulgaris ssp maritima	22	R
Amaryllidaceae	Wild Leek	Allium ampeloprasum	31	L
Apiaceae	Celery	Apium graveolens	13	Sh
	Cicely	Myrrhis odorata	–	L
	Coriander (cilantro)	Coriandrum sativum	279	S
	Dill	Anethum graveolens	43	L
	Fennel	Foeniculum vulgare	91	L
	Garden Lavage	Levisticum officinale	–	L
	Parsley	Petroselinum crispum	88	L
Asteraceae	Wild artichoke (Cardoon)	Cynara cardunculus	22	Fl
Brassicaceae	Horseradish	Armoracia rusticana	85	R
	Land cress	Barbarea verna	32	L
	Wild cabbage	Brassica oleracea	19	L
Cornaceae	Cornelian cherry	Cornus mas	–	Fr
Lamiaceae	Hyssop	Hyssopus officinalis	–	L
	Lavender	Lavandula spp	–	L
	Marjoram	Origanum majorana	270	Fl
	Sage	Salvia officinalis	315	S
	Savory	Satureja hortensis/montana	272	S
	Thyme	Thymus vulgaris ssp	276	S
Lauraceae	Bay leaf	Laurus nobilis	313	L
Oleaceae	Wild olive (Oleaster)	Olea europaea ssp europaea var sylvestris	–	Fr
Vitaceae	Wild grape	Vitis vinifera spp sylvestris	–	Fr

Mainly Northern Europe

Apiaceae	Angelica	*Angelica spp*	–	R
Brassicaceae	Rutabaga (swede)	*Brassica napus ssp rapifera*	30	R
Cannabaceae	Wild Hops	*Humulus lupulus*	5	Fr
Grossulariaceae	Blackcurrant	*Ribes nigrum*	60	Fr
	Redcurrant	*Ribes rubrum var. sylvestric*	60	Fr
	Gooseberry	*Ribes uva-crispa*	44	Fr
Myrtaceae	Sweet gale or gruit	*Myrtus gale*	–	L

NEOLITHIC INTRODUCTIONS (10,000 to 5000 years ago)

Mainly West Asia

Fabaceae	Bitter vetch	*Vicia ervilla*	–	S
	Chickpea	*Cicer arietinum*	69	S
	Fava bean	*Vicia faba*	72	S
	Lentils	*Lens spp*	116	S
	Peas	*Pisum sativum*	39	S
Linaceae	Flax	*Linum usitatissimum*	533	S
Lythraceae	Pomegranate	*Punica granatum*	84	Fr
Moraceae	Fig	*Ficus*	268	Fr
Oleaceae	Orchard Olive (oil)	*Olea europaea*	195 (880)	Fr
Poaceae	Barley	*Hordeum vulgare*	69	S
	Broomcorn or proso millet	*Panicum miliaceum*	356	S
	Einkorn wheat	*Triticum monococcum ssp monococcum*	335	S
	Emmer/durum wheat	*Triticum turgidum ssp.s*	365	S
	Bread wheat/Spelt	*Triticum aestivum ssp.s*	338	S
Rosaceae	Almond	*Prunus dulcis*	598	S
Vitaceae	Vineyard Grape	*Vitis vinifera*	69	Fr

Egypt

Arecaceae	Palm date	*Phoenix dactylifera*	280	Fr

continued

Table 7.1 Continued

Plant family	Common name	Scientific name	Energy (kcal)	Plant part
BRONZE-IRON AGE INTRODUCTIONS (5000–2500 years ago)				
Across Eurasia				
Poaceae	Rye	*Secale cereale*	343	S
Rosaceae	Orchard Pear	*Pyrus communis ssp. communis*	84	Fr
	Orchard Cherry	*Prunus avium/cerasus*	74	Fr
Mainly West Asia				
Amaryllidaceae	Garlic	*Allium sativum*	149	R
Apiaceae	Carrot	*Daucus carota*	88	R
Brassicaceae	Kale	*Brassica oleracea* grp Acephala	35	L
Cannabaceae	Hemp (seed)	*Cannabis sativa*	553	S
Iridaceae	Saffron	*Crocus sativus*	311	Fl
Poaceae	Oats	*Avena sativa*	373	S
Rosaceae	Orchard Apple	*Malus domesticus vars.*	56	Fr
	Plum/Damson/Greengage	*Prunus domestica*	45	Fr
East Asia				
Poaceae	Foxtail millet	*Setaria italica*	364	S
Mainly Southern Europe				
Amaranthaceae	Chard (cultigenic)	*Beta vulgaris ssp. vulgaris var cicla*	19	L
GRECO-ROMAN INTRODUCTIONS (2500–1600 years ago)				
Mainly East Asia				
Anacardiaceae	Pistachio	*Pistacia vera*	607	S
Cannabaceae	Hops	*Humulus lupulus*	5	Fr
Poaceae	Rice	*Oryza sativa*	110	S
Rosaceae	Apricot	*Armeniaca vulgaris*	48	Fr
	Peach	*Persica vulgaris*	38	Fr

Region / Family	Common name	Scientific name	No.	Part
Mainly West Asia				
Amaryllidaceae	Onion/Shallot/Scallion	*Allium cepa vars.*	39	R
Apiaceae	Anise	*Pimpinella anisum*	337	S
	Chervil	*Anthriscus cerefolium*	237	S
	Cumin	*Cuminum cyminum*	375	S
	Parsnip	*Pastinaca sativa*	75	R
Asparagaceae	Asparagus	*Asparagus officinalis*	22	L
Asteraceae	Endive/Radicchio/Chicory	*Cichorium intybus*	17	L
Brassicaceae	Garden cress	*Lepidium sativum*	31	L
	Watercress	*Nasturtium officinale*	11	L
	White mustard seed	*Brassica (or Sinapis) hirta*	508	S
Fagaceae	Sweet chestnut	*Castanea sativa*	245	S
Juglandaceae	Persian or "English" walnut	*Juglans regia*	654	S
Moraceae	Black mulberry	*Morus nigra*	43	Fr
Mainly South Asia				
Brassicaceae	Brown Mustard seed	*Brassica juncea*	478	S
Cucurbitaceae	Cucumber	*Cucumis sativus*	14	Fr
Malvaceae	Old-world cotton	*Gossypium arboreum/herbaceum*	–	S
Pedaliaceae	Sesame	*Sesamum indicum*	564	S
Rutaceae	Citron	*Citrus medica*		Fr
Mainly South-East Asia				
Brassicaceae	Radish	*Raphanus sativus*	16	R
Lamiaceae	Basil	*Ocimum basilicum*	16	L
Egypt				
Asteraceae	Lettuce	*Lactuca sativa*	14	L
Cucurbitaceae	Watermelon	*Citrullus lanatus*	30	Fr
Mainly Southern Europe				
Asteraceae	Globe Artichoke	*Cynara cardunculus var. scolymus*	48	Fl
Brassicaceae	Broccoli (cultigenic)	*Brassica oleracea grp Italica*	33	Fl
	Kale/Cabbage (cultigenic)	*Brassica oleracea grp Capitata*	26	L
Fabaceae	Licorice root	*Glycyrrhiza glabra*	–	R

continued

Table 7.1 Continued

Plant family	Common name	Scientific name	Energy (kcal)	Plant part
MIDDLE AGE INTRODUCTIONS (1600–600 years ago)				
Mainly East Asia				
Polygonaceae	Rhubarb	*Rheum rhabarbarum*	20	Sh
Mainly West Asia				
Amaranthaceae	Spinach	*Spinacia oleracea*	23	L
Amaryllidaceae	Chive	*Allium schoenoprasum*	30	L
Brassicaceae	Cauliflower	*Brassica oleracea grp Botrytis*	24	Fl
Mainly South Asia				
Piperaceae	Black pepper	*Piper nigrum*	251	S
Poaceae	Sugar cane	*Saccharum officinale*	399	Sh
Rutaceae	Bitter orange	*Citrus aurantium*	48	Fr
	Lemon	*Citrus lemon*	29	Fr
	Lime	*Citrus aurantiifolia*	30	Fr
	Pumelo	*Citrus maxima*	–	Fr
Solanaceae	Aubergine (eggplant)	*Solanum melongena*	35	Fr
Northern Europe				
Amaranthaceae	Beet root (cultigenic)	*Beta vulgaris ssp. vulgaris var cicla*	44	R
Brassicaceae	Brussels sprouts (cultigenic)	*Brassica oleracea grp Gemmifera*	88	L
AGE OF DISCOVERY INTRODUCTIONS (600–200 yrs ago)				
Mainly East Asia				
Lauraceae	Camphor	*Cinnamomum camphora*	–	Sh
Schisandraceae	Star anise	*Illicium verum*	–	S
Mainly South Asia				
Arecaceae	Coconut	*Cocos nucifera*	660	S
Lauraceae	Cinnamon	*Cinnamomum verum/cassia*	246	Sh
Zingiberaceae	Cardamom	*Elettaria cardamomum*	311	S
	Turmeric	*Curcuma longa*	312	S

Mainly South-East Asia				
Arecaceae	Sago	*Metroxylon sagu*	–	S
Lauraceae	Nutmeg-mace	*Myristica fragrans*	526	S
Malvaceae	Tea	*Camellia sinensis*	1	L
Myrtaceae	Cloves	*Syzygium aromaticum*	274	Fl
Zingiberaceae	Ginger	*Zingiber officinale*	335	R
Mesoamerica & Caribbean				
Fabaceae	Common and runner bean	*Phaseolus vulgaris/coccinea*	127	S
Malvaceae	Cacao (Cocoa)	*Theobroma cacao*	228	S
	New world "Mexican" cotton	*Gossypium hirsutum*	–	S
Marantaceae	Arrowroot	*Maranta arundinacea*	356	R
Myrtaceae	Allspice	*Pimenta dioica*	263	S
Orchidaceae	Vanilla	*Vanilla planifolia*	287	S
Poaceae	Maíz	*Zea mays*	361	S
Solanaceae	Bell pepper	*Capsicum annum*	19	Fr
	Chili pepper	*Capsicum spp*	39	Fr
	Tobacco	*Nicotiana tabacum*	–	L
	Tomato	*Solanum lycopersicum*	18	Fr
South America				
Bixaceae	Annato	*Bixa orellana*	0	S
Bromeliaceae	Pineapple	*Bromelia cosmosus*	49	Fr
Euphorbiaceae	Tapioca (cassava)	*Manihot esculenta*	89	R
Fabaceae	Peanut	*Arachis hypogea*	566	R
Malvaceae	New world "Pima" cotton	*Gossypium barbadense*	–	S
Rosaceae	Garden strawberry	*Fragaria x ananassa (cultigenic)*	31	Fr
Rubiaceae	Quinine	*Cinchona officinalis*	–	Sh
Solanaceae	Potato	*Solanum tuberosus*	86	R
Sub-Saharan Africa				
Rubiaceae	Coffee	*Coffea arabica/canephora*	1	S
Zingiberaceae	Grains of paradise	*Aframomum melegueta*	–	S

continued

Table 7.1 Continued

Plant family	Common name	Scientific name	Energy (kcal)	Plant part
Mainly Northern Europe				
Brassicaceae	Kohlrabi (cultigenic)	*Brassica olerace grp Gongylodes*	27	R
	Savoy cabbage (cultigenic)	*Brassica oleracea*	26	L
Mainly Southern Europe				
Amaranthaceae	Sugar beet (cultigenic)	*Beta vulgaris ssp vulgaris var Altissima*	–	R
Brassicaceae	Broccoflower (cultigenic)	*Brassica oleracea*	31	Fl
	Broccoli romanesco (cultigenic)	*Brassica oleracea grp Botrytis*	51	Fl
MODERN INTRODUCTIONS (<200 years ago)				
Mainly East Asia				
Actinidiaceae	Kiwi	*Actinidia spp*	61	Fr
Ebenaceae	Persimmon	*Diospyros spp*	69	Fr
Mainly South Asia				
Anacardiaceae	Mango	*Mangifera indica*	60	Fr
Musaceae	Banana	*Musa spp*	88	Fr
Mainly South-East Asia				
Sapindaceae	Lychee	*Litchi chinensis*	66	Fr
Mesoamerica & Caribbean				
Lamiaceae	Chia	*Salvia hispanica*	485	S
Lauraceae	Avocado	*Persea americana*	160	Fr

South America				
Amaranthaceae	Amaranth	*Amaranthus caudatus/cruentus*	102	S
	Quinoa	*Chenopodium quinoa*	120	S
Anacardiaceae	Cashew	*Anacardium occidentale*	571	S
Arecaceae	Acai	*Euterpe oleracea*	–	Fr
Caricaceae	Papaya	*Carica papaya*	43	Fr
Erythroxylaceae	Coca	*Erythroxylum coca/novogranatense vars.*	–	L
Lecythidaceae	Brazil nut	*Bertholletia excelsa*	660	S
Passifloraceae	Passion fruit	*Passiflora spp*	96	Fr
Sapindaceae	Guarana	*Paullinia*	–	S
Solanaceae	Physalis	*Physalis peruviana*	–	Fr
North America				
Asteraceae	Sunflower	*Helianthus spp*	582	S
Ericaceae	Blueberry	*Vaccinium spp*	57	Fr
Juglandaceae	Pecan nut	*Carya illinoinensis*	710	S
Sub-Saharan Africa				
Arecaceae	Palm oil	*Elaeis guineensis*	885	S
Poaceae	Sorghum	*Sorghum bicolor*	359	S

Note
A dash indicates no data.

Most evidence points to an impoverished natural pool of agricultural wild types across prehistoric Europe. Archaeobotanical findings suggest only hazelnuts, beet root and kale/cabbage had progenitor wild-type varieties growing in northern and central Europe during Neolithic times, between 10,000 and 4,000 years ago (Colledge and Conolly 2007). Wild types of apple (*Malus sylvestris*), sloe (a type of plum, *Prunus speciosa*) and pear (*Pyrus* spp) have also been discovered in prehistoric midden remains at various sites in western Europe, but orchard varieties as we know them today evolved from medieval imports and have been variegated from species sourced outside the region. These and other domesticated orchard fruits depend on grafting to achieve their reproductive potential and this technique, again, arrived from west-central Asia and was only first well established in Greco-Roman times (Mudge *et al.* 2009). Radiocarbon dating of charred seed materials supports a view that most crops species first entered Europe through the Mediterranean and subsequently migrated north-westward (Colledge and Conolly 2007). In terms of GDD, this is expected – crops originated in central Asia and the Mediterranean because these areas have longer growing seasons and then moved towards higher latitude regions where GDD is much lower. Fewer crops are able to reach maturation within these shorter periods and the risk of crop failure due to sudden weather changes during these narrow "growth windows" is much higher as a consequence.

But in relation to precipitation, this pattern of migration is counter-intuitive. Why would dry regions act as incubators of source crop plants and regions with much greater moisture availability prove to be recipients? Most people would view arid regions as a challenge to human sustenance, requiring external inputs, not as a source of it. The answer rests, paradoxically, with the impact that increased rainfall has on the structure of climax vegetation. Climax vegetation is an ecological term used to describe the plant community that eventually self-assembles under prevailing environmental conditions. Temperature and moisture are the primary determinants of climax vegetation although soils, episodic disturbances – such as hurricanes or wildfires – and barriers to seed dispersal can also have a hand in shaping the climax vegetation of any region. In the case of Neolithic Europe, this amounted to seasonal broadleaved birch and alder forests in the west, gradually inter-mixing with coniferous forests in the north and east. In southern Europe and across the Mediterranean into the Levant and towards the Persian Gulf, the climax vegetation becomes more open and shorter in stature, disappearing altogether over most parts of the Arabian Peninsula. Although the Sahara region received somewhat greater rainfall as northern glacial masses retreated at the beginning of the Neolithic *sensu lato*, palaeoenvironmental indicators suggest that this condition was temporary and the deep aridity that characterizes the region today was in development at least 4,000 years ago (Kröpelin *et al.* 2008).

This transition from dense, closed-canopied forests to open, scattered grasslands and desert played a key role in facilitating plant domestication and incipient agriculture. In central and north Europe, vast stretches of contiguous forest provided fewer ecological opportunities for the diminutive, but rapidly maturing, grasses and legumes that would constitute the founder crops of European agriculture. These founder crops – emmer and einkorn wheat, barley, chickpeas, lentils, peas and bitter vetch (Zohary *et al.* 2012) – are characterized by ecologists as disturbance-dependent or ruderal species if found within a contiguous forest ecosystem. They are found in patches of recently disturbed forest, after fires, or along waterways and coastlines or anywhere closed-canopy conditions could not establish or be maintained. These are not ideal conditions for progenitor staple food crops, mainly because the high solar irradiance they require is inflected by surrounding forest cover. But across much of the region south of Europe the progenitor species of the founder crops would have constituted part of the climax vegetation, an environment ideally suited to ruderal species. Ruderal species exhibit certain ecological attributes, such as long-lasting seeds that produce rapidly growing plants of relatively small structure that reproduce quickly – often only once or twice – and then die. These attributes allow species to prosper in environments where moisture and nutrient availability is brief and less predictable, but sunlight is readily available. They are opportunists, able to make use of ephemeral environmental conditions. They require relatively little time from germination to seed production and often produce many seeds that account for a relatively large fraction of the total plant weight. In the case of each of the seven founder crop species, seeds are the harvested portion of the plant and would have been the primary focus of breeding. The longevity of their seeds made ruderal species ideally designed for human use, particularly during the long winter hiatus that offered very little in the way of fresh plant material, either for food or the seed needed to sow crops in the following year. But some of the traits that make ruderal plants such great opportunists also impaired the size and predictability of seed set and these were bred out of the domesticated varieties leaving them more vulnerable to disease and unseasonal changes in weather.

The situation for trees is very different. Their life histories are more ponderous. They take considerably longer to reach reproductive maturity and the seed crop accounts for a much smaller fraction of total plant weight, although a single tree can sometimes reproduce annually or biennially over decades or even centuries. After reaching adulthood, trees are advantaged in two important ways. First, they are far more resistant to bouts of frost, drought and disease that chronically led to catastrophic failure in annual crops. Journal entries during periods of food scarcity describe how bouts of unseasonably cold temperature and both flood and drought were the primary drivers of crop failure. Narratives from the time speak of some communities across Europe resorting to making bread from processed acorns, telling us that although annual crops failed, tree seed crops probably did not (e.g. Spain, Grove and Rackham 2001). Second, trees are better at capturing nutrients and moisture due to their extensive root systems, making them less vulnerable to soil exhaustion. Forests internally recycle nutrients through a process of release (through dead leaves, branches, flowers and fruit), decomposition, mineralization and re-uptake. In some instances, this recycling can be hyper-efficient with very little of the nutrient base conveyed outside the forest ecosystem. Many agricultural crops require significant nutrients to support their growth over a very short season. But unlike natural forests, base agricultural crops are harvested annually and the absorbed nutrients are exported to humans with the bulk lost from the agro-ecosystem. Exposed soils after harvest also make the agricultural field more vulnerable to erosion and nutrient leaching. Nutrient depletion during medieval times became an important limiting factor to improvement of crop productivity and increasing food security (Newman and Harvey 1997). Phosphorus depletion may have been of particular importance where levels could not be replenished naturally through weathering of phosphate rock. Many of the leguminous crops, such as peas and vetch, fixed nitrogen through an association with rhizobia – symbiotic bacteria that invade the roots of many plants – but these only came into play when farming practices were modified to rotate legumes with cereals and pasture during the late Middle Ages. The latter rotation, and the manuring and marling of fields to replenish nutrients taken out through the harvest became a critical lynchpin in the management of crop yields prior to the industrial production of agro-chemicals (Glennie 1988). Most plants also form symbiotic relationships with some type of mycorrhizal fungi. These fungi – including the highly prized truffles and some morels – improve mineral uptake by extending the root volume of plants and accessing forms of nutrients in the soil, such as phosphate, that are otherwise not easily taken up by plant roots. Almost all trees in Europe have some form of mycorrhizal association and many important crops, such as wheat, oats and barley, do as well. But the brassicas – perhaps the largest and most variegated crop group native to Europe – do not form these relationships (Brundrett 2009).

Domesticated livestock arrived with the basket of founder crops from west and central Asia and became a central part of the early agricultural "package" that we now know as traditional European farming. Cows, goats, sheep and chickens find their maternal origins in Asia and the Middle East, entering Europe via Greece and Egypt (e.g. Perry-Gal *et al.* 2015). Current genotypic studies suggest only the domesticated pig among the "big five" livestock has evolved from ancestral stock native to Europe, and this possibly as an admixture of native and Asian varieties (Larson *et al.* 2007). For a host of reasons, livestock offered a particularly useful source of nutrition to northern Europeans, as they continuously struggled with crop failure. One of the principal roles of livestock arriving with founder crops involved their use as a source of power. Pulling ploughs or running mills with working animals allowed farmers to be more ambitious in their agricultural pursuits while undertaking other non-field activities with their increased "spare" time. The efficiency of using animals was further improved over the course of the Middle Ages as farmers began to realize the much greater plough work they could achieve by substituting oxen with horses. Horses could do more work and consumed less fodder and by the end of the fifteenth century in England, the stocking densities of working horses had nearly doubled, while those of oxen had declined by a third (Broadberry *et al.* 2008). Second, non-working livestock offered a means of converting largely unpalatable parts of the region's NPP, such as inedible pasture grasses (cattle, sheep), seedmast (pigs), and insects (chickens, pigs) into a highly accessible source of nutrients and calories. For example, pigs – the quintessential animal of the poor – continued to be herded, or pannaged, into hardwood forests across rural Europe, to feed on acorns, beechmast and chestnuts. In some regions, this continued well into the nineteenth century (Witney 1990; Szabó 2013). Third, livestock offered a non-seasonal source of food that was unavailable through crop production and allowed people with little access to the bioproductivity outside their region to survive the winter shutdown in plant growth. Finally, livestock formed a critical part of nutrient recycling within the agro-ecosystem. Their manure remained a highly prized and highly necessary soil ameliorating agent and assisted in maintaining levels of agricultural productivity throughout medieval Europe.

All told, there were compelling reasons why European agriculture was both botanically hypodiverse and, by association, heavily dependent on livestock. The native forests were more diverse than the climax vegetation across the Levant and North Africa, but included very few species with fleshy fruits or seeds that were readily consumable. The exception was hazelnut – its consumable nut, useful wood and capacity to coppice made it almost as important as many founder crops to early European societies and its utilitarian value may very well account for its ubiquitous distribution across Europe today. The few native species with consumable fruits – wild plum, pear and cherry in the north and the oleaster and wild fig in the south – eventually mixed with extra-regional varieties to form the foundation of European arboriculture. To this small list of native fruit trees, we can add raspberry, blackberry and *Ribes*, but beyond this handful of woody, perennial plants producing edible fruits and nuts, there was little on offer from the climax forests of Europe (see Table 7.1).

By comparison to the eight trees and shrubs producing edible fruits and nuts across Europe, agricultural use of tree crops outside the region was much greater. By the sixteenth century, an additional 20 tree and shrubs from various parts of Asia had been added to the European panoply of food plants. This included modern orchard varieties of the apple, plum, cherry, olive, peach and apricot, as well as palm dates, almonds, chestnuts, walnuts, lemons, oranges and cherries. In the Americas, the number of woody plants being domesticated, or actively cultivated, by indigenous societies leading up to the Iberian arrival was extraordinary compared to those in Europe. Clement (1999) identified more than 77 trees and shrubs that provided edible fruits or seeds and were probably being actively managed by indigenous agriculturalists at the time. I know first-hand the complex systems of field-fallow management practices employed by these agriculturalists in the Amazon today, including a large number of the species identified by Clement (e.g. Hammond *et al.* 1995). Many of these, of course, became common food items in Europe as part of the Columbian Exchange, including cocoa, chilli, tomato, Brazil nut and cashew.

We can see from Table 7.1 that the ecological problem with Europe's vegetation was that it consisted of few plants with consumable seeds or nuts. Leaves, shoots, roots and fruits made up the overwhelming bulk of plant-based foods from native species and this did not change until seed crops introduced from Asia and the tropics changed both European agriculture and diets. The biological undercurrent that makes this important has to do with caloric intensity. Seeds on an equal weight basis contain twice as much energy as other plant parts and in some comparisons, more than five times the energetic reward. Beet roots, kales, cabbages and herbs alone were not fit-for-purpose in meeting the energetic demands of humans, nor were the majority of small fruit-bearing trees and shrubs. There are exceptions, and it is not happenstance that olive oil and hazelnuts – both high in energy – have become "legacy" food ingredients across different parts of Europe. The average caloric advantage of consuming seeds over other plant parts in Europe is illustrated in Table 7.2 but superior shelf-life and a predisposal to a broad spectrum of cooking techniques also made seed-based foods more efficient in satiating daily energetic needs.

European forests also differed from those in the tropics in their environmental histories, the fundamental impact this had on the pool of

Table 7.2 The relative energetic value of consuming various crop plant parts based on the caloric content of food plants depicted in Table 7.1

Plant part consumed	Energetic content (avg kcal per gram)	No of species	
		All	Native
Seeds (S)	360.8	52	8
Flowers (Fl)	198.6	11	4
Shoots (Sh)	169.7	4	1
Roots (R)	133.9	15	4
Leaves (L)	68.8	30	22
Fruits (Fr)	63.4	42	12

ancient plant types and how they evolved into a modern community of forest species. Europe was more profoundly shaped by the last Ice Age than any other heavily populated region and the vegetation pool still reflects the influence of widespread glaciation. This difference is most cogently explained by the taxonomical relatedness of native plants within Europe and other regions and the level of overlap between these regions. The most striking difference between the taxonomy of European forests and those found in the tropics and along the other temperate coastlines is the virtual absence of any species associated with "primitive" or "basal" groups – or clades – of flowering plants. These clades – together sometimes informally referred to as paleodicots – represent branches of the evolutionary tree that separated or became isolated from the main progenitor group at an earlier point in our planet's botanical history. Climate change and geological processes are typically the main factors responsible for separating and isolating ancestral plant populations, leading to new species. Most of the plant families attached to these early offshoots are tropical.

Why does the distribution of primitive flowering plants matter to globalization? Contained within this group is the "holy trinity" of plants that would ultimately form the bulk of the tropical spice trade: *Piper nigrum* (botanical family Piperaceae), the black and white pepper plant; *Myristica fragrans* (Myristicaceae), the source of nutmeg and mace; and *Cinnamomum vera* and *C. cassia* (Lauraceae), the sources of cinnamon. These were the Iberian targets that prompted their initial navigational exploits at the end of the fifteenth century. Prior to the onset of trade between Europe and the tropics, most Europeans had little contact with plants from these "primitive" groups. Black pepper had made its way through Middle Eastern markets from India since Roman times and the relatively small quantities were traded just enough to pique the interest of the wealthier classes in Europe. But apart from this important link, there was little else. Only six species, one tree – *Laurus nobilis* (Lauraceae) – and five species of low-lying herbaceous plants in the nephrotoxic birthwort family – the Aristolochiaceae – grow naturally in the region, and primarily along the southernmost margins of the Mediterranean and in the Balkans. This compares to thousands of known species in this group of primitive angiosperms found across the tropics (Figure 7.3). However, this enormous gulf in representation is not simply a case of extreme differences in latitude. Many "primitive" flowering plants species are found in the temperate forests of North America and China, including sassafras (*Sassafras albidum*, *S. tzumu*, Lauraceae), the paw-paw (*Asimina triloba*, Annonaceae) and tulip tree (*Liriodendron tulipifera*, Magnoliaceae) across much of the eastern United States. The magnolias, a large genus containing more than 200 species of trees and shrubs, are represented along the entire length of the Americas from Canada to the Amazon and in East Asia growing naturally from the Korean Peninsula southward to Borneo and Sumatra. Many

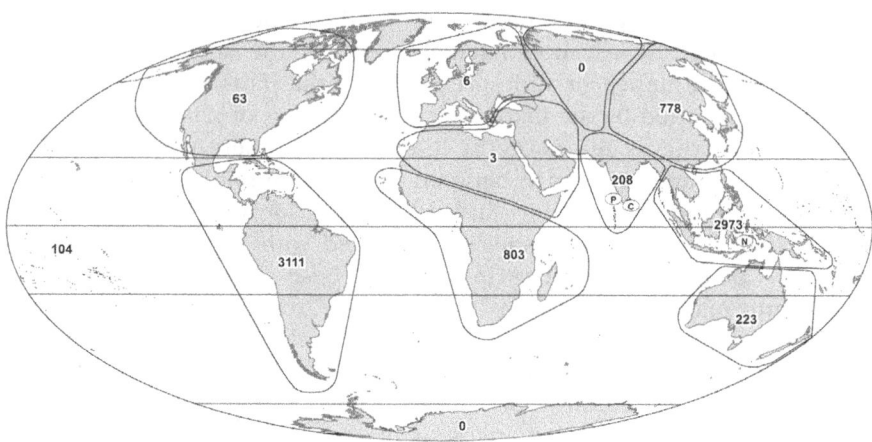

Figure 7.3 A general estimate of the known number of primitive "magnoliid" flowering plant species across various regions of our planet.

other species of primitive angiosperms, some from small botanical families such as the Saururaceae and Calycanthaceae are also found in both tropical and temperate locations in the Americas and Asia, but not in Europe or the adjoining regions southward (Figure 7.3) The tropical spice trade grew in part from an absence of these primitive plants and the class of phytochemicals, particularly eugenol, cinnamaldehyde and piperine, that were not present in the native vegetation of Europe. Bay leaf was the only indigenous source of eugenol and it is no accident that it became an important food seasoning within the toolkit of European cookery, but more later about the importance of tropical phytochemicals and their role in globalization and the tropical trade.

Native European forests offered little in terms of sustenance, dominated natural growth and suppressed the evolution of wild-type varieties of cereals and pulses. As mentioned earlier, the life histories of plants making up the underlying climax vegetation were ecologically incompatible with those of the introduced agricultural crops. This incompatibility was due to the elevated bioproductivity of Europe compared to west Asia that, paradoxically, had a stultifying effect on the development of progenitor food crops because it supported an expansive closed-canopied forest ecosystem dominated by species that yielded little nutrition. More than 90 per cent of the agricultural plants and animals utilized in Europe up to the time Columbus arrived in the Americas had been imported at some point in time from central and west Asia. Many of these, such as chicken or peaches, followed the very same path as the plague and influenza – originating in East Asia and arriving in Europe after spending some time in a west Asian "agricultural reservoir" where they were further improved and refined.

Most of the native food plants that are native to Europe are not endemic to the region either. For example, the hazelnut is naturally distributed through Turkey and Iran, and many similar (congeneric) species are indigenous to North America, including an American hazelnut (*Corylus americana*) and the more common filbert (*C. maxima*), as well as wild raspberry and blackberry species. Mints, mustards and most of the other 25 weedy, herbaceous plants that make up half of food plants considered native to Europe evolved along the margins of southern Europe and none were unique to the region, being widely distributed across Eurasia and, rarely, the Americas as well (e.g. the dandelion, *Taraxicum* spp). Consequently, it is fair to say that Europe has always been a recipient region in terms of agricultural resources, at least since the beginning of the Neolithic period, some 10,000 years ago. Perhaps only the kales, beetroots and their constellation of allied varieties and cultigens stand out as principally European in origin. This important pool of cabbages, broccolis, cress, cauliflower, rutabaga and sugar beet could be considered Europe's indigenous contribution to global agricultural resources, but very little else.

As Diamond (1997) points out, the movement of crops that have evolved at similar lines of latitude was ecologically more straightforward. Movement across the Eurasian continent from East Asia to Europe was not blocked by oceans or latitudinal changes in climate although environmental conditions across central Eurasia certainly remained very different – as they are today – from those prevailing in western Europe due to the influence of this enormous land mass in propagating more extreme seasonal temperature changes in the central regions. But the evolution of plants that were closely related was far more likely across Eurasia than among the different continents isolated from one another by the oceans or extreme differences in latitude. The problem for early European agriculture was that these transfers from temperate Asia were not very diversified. If we examine the food plant stock available in Europe before establishing direct contact with the tropics, what we see is a group contained within a surprisingly small number of taxa (Table 7.1). Nearly 80 per cent of the fifty-odd (wild-type) food plants native to Europe are drawn from only five plant families: the Apiaceae, Asteraceae, Brassicaceae, Lamiaceae and Rosaceae. Neolithic introductions from west Asia added nine food plants in two new, major plant families to the diets of Europeans: wheat, barley and millet in the Poaceae and various legumes from the Fabaceae. But over the subsequent 3,500 years, these seven botanical families continued to account for two-thirds of the new food plants introduced to Europe up to the collapse of the Roman Empire (Table 7.1). In the aftermath of Rome, the fragmentation of its former provinces into kingdoms extinguished many of its former technological and cultural advancements. But it also led to a revival of imperial power across Turkey, the Levant and Persia and the establishment of city-states across Italy dedicated to and dependent upon broadcasting trade and commerce as far as could be achieved. These commercial links invariably built upon those established by the Romans, but, more importantly, began to establish new contacts with tropical Asia via Arab and Persian merchants. A proxy trade with the tropics via West Asian markets changed the botanical affiliations of introduced food plants. Although few species arrived during the medieval period by comparison to Greco-Roman times, only a quarter of the new species being added to Europe's pool of food plants belonged to the seven ancient plant families. Discovery of maritime links to the tropics further diluted the dominance of these groups. By the 1800s, only a fifth of new food plants being consumed in Europe belonged to these families – a trend that has continued up to today (Table 7.1).

Of course, accessioning new plant products from the tropics directly into European agriculture would by comparison to those from West Asia prove far more challenging – nay impossible in many instances – and spark the realization that to access the immense wealth of our planet's biological diversity, Europe would need to expand to the tropics. The consequence was a doubling of the number of foods and flavourings that were available originally to Europeans through their native pool of flora (Table 7.1), transforming one of the world's most hypodiverse, isolated food supplies into one of its most hyper-diverse within the space of three centuries. By the 1800s, the tropics were contributing food products from at least another 17 plant families unrepresented in Europe, as well as four of those already present. Outsourcing both natural productivity and diversity to the tropics led to a boom in Europe, both in terms of the nutritional and biochemical changes to everyday life of workers and in the expansionary impact it had on trade, taxes, credit and economic growth.

To feed, is to make, is to serve

There has been an underlying periodicity to the way societies have evolved based on acquisition, adoption and diversification of new approaches to work. We often refer to the transition between these periods as "revolutions" – agricultural, industrial, technological, digital. It goes like this – we started with a need to feed communities and developed a system of increasing food production to meet this goal. Having built this principal foundation, these communities coalesced and some found themselves with more time to spend beyond meeting their daily caloric needs. We then stepped into developing increasingly sophisticated tools to increase economic productivity at larger and larger scales of production. Having mastered the basic principles of industrial activity, we then moved on to diversify and expand services that further increase efficiencies and capabilities of industry and society. It is this process, feed first, make second, serve third, that the discovery of the tropics enabled and accelerated. Of course change is rarely as punctuated or segmented as the above description might imply. But prior to the onset of global trade, feeding the population through domestic agriculture was tantamount to economic growth in Europe for two significant reasons.

First, it fostered higher survivorship, particularly among children. Survivorship analysis tells us that the likelihood of death during an individual's life is greatest at its beginning and its end. It forms a bathtub shaped curve where the bottom of the tub depicts the risk to adults throughout their lives from disease, illness, accidents and misadventure. On aggregate, this remains very low for adults relative to the impact that these factors have on newborn children and the elderly. Improving nutrition is universally accepted as a principle precursor to good health and, by referral, as a basis for surviving early infections and living longer. Medieval societies in Europe continuously struggled to maintain healthy nutritional standards. Crop failures, land tenure systems and market prices often worked in unison to create a spiral of dearth that mainly affected the lowest economic classes – the classes that produced through labour, not capital. This in turn made them more susceptible to the various diseases that afflicted populations at the time. These risk factors – weather, crop yield, tenure and prices – would ultimately find their way first into Adam Smith's (1776) views on commodity production, labour and economic rents and 20 years later into Thomas Malthus' (1798) treatise on the relationship between population growth and agricultural production. While Malthus' views are often much maligned today, population size and poor agricultural production appear to better explain the frequency of famine and starvation in Europe leading up to the end of the seventeenth century than factors influencing the distribution of grain itself (Alfani and Ó Gráda 2018). This is in contrast to a modern agriculture system technically capable (by some estimates, just barely) of meeting the needs of our planetary population yet coincides with outbreaks of starvation principally due to distribution barriers. Crop failure in pre-Enlightenment Europe simply led to increased mortality when populations grew too large to be sustained through periods of dearth. Growing and sustaining a nation's population was and is fundamental in determining the size of the economy for a host of reasons linked to both production and consumption. As I will discuss in the next chapter, the size of any modern nation's population and economy are strongly inter-related and economic size plays a critical, reflexive role in ensuring the sustainability of food supply and environmental use, but trade and globalization does not necessarily improve food security for all nor engender the same economic outcomes.

Second, agricultural output stimulated economic growth because it acted as the primary energy system, engendering a greater capability for work among the working-age population of people and draft animals. One of the greatest social problems in Europe leading up to nineteenth century was a chronic inability to meet the daily caloric requirements of most adults. Although not all historians agree on the nutritional intake of medieval populations, it is generally thought that in pre-plague, medieval England, agricultural workers required between 3,800 and 4,000 kilocalories per day. The production estimates that have been produced based on tithes and rent records from the period suggest that per capita supply of grain amounted to less than a third of this amount. Energy from meat, cheese and butter would have augmented grain-based sources (including beer and ale), but it is unlikely that most people were able to achieve on a daily basis the ~ 4,000 kcal needed to support a healthy rural labour force. The significant shortfall of available energy from wheat, oats and barley highlights the razor-thin margin in meeting basic requirements from agriculture in medieval England, particularly where this could not be made up through meat and dairy products that were more expensive. Crop failure due to poor weather under these precarious caloric circumstances would often rapidly lead to famine (Healey 2011). Even by 1800, the grain supply was only able to deliver around 1,600 kilocalories per day, although productivity had improved substantively and the number of rural labourers had halved since 1300 (Broadberry *et al.* 2011). Ensuring that agricultural workers and their families were reasonably fed was a core precursor to increasing agricultural productivity. A large, underemployed labour pool beset by illness and incapacity due to poor nutrition fundamentally undermined a drive to increase the amount of grain and livestock produced. Campbell (1988) went as far as to state that "in all pre-industrial economies the limits to economic growth were set by the productivity of labour in agriculture". This was invariably so, since the overwhelming majority in Europe were employed in the agricultural sector and the relationship between caloric intake, weather-induced crop failure and productivity would drive recurrent bouts of famine and stymie domestic economic growth in most European medieval societies.

This is very different from the situation today, where domestic agriculture accounts for less than 2 per cent of economic activity in Western Europe, yet caloric intake has never been as high. Food security and diversity have been significantly enhanced by technologies that improve yields and allow food to be transported over vast distances without spoilage. The rapid growth in other sectors of the economy that make goods and deliver services has also acted to shrink agriculture's economic share, although many of these too are attached in some way to the distribution of food and drink. But this modern condition was achieved only recently, through rapid improvements since the 1950s in crop yields, agricultural productivity and transportation technology. Even in the late 1800s, some European countries such as Great Britain, Spain and Portugal were still failing to grow domestic agricultural output at a rate higher than the growth in their populations (Mitchell 1975; Mitchell and Deane 1962) and some have questioned whether Europe underwent an agricultural revolution at all based on lacklustre growth in agricultural productivity between 1700 and 1860 (Clark 1999).

As I read through the plethora of historical and economic studies on medieval and pre-twentieth-century patterns of agricultural change in Europe, I realize that the looming issue of the tropical trade is often altogether absent. It is as though agricultural, industrial and economic growth in Europe was as isolated as the region was biologically or as though the tidal wave of agricultural production arriving from the tropics had no or little bearing on the (mis)fortunes of European agriculture. While this is obviously true for most of medieval Europe (except the trading Italian city-states), we know this is not so from the brief history of events shaping sugar production presented in the previous chapter, and the manifest changes that took place in maritime trade, taxation and agriculture as a result of cane sugar's increasing influence between 1600 and 1850. Cane sugar manifestly changed the diets of most Europeans, augmenting the caloric intake of labourers as the price declined with an expanding supply sourced in the West and East Indies. Based on average consumption rates per person, as illustrated in Figure 6.2, cane sugar was adding an estimated 22 kilocalories to the average daily intake of British residents by 1700, growing to nearly 100 kcal by 1800 and 130 kcal by 1850. These subsidies from cane sugar imports, amounted to an estimated 5–8 per cent of the average daily energy supplied by domestic production of wheat and other cereals.

The introduction of maize also quietly altered the energetic consequences of crop cultivation across Europe. Early expeditions to the Caribbean and North America returned with maize seeds and the indigenous knowledge needed to cultivate the crop in temperate climates. This information was rapidly integrated into early herbal compendiums and books describing agricultural techniques (Finan 1948). Genomic analyses indicate that several maize varieties were introduced from both the tropical and temperate regions of the Americas in the sixteenth century (Rebourg *et al.* 2003; Tenaillon and Charcosset 2011). These in turn spawned the modern land races of maize that are now one of the most important crop species (by weight) grown in eastern and southern Europe. Less than 50 years after Columbus' arrival in the Caribbean, maize was being grown in Germany and in many parts of southern Europe. By the early 1600s, the crop had been introduced to most of continental Europe. Although never rivalling the production of traditional staple crops, the adoption of maize would have slowly boosted the energetic returns of European agriculture. Modern varieties return between 8 and 10 per cent higher energetic content than the other major grains grown in Europe, including wheat, rye and barley.[6] If this difference has remained constant since its introduction, then maize would have added to the caloric return of farming where it was adopted, particularly through its use as a livestock feed. Sugar cane and maize were two important contributors to the process of achieving greater food security across Europe, but there were others. The potato also became an integral component of the European diet after its arrival in the seventeenth century. Pulled from the Andean highlands, the starchy relative of the chilli pepper was well adapted to the cooler climate of Europe. It also performed well in many of the landforms not suited to more demanding cereal crops. But unlike potatoes – that only return 20 to 25 per cent of the energetic content found in cereals and were typically propagated from clonal stock rather than seed – maize and cane sugar were energetically superior to the main plant-based sources of calories that had arrived thousands of years earlier from West Asia.

Stimulating trade

The energetic reward that arrived with maize and cane sugar contributed to an improved diet of European labourers either directly or through their fortification of other plant and animal products. It may sound particularly odd to speak of sugar in this way, but chronic under-consumption of calories was a major problem confronting European countries leading up to the twentieth century, whether the public institutions of the time chose to recognize it as such or not. Labour productivity has always been a critical factor in achieving economic success and any agents, conditions or devices that improve the amount of work achieved within a given period of time are by proxy also important to national economic growth.

Perhaps the most transformational agent that arrived as a result of the accelerated trade between Europe and the tropics in the fifteenth century was the widespread introduction of a group of biomolecules that had no energetic value and contributed very little to human nutrition. We refer collectively to these as stimulants and they arrived in Europe primarily through dried bulk seeds and leaves of three tropical products: coffee, tea and cocoa. Stimulants act to increase the activity of the central nervous system and the general action of caffeine and other methylxanthines contained in these plants is to increase alertness and wakefulness, improve concentration and alleviate fatigue when consumed in reasonable quantities. Caffeine creates these responses by blocking the action of adenosine on a special class of receptors that are responsible for the modulation of neurotransmitters, such as dopamine and glutamate that act as chemical bridges between neurons in the brain. The gradual accumulation of adenosine along these receptors is believed to induce drowsiness and caffeine acts to retard this build-up. There are many other chemicals that induce a stimulative effect on the human nervous system, including those delivered through the ingestion of other plant bioproducts, such as cocaine, ephedrine, nicotine and cathinone. But these products are not part of the methylxanthine group of purine alkaloids and have an additional impact on the release of dopamine and other "pleasure" biochemicals that can lead to a severe physical and/or psychological dependency. By contrast, the action of caffeine and other methylxanthines on adenosine accumulation is temporary and reversible without the severe physiological responses that characterize long-term use of other unrelated classes of alkaloid stimulants. In other words, when you stop drinking coffee or tea, the binding effect of caffeine on your neural receptors dissipates and the accumulation of adenosine continues, sometimes with a rapidity that leads to an "after-caffeine slump".

Today, caffeine is the most consumed psychoactive drug globally, but in medieval Europe labourers were not exposed to caffeine or any other biomolecules that induced a stimulating effect. Only a single native plant in Europe, the familiar European holly (*Ilex aquifolium*), contains any measurable amount of caffeine and this is concentrated in the bright red berries that are generally considered toxic to humans. In Paraguay, the leaves of the congeneric plant *Ilex paraguariensis* are still used to make the traditional caffeine-laced tea, called *mate*, that is consumed across much of southern South America. Early European chroniclers also recorded the use of another native *Ilex* species to make a medicinal tea, *yaupon*, in the south-eastern United States but there is no historical evidence suggesting that Europeans produced a similar beverage from their own holly plant. It is difficult to imagine daily life without coffee or tea (at least for me), but to think that generations across Europe did so until the 1600s is extraordinary. This is even more extraordinary if we consider that the bulk of the human population at the time was using or at least living coincident with plants containing stimulants. Before the European Age of Discovery, the Chinese and their neighbours in South-East Asia were drinking green tea and trading this along the Silk Road and Tea Horse Road across central and West Asia. Arabs, Turks and Persians were drinking coffee from Mocha in Yemen before the fifteenth century as well as Umharic-speaking peoples of the Ethiopian region and other parts of north-east Africa. Chewing the seeds of the kola fruit (*Cola* spp) and the leaves of *khat* – the cathinone-containing plant *Catha edulis* – are both considered practices that find their origins well into the Bronze and Iron Ages of the northern and West African regions where the plants grow naturally (Lovejoy 1980). Similarly, *coca, yerba mate, guayusa* (another *Ilex* holly found in the Amazon), *guarana* and various species of mildly caffeinated relatives of cacao (*cupuaçu*) formed integral parts of the lives of South American indigenous people while tobacco was smoked and traded across most of the Americas (Wilbert 1987). Cacao played a similar but more complex role across most of Central America and Mexico (Coe and Coe 1996). Across the island archipelago of South-East Asia and Melanesia the betel nut has been chewed traditionally as a packet or plug in association with the betel leaf (*Piper betle*, a relative of black pepper), calcium hydroxide and other spices since the Asian Neolithic. During the Bronze–Iron Age it is believed to have made its way along regional trade routes to South Asia as far north as Kashmir (Zumbroich 2007).

Mapping out the global distribution of these various plants at the time of Columbus' voyage identifies a series of regions naturally devoid of stimulants (Figure 7.4).[7] What I notice immediately is the absence of any

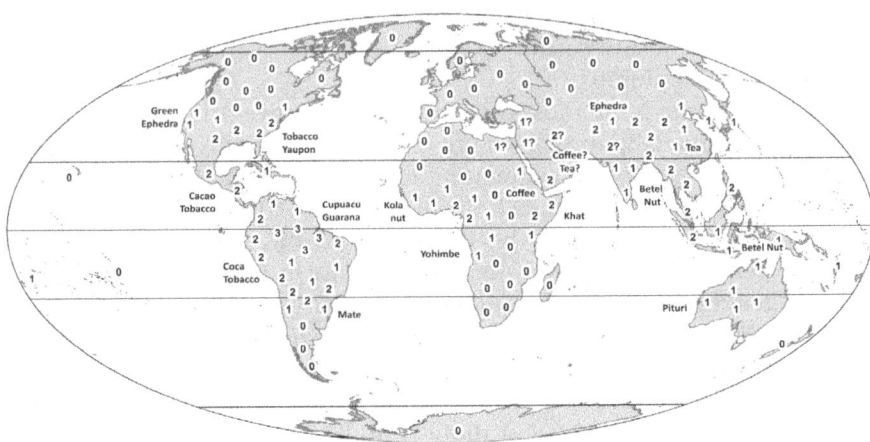

Figure 7.4 Regional distribution of various stimulant-containing plants prior to onset of European expansion and globalization of trade, *c.*1500 AD.

naturally occurring sources of these neuro-stimulants in Europe. Comparing the presence of these psychoactive compounds to the distribution of palaeo-angiospermous magnoliids in Figure 7.3 also shows an interesting congruence between the distributions of these two groups. Neither appears to have successfully colonized the high latitudes of Europe, North America and Asia north of 40–45 degrees. The only other regions that appear to not have a widely consumed neuro-stimulant available naturally (i.e. not through long-distance trade) are the vast Saharan area across North Africa; the area of south-east Africa that today includes the nations of South Africa, Namibia, Lesotho, Malawi, Zimbabwe, Botswana and Madagascar; and the islands of New Zealand.

The plant neuro-stimulants depicted in Figure 7.4 are not equal in their impacts on humans of course. They are all alkaloids or alkaloid derivatives but belong to a variety of classes within this group. Methylxanthines – the group of molecules that includes caffeine, theobromine and theophylline – are generally characterized as purine alkaloids. Most other stimulants fall into other alkaloid classes, including tropanes (cocaine), pyridines (nicotine), phenethylamines (ephedrine), monoamines (cathinone) and indoles (yohimbine). These latter substances are considered significantly more noxious to human health when consumed in large amounts over prolonged periods than the methylxanthines. In every instance, their consumption originated from practices associated with traditional indigenous cultures and the amount and frequency of use is believed to have been significantly lower than in globalized, urbane societies where cultural context has been stripped from their pharmacological effects on human health. Nicotine and cocaine are perhaps two of the best known examples of traditional neuro-stimulants being adopted within a globalized system where the original basis of adoption by human societies has been discarded. For example, coca leaves have been chewed by Andean civilizations for millennia, but ingestion of cocaine through the traditional use route is a small fraction of the amount introduced through the powder concentrate that is now distributed globally. Some studies have indicated that cocaine at small quantities may be of positive benefit to life at extreme altitudes, such as those experienced in the Altiplano regions of Peru and Bolivia, although this is not conclusive (Bolton 1979; Bray *et al.* 1983).

The introduction of tropical neuro-stimulants in the sixteenth and seventeenth centuries certainly changed the psychoactive foundation of cultural traditions and work practices of Europe. During medieval times, alcohol was the primary psychoactive drug consumed on a daily basis. The physiological impact of beer, ale, wine, mead and spirit consumption is largely antagonistic to work since its primary effect on neuro-transmission is sedatory rather than stimulatory. It achieves this by reinforcing the effects of the main neuro-transmitter responsible for relaxation of our brains and muscles – gamma-aminobutyric acid, or GABA. Records from the time suggest that alcohol was consumed at every meal, but that the alcohol content of these "small beers" – at 1 to 2 per cent – was less than half of the alcohol found in most modern ales or lagers. Stronger ales and beers were made, but these were typically more expensive and reserved for wealthier families or consumption in social situations, not as part of the daily work pattern. Nonetheless, most labourers at the time were probably under mild sedation throughout most of their working lives. The rapid introduction of a broad class of sedatory antagonists such as the methylxanthines could only have had a revolutionary impact on the way in which work was carried out as coffee and tea became the drinks of choice during working hours across much of Europe, augmented by increasing consumption of chocolate and tobacco. The latter of course is the primary source of nicotine, an alkaloid that unusually can act as both a stimulant and a sedative, depending on the amount consumed.

To trade, is to tax, is to credit, is to grow

Colonization of the tropics allowed Europeans to develop a new market in neuro-stimulants that previously were unknown to them. The pleasurable effects of this novel class of biochemicals meant that the sedative impact of alcohol could be complemented, if not challenged, by an entirely new market sourced primarily from trade with the tropics. Combined with the complementary role of sugar, the novel supply of new spices and expansion of staple food choices led by maize and potatoes, the tropics incrementally improved food diversity and security in Western Europe. The caloric subsidy from the tropics delivered between the seventeenth and nineteenth centuries was non-trivial, but certainly not of a magnitude capable of sustaining population growth at the levels witnessed over this period or altering the balance of nutrition. Instead, the tropics provided a new class of tradeable products that increased the options for nations seeking to balance out their staple grain supplies and needs through more traditional, regional sources. The role of neuro-stimulant consumption – as well as other products from the tropics – in delivering a new dynamism to European commerce did not happen overnight. The supply of natural products arriving from Asia and the Americas grew incrementally and at different rates across Europe. For example, tea had already been established as a part of everyday life for the upper class in Portugal more than a century before its introduction to Great Britain in 1662 by Catherine, the Portuguese queen consort of Charles II, but Britain would overwhelmingly become the larger consumer of tea by the end of the eighteenth century. Conversely, the German and Dutch trade in neuro-stimulants was increasingly driven by coffee, particularly as production in Java began to rival the traditional coffee supply from Mocha and re-exportation to Germany became a lucrative business for British merchants (Smith 1996). The trend in consumption of coffee and tea by the British changed in the wake of various events that began to favour tea over coffee. Alterations to tariff duties on tea and coffee during the eighteenth century made tea more profitable for import and coffee more favourable to re-export to Europe. A disruption of supply chains brought on by war with Spain from 1738–1745 also directly impacted coffee consumption in Britain, while the growing clout of sugar-growers seeking to convert coffee plantations in the West Indies and the East India Company's successful establishment of trade ties with the Chinese tea market pushed the British neuro-stimulant market towards Asian tea (Smith 1996). A reversal in the relative growth of coffee versus tea consumption would not occur again until the Second World War and the subsequent independence of former colonies, particularly India. Since then, average coffee consumption per capita has grown to exceed that of tea for the first time in nearly 300 years (Figure 7.5). In terms of average daily caffeine intake, coffee acted as the main source up to the 1750s. Thereafter, the main source of caffeine alternated between tea and coffee – from 1750 to 1820 tea became the main source and then coffee again briefly from 1830 to 1860 and tea for over a century from the 1860s to the 1970s. The end of empire at that time witnessed a precipitous decline in tea consumption and coffee once again became the main daily source of caffeine. Current levels of consumption indicate that coffee accounts for a four-fold intake relative to tea, a historic high (Figure 7.5). But from the mid-seventeenth century up to the take-off in tea consumption in the early 1840s through to the coffee revival in the 1960s, the average daily caffeine intake in Great Britain had increased nearly 10,000-fold. In reality, this would have been much greater for some and very little for others, but from a society with little exposure to neuro-stimulants

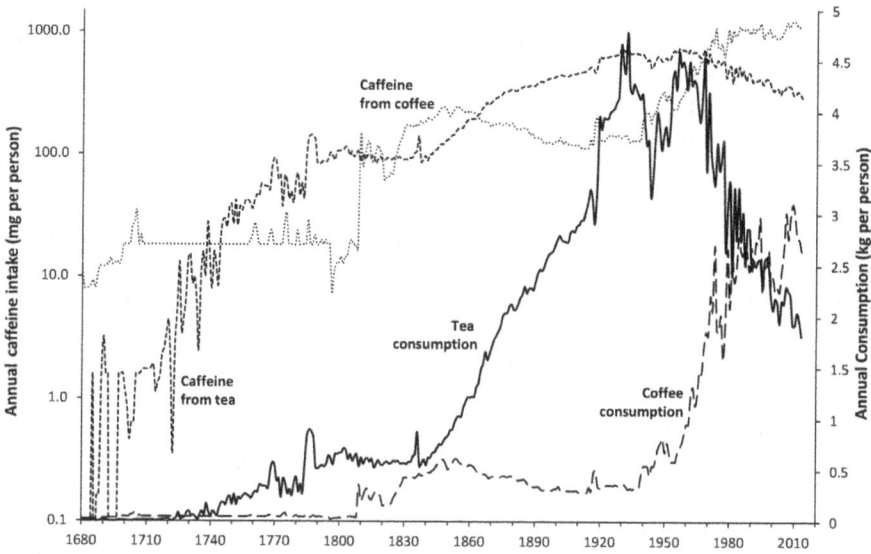

Figure 7.5 Changing consumption of tea and coffee in Great Britain from introduction to current times.

before the seventeenth century, the rise of caffeine, theobromine and nicotine in Europe was breathtaking.

It was not simply the putative effects on work productivity that made the introduction of tropical foods and stimulants arguably the most important event in the European drive towards globalization. It also stimulated an enormous increase in tax receipts and as a consequence of this expansion of public revenue, the availability of credit. While customs and excise taxation had commenced in the 1300s under systems of poundage (dry goods) and tunnage (wine) in England, tropical products only began to influence public finance with the levying of the first explicit duties on sugar and tobacco in 1685. From that point on, the raising of funds from tropical commodities through additional and special tariffs became a regular component of British crown finance. For example, consideration of an additional duty on cane sugar in 1744 was proposed in Parliament "to produce 80,000*l.* a year, sufficient to serve as a fund to borrow 1,800,000*l.* at 3 per cent" (Dowell 1888). Although this failed to pass into law in the wake of the farcical impositions of the 1733 Sugar Act and pressure from "plantation class" representation within the assembly, controlling supply of tropical products through colonial systems ensured that customs duties, excise taxes and sellers licences became a prodigious source of revenue to the British Parliament and Crown. Abbé Raynal (1776) describes the growing contribution of Chinese tea to British public finance:

This custom of the English, joined to the great demand for tea in their settlements, made them, towards the end of the last [seventeenth] century masters of almost all the trade carried on between China and Europe. The heavy duties laid by the British government on that foreign production, at last made other nations, and France in particular, sensible of the advantages of commerce.

(A. Raynal, Vol. II, Book V)

Apart from excise taxes on alcohol production and consumption, tropical imports of sugar, coffee, tea and cocoa, as well as Virginia tobacco were underpinning an increasing portion of this "loan security" from the late 1600s. By the second decade of the nineteenth century, duties from tropical product importation and consumption had ballooned into one of the most significant components of the public revenue, exceeding sums generated from every domestic source excepting excise taxes on alcohol (Powell 1825). Taxes from neuro-stimulants at this time were accounting for nearly 15 per cent of total public revenue from direct and indirect taxation.[8] As Ashworth (2003) notes for Great Britain, "addictive items proved by far the best security for loans [and] in this sense played a vital role in forging the English/British state". The very same could certainly be said of the Dutch Republic, where the taxation of trade and consumption in the wake of their successes in the East Indies tea and coffee trade were in place well in advance of the various indirect approaches later adopted by the British. But the Dutch also later became the first nation to recognize the role of reducing tariffs and liberalizing trade in stimulating further economic growth, albeit in part under duress imposed by the actions of the British through the Navigation Acts and the French programme of *adroits* and trade terms known loosely as "Colbertism". Throughout the nineteenth century, Dutch government income from customs duties remained one of the lowest in Europe – the proportion of revenue from this source was nearly half that of Great Britain – although this was generally offset by the imposition of much higher excise rates (Mitchell 1975).

The growth of the tropical trade – particularly within the colonial system – was a boon for public finance in Europe, at least on the surface. But the opportunity to impose customs and excise on a vast new range of tropical products brought with it the need to protect an even wider geography of trade interests. At the end of the eighteenth century, the size of the British merchant fleet had grown to nearly 11,000 ships with a 1.1 million ton capacity being operated by 87,000 men. A full third of the British army was now stationed in various tropical colonies with the Navy consisting of 115 battle ships and nearly 39,000 sailors (Dowell 1888). The cost of maintaining this considerable force had been precipitating a rapid rise in government debt for nearly a century but military expenditure was exceedingly high for most European nations at the time. For the British, it was exceptional. By the end of the eighteenth century and with little respite from constant conflict with the French over colonial control and maritime supremacy, the expenditure on naval, army and ordinance was nearly six times the amount spent on other government functions (the Civil List) (Anderson 1764). I estimate these expenditures at 9.2 per cent of national GDP – an amount only exceeded by "defence" spending during the Civil War (11 per cent of GDP), the First and Second World Wars (50 and 60 per cent) and the Korean War (11 per cent). Only interest payments on outstanding debt were greater at the time – pointing to creditworthiness as the true source of economic and geopolitical supremacy.

Generating revenue from tropical products became a fundamental part of servicing this debt while simultaneously underpinning the creditworthiness of public finance. The expansion of credit that accompanied colonial expansion in turn was a fundamental part of the European transition towards industrialization since it primed a financial system capable of handling increasingly complex capital transactions. The expansion of "sober" credit – unlike the scheming employed through various "bubbles" – ultimately underpinned the accrual of wealth and this in turn was catalyzed by the massive increase in tradeable goods that accompanied European entry into the tropics. The vast amount of government debt available at the time, the clever provision of transferability of this debt, its consolidation through the Bank of England and most importantly, the extinction of royal defaulting on debt re-payments and interest payment "holidays" created a bedrock alternative to other, riskier investments at times of economic uncertainty. To achieve this required not only institutional reform, but a deep and diverse base for taxation that, at the time, only the trade and plantation production of tropical products could provide.

Productivity's promise

Disliking or even refuting the notion of European global dominance up to the twentieth century has become something of a cause célèbre. But whether we like it or not, evidence clearly follows a path to the irrefutable conclusion that European society, and by proxy those that have adopted the global system they engendered, underwent an economic evolution that transcends most other human developments. As we have seen, this was certainly not due to any inherently superior devices, technologies, immunities, religious beliefs or agricultural resources. Most of these were either borrowed or acquired from Asia or the Middle East. Many authors have focused on these and other factors – institutions, rule of law, etc. – as the strengths that led to European dominance and globalization. But the evidence points to a different driver of European expansion – that the region and its population were inherently handicapped by bioproductive isolation and its ancillary low biological diversity. Native forests lacked the sort of diversity that makes trees a vital source of food and nutrition in the tropics. The biochemical make-up of the native flora lacked many of the groups that became globally dominant, such as neuro-stimulating alkaloids and a wide range of phenylpropenes such as eugenol, safrole and anethole that were absent or otherwise restricted to a handful of small indigenous herbs. And as we have seen in the previous chapter, the region also lacked an abundant source of sucrose until the nineteenth century. If we consider that trade and commerce leading up to the mid-1800s were dominated by the exchange of these and other biological products rather than those born from minerals and fossilized deposits, it is clear that Europeans did not accelerate globalization from an initial position of strength, but from relative weakness. Certainly the financial and commercial institutions that were created in the Italian city-states, the Dutch Republic and England in particular played a formative role in delivering a new economic system that eventually integrated trade, capital and credit at a scale of economic expansion never witnessed before. This process – as in the case of the troublemaking South Sea Bubble, the widespread dispersion of appropriated crop varieties, the introduction of deadly diseases or the institutionalization of slavery as a factor of production, among many – was not without its great detractions. Few transformations are. But European progress during the Age of Discovery perhaps is best viewed as a sweeping new chapter in an inevitable march towards globalization that was already under way, as evidenced by the movement of crops, diseases, people and ideas across the Eurasian continent. Europeans were simply seeking a way to catch up due to the bioproductive constraints of their continent, utilizing the technological means and socio-cultural impetus available to them and discovering productivity's promise – the innate currency of human development – in the tropics.

Yet, while European institutions imbibed trade from colonial possessions in the tropics and reaped many of the nutritional, economic and social benefits, what kind of legacy, if any, has been left and what impact has this had on the societies and natural environment of post-colonial tropical nations? In particular, how have the most bioproductive countries in the tropics – those with tropical forests – fared in the aftermath of the world's first wave of globalization? How has this legacy impacted their standing in the new wave of globalization? These are the topics of the next chapters.

Notes

1 NewsDay, 23 Feb 2011, *Ndebele–Shona relations: recipe for disaster.*
2 Some economically useful metal deposits, such as aluminium-bearing bauxite, are linked to bioclimatic conditions.
3 Growing degree days (GDD) are the accumulation of heat units over a period of time, usually a year. The term describes the physiological growing season of plants based on a timeline of daily temperature change, absent the effects of

moisture and nutrient limitations. GDD are determined by subtracting a baseline temperature from the square of the sum of daily maximum and minimum temperatures. Crops have varying tolerances and thus different base temperatures at which growth is curtailed. For wheat, it might be 0°C and for maize 10°C. Here, a 5°C base was used to compare different GDD. When daily minimum temperatures fell below 5°C, they were re-set to the baseline in the calculation (McMaster and Wilhelm 1997).

4 Bound westward by –26 degrees longitude and eastward by +70 degrees longitude with 0 degrees longitude running through Greenwich, UK and northward to the Arctic Circle and southward to the equator. This mask was then used to extract regional data for the various coverages depicted in Figure 7.1. Datum and projections as described in Chapter 2. Single degree latitudinal averages were then calculated from the raster data using a zonal classification grid. By taking averages, pockets of very high and low values are modulated. Data sources: GDD: CRU-UEA, New *et al.* (1999); Precipitation: Worldclim2, Fick and Hijmans (2017), Harris and Jones (2017); Crop yields: Monfreda *et al.* 2008.

5 Food plants and the timing or their arrival in Europe compiled from Jennings and Cock (1977), Vavilov (1992), Wiersema and Leon (1999), Heywood (1999), Hancock (2004), Prance and Nesbitt (2005), Colledge and Conolly (2007), and Zohary *et al.* (2012). Energetic contents from the USDA National Nutrient Database (ndb.nal.usda.gov/ndb).

6 The energetic content of maize and other modern foodstuffs used in the calculations are derived from the USDA National Nutrient Database. These are values for modern crop varieties and the energetic content of medieval crops may have been significantly different. The assumption here is that changes in crop nutritional quality have remained similar across the major food crops grown in Europe.

7 The generalized presence/absence of neuro-stimulants is based on the known, natural distribution of the source plant and a record of human use and consumption. Plants containing neuro-stimulants but not used outside single communities/tribes or restricted to rare or infrequent ceremonial or medicinal uses are not included here.

8 From statistics in Powell (1825). This amount includes income from excise and customs duties on tobacco, coffee, tea, cocoa and chocolate. Total public revenue includes all sources of income listed, exclusive of borrowings.

Part III

Modern consequences

8   Smaller, younger, poorer

High in a remote part of the Maya Mountains in Belize away from the usual tourist destinations, I am told that we need to be on alert for illegal loggers and "*xateros*". My Belizean colleague tells me that they are typically well armed and he has brought along a couple of members of the Belizean Defence Force just in case. *Xateros* are collectors of the smooth, green fronds of the *xaté* palm (two species in the highly endangered genus *Chamaedorea*) that grows here in abundance. The young fronds of this small, understorey palm have become the source of a floriculture trade that rocketed from a small specialist supplier to Europe and North America in the 1950s to one that is now worth more than US$3 million per year and provides many jobs and a source of income to residents of the region (Bridgewater *et al.* 2006, 2007). The delicate fronds commonly accompany other tropical flowers in grand floral displays, but are particularly in demand during the Easter period. The palm produces fronds seasonally and at a steady, but rather low, rate, although it can vary with conditions from year to year. Once extraction adheres to the growth rate, however, harvests can be sustained indefinitely. The volume of fronds produced can be expanded by encouraging a larger population of the *xaté* palms to take hold in the understorey or through enrichment planting. But over-harvesting through the activity of illegal *xateros* threatens the long-term survival of this extractive industry. The objective of these "palm-poachers" is to take as many saleable fronds as they can find, often leading to the complete defoliation of well-established adult plants. Due to the slow growth conditions of the palm's favoured understorey environment, these are not quickly replaced and most of those defoliated in this way soon perish. The impact of these losses on the population dynamics of the species can be devastating and the *xaté* palm is now rare over large tracts of forest that only a few decades ago were heavily carpeted with the vibrant green "fish-tail" leaves of young palms. The ultimate outcome is the loss of an important local industry that provides a critical source of income to some of the poorest rural households and an increase in violence associated with illegal elements in the trade. Illegal logging, hunting and the collection of live specimens for the pet trade have

followed, depleting some of Belize's largest national parks of the very elements that help to sustain its relatively healthy tourism economy and
beneath all of this is the hidden spectre of a vibrant and violent trans-
shipment of illicit drugs from South America via Belize to Mexico and the
United States.

The most interesting aspect of the *xaté* economy, however, is not the
interplay of the various roles and rewards within the industry, or its significance to local livelihoods, although these are all important in other ways.
Rather, it's that none of these are Belizean. The illegal *xateros*, the local
livelihoods, the processing plants and the international trade are all situated across the border in Guatemala. In fact up until very recently, the
only Belizean in this multi-million dollar imbroglio was the palm. Flying
over the nearby border between Belize and Guatemala quickly reveals one
of the main reasons why the *xaté* of Belize is being used as if it is Guatemalan – there is very little intact forest remaining on the Guatemalan side of
the border. In fact, illegally harvesting *xaté* in the protected areas of the
Maya Mountains is not the only incursion. Satellite imagery shows large
swathes of Belizean forest removed along the "official" border for *milpa*
agriculture – the traditional subsistence farming technique of the Maya
consisting of maize, beans and squash that pre-date the arrival of Columbus and the Spanish *conquistadores* in this part of the New World tropics.
Again, it is Guatemalans that have extended their patchwork of *parcelas*, or
fields, over the border. A boundary that appears to have very little relationship with the normal function an international border is expected to
perform. In fact, the official Guatemalan stance on the border since 1940
is that it does not exist due to unfulfilled obligations dating back to a
treaty signed prior to Belize's independence in 1981 and this view remains
in place today.[1]

The convolution that is Belize's *xaté* economy with all that it entails –
illegal border crossings, sporadic violence, increasing social dissent and
widespread environmental degradation – epitomizes the modern condition of many tropical nations. Similar scenarios could just as easily depict
the harvest of bush-meat or fish, the mining of precious gemstones, or the
wholesale extraction of rare tropical hardwoods, natural rubber, palms,
orchids or live birds for the pet trade. Situations where resources are purloined without consideration of national sovereignty or the usufruct rights
of a nation's citizens, or without any intention of recycling the proceeds of
extraction towards a nation's sustainable development, have plagued most
countries in the tropics since their earliest days of nationhood. Tropical
nations, most with hyper-productive natural landscapes and relatively
untapped mineral assets, have become particularly prone to conflicts –
both external and internal – over how natural resources are extracted,
who stands to benefit and who stand to suffer from the use of a nation's
most important asset class. In both cases the integrity of the government
and its relationship with its people is often brutally tested over the

allocation of land and natural resources. When the Belizeans got word that the British and Belizean leadership were going to accede to more of the Guatemalans' demands than they were willing to accept during the 1981 Heads of Agreement negotiations, they rioted violently. This disconnect between the biophysical and political geography of the Guatemala–Belize border region continues to swell social uncertainty that is not going to dissipate anytime soon.

Map lines

The physical geography of our planet is a map of nature – a map of the changing relationship between the condition of land, water and air that is shaped by the interaction between the three major sources of energy available on Earth. Its internal dynamo – a force that moves continents, makes mountains, gives shape to our atmosphere and spins our planet: the force of gravity – produces tides and the planetary motion that creates the seasons; and the flux of solar radiation that nurtures life creates the winds and drives surface currents in the oceans that collectively weather away the past products of the internal dynamo.

The political geography of our planet is a map of nations – a map of the changing relationships between culture, governance, ownership, commerce and technology that have led to a puzzle-piece assembly of discrete national living spaces. Of the 510 million square kilometres of land on our planet, only the continent of Antarctica – a land mass that remained unseen until 1820 – stands as a piece that does not fit in the puzzle we have created, subject to a treaty of mutual exclusion among nations that does not allow any nation to exploit the continent for its own national purposes – for now.

The Australian-British writer and historian, Paul Carter, wrote of the distinction between physical and political geographies as a way to understand the anatomy of Britain's colonization of his adopted homeland. Carter's spatial history of Australia brings to bear the important role of the earliest European explorers and surveyors in paving the path to ownership. They did this, arguably without direct intent, by converting the physical features of an unknown landscape into recognizable ones through a new nomenclature that assigned names to features based on the specific travails and successes of the expedition at the time of encounter or paid homage to historic British figures – Wellington, Nelson, King George – that had nothing to do with Australia. They assigned names based on the availability or scarcity of water and other resources pre-supposed from the understood relationships between topographic features in Europe. If a mountain did not have a stream flowing down its flanks, it became "Mount Deception" or if there was an area of particularly parched landscape without a feature, the highest point was given a name, such as "Mount Misery" (Carter 1987). Findings from the many scientific activities that accompanied expeditions were employed in name-making to add a sense of serious purpose to the landscape. Familiarity was spread to the plants and animals attached to these features. Trees seen growing on hills became oaks, birds seen fluttering along a stream became robins.[2] In this way, the physical geography of Australia had already been placed in a distinctly British cultural context by the time immigration from Europe accelerated on the back of the 1850s gold rush. Maps pieced together from the various surveys documented this new political landscape, rationalizing an unfamiliar natural world to colonists that had no history with an environment that was far removed and physically and biologically very different from the British Isles. Plans could be more readily pitched to potential investors and ventures more easily organized based on this new, understood geography. Ownership could be assigned and land management units could be created using recognizable landmarks as a device to quickly draw up boundaries in the absence of more precise and time-consuming cadastral surveys. As quickly as the potato, tomato and chocolate became European, Australia became British.

Across the Pacific, a similar process had already been employed in delimiting Spanish and British claims along the north-east coast of South America. Half-way through the 1498 voyage of Columbus – his third to the New World – he and his crew became the first known Europeans to set sight on the South American continent as their ships entered the Gulf of Paria. Wedged between the island of Trinidad and the northern edge of the vast complex of palm and mangrove-lined *caños* that form Venezuela's Orinoco delta, the gulf provided the shallow and calm waters from which Columbus began the first empirical mapping of the continental coast. He took it upon himself, like Cook in Australia and the exploring British surveyors that followed, to begin selectively naming the various natural features he encountered. The sixteenth century friar, Bartolomé de las Casas, described Columbus' serial effort around the Gulf of Paria:

> He saw that the coast was filled with good harbors and a very high land; by that lower coast he saw many islands toward the north and many capes on the mainland, to all of which he gave names: to one, Cabo de Conchas; to another, Cabo Luengo; to another, Cabo de Sabor; to another, Cabo Rico. A high and very beautiful land.... On going out of the mouth, he saw an island to the north, which might be 26 leagues from the north, and named it La Isla de la Asuncion; he saw another island and named it La Concepcion, and three other small islands together he called Los Testigos. They are called this today. Another near them he called El Romero, and three other little small islands he called Las Guardias. Afterwards he arrived near the Isla Margarita, and called it Margarita, and another near it he named El Martinet.
>
> (B. de las Casas (from Thacher 1903))

Thus, on his third voyage, Columbus transformed from the daring early explorer of his 1492 voyage to a systematic – almost manic – surveyor, or in the words of Carter "a harbinger of civilization ... preparing the path for orderly colonization" (Carter 1987). Every physiographic feature required a relationship – with the trials and rewards of his experience on the day of naming, with historic Spanish icons and, perhaps most importantly, with his own piety. Cabo de Conchas points to a feast of conch (a gastronomic delicacy in the region), Margarita Island was named for the rich reward of its pearl beds (margarita is Latin for pearl), and Boca de Drago signals the dangers awaiting vessels passing through the "mouth of the dragon", while the island Isabela pays respect to its eponymous queen and the islands of Asuncion and Concepcion pay homage to the guiding hand of Providence. These "new" waypoints were now culturally interpretable to the Spanish that would follow. They became familiar without being known.

Columbus' record of his encounter with the native inhabitants of the area would only act to further agitate European would-be adventurers into a frenzied state of anticipation. Again, de las Casas describes the initial encounters:

> Afterwards in the afternoon there came more from all the territory, many of whom wore at the neck pieces of gold of the size of horseshoes. It appeared that they had a great deal of it: but they gave it all for hawk's bells and he did not take it ... yet he had some specimens from them and it was of very poor quality and appeared gilded anew. They said, as well as he could understand by signs, that there were some islands there where there was much of that gold, but that the people were cannibals.... The Christians saw one Indian with a grain of gold as large as an apple. Another time there came an infinite number of canoes loaded with people, and all wore gold and necklaces, and beads of infinite kinds...
>
> (Thacher 1903)

His travels along the mainland were cut short due to illness, but upon his return to Spain word quickly spread of his findings. Columbus would not set eyes on the continent again, although he would make one final voyage through the Caribbean. His brief encounter with indigenous inhabitants in the region, however, planted a seed in the minds of his fellow adventurers of an epic myth that would catalyze the European pre-occupation with the region for centuries – El Dorado.

Within a few years, a new cohort of Spanish adventurers had established nascent colonies on the islands of Margarita and Trinidad. These would be the launching point for numerous exploratory surveys up the Orinoco River and along the coastline and waterways of the Guiana coast. Alonso de Ojeda, Vicente Yáñez Pinzón, Diego de Ordaz, Pedro Maraver de Silva,

Antonio de Berrio – these names may not resonate like those of Columbus, Cortés, Pizarro or Quesada, but these *doradistas* were convinced that the great finds of their celebrity predecessors – vast hoards of gold and silver worked by the Aztecs, Incas and Chibchas – were also to be found in the Guiana Highlands. El Dorado was there, in his city upon the lake, waiting to be met by the adventurer willing to make the journey. One such adventurer was the Englishman, Sir Walter Ralegh, although he proved to be something more than a *doradista* and his efforts to lay claim to the legend of El Dorado, rather than the gold itself, in part as a ploy to stay an inevitable execution back home, would inevitably play a role in the eventual map lines that would divide up the region between British, Spanish and Portuguese interests.

Ralegh, like most Europeans that preceded and followed, set sail from the Canary Islands in 1595. He arrived in Trinidad filled with the same purpose as the Spaniards – to carry out a series of expeditions, or *entradas*, in order to pin down the precise location of Manoa and convey this information to his sovereign and patron, Elizabeth I, which he did with great aplomb in his travelogue *The Discoverie of the Large, Rich and Bewtiful Empyre of Guiana*, published in 1596. It is generally accepted that Ralegh and his companions visited some of the locations that he mentions in his report, spreading the news of the Virgin Queen as arch-enemy of the Spaniards and saviour to the indigenous tribes-people of the region (Whitehead 1987). But the remaining prose, a mixture of gossip and outright fiction mingled with a rudimentary evaluation of mineral resources, appears designed purely to entice those in England as to the prospects of the region. At the centre of this enticement stood El Dorado. Ralegh would draft a rudimentary map of the region, formed from documents purloined from the Spanish, aided by local indigenous informants and the general observations made during his own, modest efforts to explore the lower Orinoco and its tributaries that were thought to be the gateway to Manoa. The city on the lake, impossibly out of scale and with far greater detail than most other elements featured on the map, provided the landmark needed to begin rationalizing ownership of the land, or as D. Graham Burnett suggests "…served as the gravitational center of [the future] British Guiana" (Burnett 2000). Ralegh's reprieve at the hands of his patron Queen Elizabeth would not last long after her death, despite convincing James I to allow him a second voyage in pursuit of riches in the New World previously granted to him under Elizabeth's royal charter. Having failed to produce the treasure estimated in *The Discoverie*, he turned to privateering and upon his return to England, out of favour and without royal support,[3] was sentenced and beheaded as a "diplomatic gesture" to ease Spanish rebuke of his actions against their New World fleet. His efforts, however, would transform El Dorado from a tall-tale at the end of a chain of local whispers to a geographic feature – a city, Manoa on a lake, Parima – recognized on virtually every map of the region produced from the sixteenth to early nineteenth centuries. The role of the Spanish *doradistas* – Antonio de Barrio in particular – was lost. The great Trinidadian author and historian V.S. Naipaul (1969) wrote: "All that was known of Berrio was what Ralegh had written. When Berrio's papers were recovered, three hundred years later, the Spanish Empire was over and the El Dorado legend was fixed: it was Ralegh's."

When Robert Schomburgk was commissioned by the Crown in 1835 to undertake a boundary survey to formalize Britain's colonial interest in Guiana, he relied heavily on Ralegh's early explorations as the *sine qua non* in legitimizing the international boundaries. After more than 200 years had passed, Ralegh's *Discoverie* had been transformed under the mass of an expanding empire from the fictional tale that it clearly was into a document of legitimate geographic exploration that had simply feigned the fantasy of El Dorado in order to paste over the real hardships facing would-be colonists invariably critical to the success of any project in the region (Burnett 2000). A new genre of travel writers rapidly consumed Ralegh's *Discoverie* adding their own embellishments. Highly regarded exploring geographers, such as Alexander von Humboldt and his travel companion botanist Aimé Bonpland, added authenticity to elements featured in the account. Ralegh's caricature of the region was by this time considered an important benchmark geography of the area. The maps drawn up by Schomburgk on the back of the British Guiana surveys would act as the formal document in delimiting the boundaries of the crown colony from its neighbours (Schomburgk 1848). Known as the "Schomburgk Line", it relied heavily on natural features to provide the points through which map lines were eventually drawn. However, beset with the difficulties that travel into the remote parts of the region almost always entails, the survey team would not set foot on much of the forested land falling along the supposed boundary.

The cartographic dissection of Ralegh's Guiana by Schomburgk would be open to a stinging exchange between Venezuela and Britain three decades later – a disagreement that ensnared the attention of a United States that was at the time also rapidly growing in power and regional influence. A boundary tribunal was promptly commissioned with the task of determining the authenticity of claims to the area. The tribunal concluded little, almost falling into farce as it attempted to identify the true boundary but without any intention of visiting the region to test the veracity of Schomburgk's efforts. Maps were to be used to evaluate the accuracy of other maps. Consequently, a bit more than a century and a half later, the boundary between Guyana and Venezuela is still being hotly contested. Many parts of the "boundary" have yet to be properly surveyed and for the most part remain unmarked. Ironically, the renewed claims by Venezuela have become more vociferous since the value of the mineral resources found in the region began to rise rapidly in the international markets since the turn of the millennium. It is in many ways a modern-day redux of the sixteenth century gold-fever that created the myth of El Dorado, a term that is still broadcast all too frequently as a portent of the great potential that awaits in the region despite centuries of evidence to the contrary.

Hero, providence, guide

Throughout the 400 years that followed Columbus' third voyage, European explorers and the hastily assembled jumble of cartographers that feverishly followed their exploits went busily about their business creating a new political geography of the tropics. It is surprising how many colonial borders, such as those dividing Belize and Guatemala, were born from simple map lines. Often these lines were drawn with little more in mind than slicing a pie that they had no hand in making. They would consume the pie, and attempt to obtain as large a slice as was possible. Boundaries informed the extent of possession when negotiating in the meeting rooms of Europe, but again these were often at best based on sporadic survey points collected by expeditions that frequently failed to visit the boundary lands used to construct the new political geography. Absent the technological advances that surveyors have at their disposal today, it is fully understandable how Ralegh, von Humboldt, even Schomburgk – whose expedition was specifically commissioned to delimit the borders – failed to visit many of the areas they used to draft their maps. But these failures were rarely revealed in a growing celebrity of the adventurer-explorer class, where success was measured only by the novelty, multitude and hardship of their discoveries and the anecdotal narrative that followed.

Quantifying their success was largely based on naming landmarks or collecting specimens. As Carter (1987) remarked on the approach of Cook's intrepid botanist, Joseph Banks, to the novelty of the Australian flora:

> whether the flora of Botany Bay was typical or special, whether it was insular or coastal, was unimportant to him. For Banks, it was the aggregate of objects that counted, not their true relations ... For Banks, names enjoyed a simple Linnaean relationship with the object they denoted. They gave the illusion of knowing, under the guise of naming.

Over the course of Schomburgk's survey along the rivers of the Guiana region, the ritual of naming followed in the footsteps of Cook and others, just as they had sowed European monikers into the land and culture of Australia 200 years earlier. Schomburgk recounts one of these opportunities when encountering a large cataract along the upper Takatu River in Guyana (Riviére 2006):

All the Indians who composed our crew, as well as those about Annay, concurred in declaring that never before had a White Man been at the Cataract. I considered myself authorized to give it a name; the greatest Cataract was therefore called after His Gracious Majesty, the Patron of the Royal Geographic Society, and to go through this ceremony in due form, we broke in our store of reserve, for one of the last bottles of wine, we had treasured up for months…. From one of the higher rocks flowed the red juice of the grape mingling with the white foaming torrent, and while naming it King William's Cataract, Great Britain's Union was unfolded, and a salute fired according to our means.

(R.S.)

Schomburgk's description tells us a great deal about the underlying factors facilitating the establishment of new political geographies during the period of colonial development. He feels entitled to name the cataract, as the first White Man to visit, rather than caring to ask its name from the local Makushi people. God, or Providence, would seem to have brought him here for this important event. He garlands the naming with as much pomp and pageantry that could be assembled under such circumstance. This is Schomburgk as self-made hero, elevating himself by recognizing the king and making a last sacrifice of wine in his patron's honour. All of this is being carried out under the curious and watchful eyes of the indigenous guides that made his arrival at the cataracts entirely possible.

Carter highlights these very same elements as driving the preparation of land for colonization in Australia – the exploring surveyors' invocation of Providential support for both cause and consequence, the promotion of themselves as heroic figures tasked with providing a transliterated narrative of the landscape they encountered in the face of overwhelming physical and mental adversity, and relegating the critical role of their indigenous companions to a literary aside. He and others, such as Burnett and Whitehead, argue in different ways that it was the narcissism of early contact that served to jump-start the social, political and economic machine back home by popularizing the actions needed to transform *terrae incognitae* from Aboriginal-inhabited landscapes to British-controlled colonies. The ultimate objectives, of course – to establish ownership over lands with the means and muster to profitably harness the bioproductivity and natural resources of the tropics towards economic expansion in Europe – were more pragmatic than providential. To achieve these objectives, however, European nations required not only means and muster, but compromise. Not compromise with the indigenous inhabitants, but with each other; and it was this almost casual agreement amongst gentlemen in the map-rooms of Europe that laid the framework for the partitioning of the tropics.

Mercantilist ways

The delimitation of boundaries and appropriation of the tropics into a patchwork of colonies established a framework for European powers to engage in a multi-centennial game of one-upmanship. Colonies became assets that could be seized, exchanged or fractionated among the holders. The British and Dutch exchanged control over Guyana no fewer than six times before control was finally settled. With French Guiana, it was eight times between four different countries (Hammond 2005). For Suriname, it was also six – the most famous exchange involving the Dutch submission of the swamp-laden Manhattan Island (New Amsterdam) to the British (who had already seized the town) in exchange for the sugar-growing coastlines of the South American colony and Run Island, the sole remaining possession at the heart of the highly profitable Indonesian nutmeg trade that they did not already control. Tropical islands, due to their relatively small size and predisposal to naval blockades, were frequently being used as collateral for exchange. The island of Guadeloupe was a sugar-growing powerhouse in the eighteenth century, producing more sugar than the British Caribbean colonies combined. It was fought over ruthlessly by the French and British for nearly a century. Many of these hand-overs were carried out as part of much broader peace-treaty agreements signed between European countries frequently engaged in direct conflict. At the Treaty of Paris in 1763 Guadeloupe was handed back to France. The price of this exchange – Canada. But the island was again captured by the British in 1810, handed over to Sweden as part of another agreement and then subsequently returned to France under the terms of yet another treaty. A further treaty between Britain and France in 1815 finally put to rest a century of horse-trading over Guadeloupe. This pattern of capture–negotiate–exchange became particularly important – perhaps more so than first contact and colonization – since it is the litany of direct conflict between European nations, not the colonization of the region per se, that has ultimately shaped the modern political geography of the tropics.

The endless stream of conflicts that dominated the history of Europe from the sixteenth to mid-twentieth century provided a surfeit of opportunity to re-balance and expand colonial property portfolios without the harsh and expensive leg-work often needed to initially wrest control from organized indigenous forces, particularly in the case of the African and Asian tropics. Mercantilism was the hidden driver of these exchanges. The desire to accumulate wealth by selling more to other nations than was bought from them became an important reason to expand colonial holdings. As we have seen in the preceding chapters, European nations had few "unique" products to exchange – almost everything grown or raised in Britain could likewise be pursued in Holland, France or Germany. Only a handful of products distinguished northern Europe from the Mediterranean – herring, cod, rye, buckwheat, olives, dates and citrus. The tropics offered foods, spices and neuro-stimulants that could not be produced in Europe, as well as a large number of biological dyes that brought new and cheaper colour varieties to the textile industries. Expanding control over tropical colonies squeezed other European nations with fewer such holdings.

As a consequence, military action and economic subterfuge were often the only options in alleviating economic stresses precipitated by the withdrawal of supply by a nation for political and economic reasons. We saw this in the case of "naval stores" and the need to supply these from the colonies to offset uncertain supplies from Scandinavia during conflict. This form of supply constraint was a dangerous game. But it was an effective tool that allowed competing empires to stay in the game and advance their domestic economies – for a while. The European powers lost their grip after the devastating impact of the World Wars, but continued to shape the economic future of the post-colonial geography of the tropics through other means. The United States and later Japan joined the post-colonial scramble to establish vice-like economic connections to the bioproductivity and mineral resources of the tropics, often with alarming social, economic and environmental consequences, as we have seen from the history of the banana. Most recently, China has re-defined this scramble. Its transformation from former colony into major economic power has re-invigorated the mercantilist model. The acquisition of land and resource-use rights by Chinese companies in virtually every tropical country can only be described as breathtaking in comparison to the relatively listless experiences of its European, American and Japanese predecessors. And there have never been as many nations to engage in the tropics as there are in the first quartile of the new millennium, due to the legacy of acquiring control over the bioproductivity of the tropics.

The size and number of nations

By the time decolonization had reached its conclusion in the late twentieth century, 98 tropical nations (88 of these with tropical forests) had been added to the UN register.[4] Another seven would be added as the four self-determined territories seceded from their parent countries and three Pacific island states were released from their post-war UN Trusteeship under the United States. Twenty-six tropical territories, some self-governing, remain affiliated with former colonial parents. All but one of these, French Guiana (with *département* status), are islands.[5] This tidal wave of independence tripled the number of nations over the course of the twentieth century. In 1900, there were fewer than 65 self-governing countries in the world, compared to over 195 recognized today. The progressive fractionation of imperial territory into smaller and smaller administrative units and the subsequent conflicts and divisions that led to cessation of territory from many newly independent nations primed the tropics for a modern political geography of many nations. Figure 8.1 quickly illustrates this outcome. It describes a general increase in the number of nations at each degree latitude as we move from the poles towards a peak around 4 and 10 degrees north of the equator. This rise in the number of nations towards the tropics runs counter to the bulk of global land area stationed north of the tropics between 30 and 70 degrees latitude (see Figure 1.5).[6]

Empire and post-independence conflict catalyzed fragmentation, but they are not the sole primers of this proliferation. The incongruous distribution of modern physical and political geographies on our planet tells us

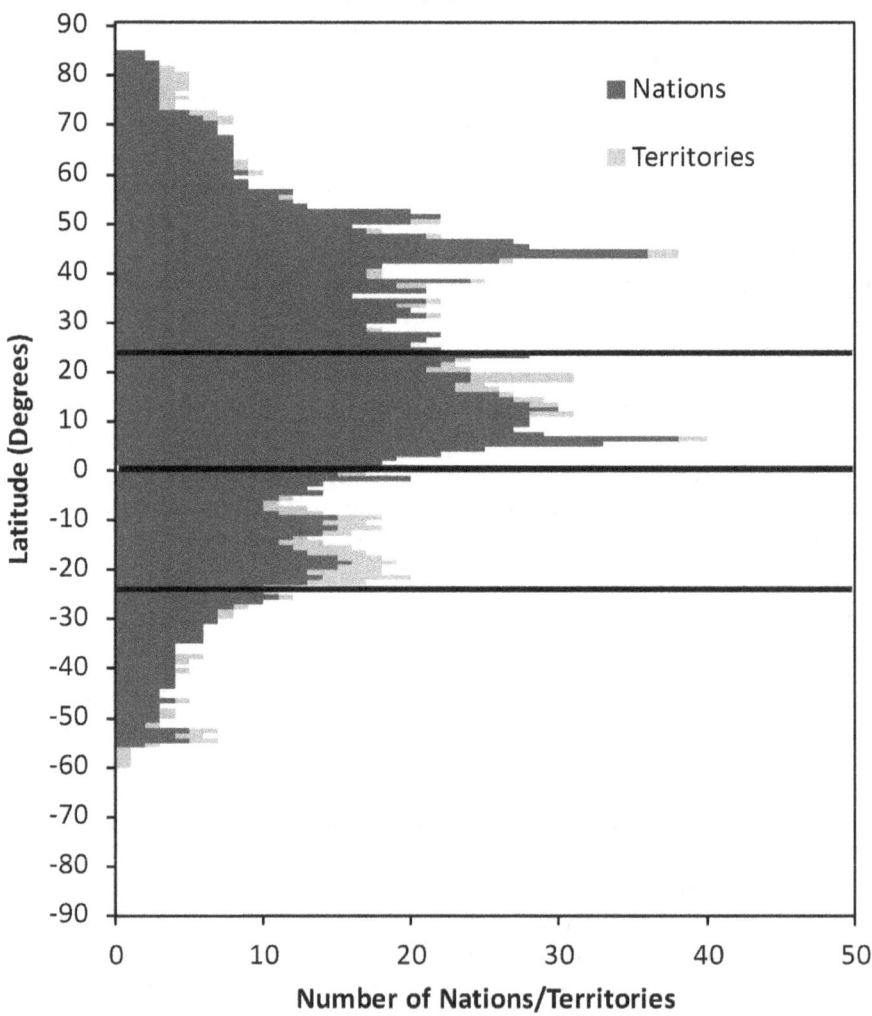

Figure 8.1 The tropics contain "peak nations" due to former colonial fragmentation and a natural abundance of islands.

that geopolitical forces also need to be viewed in light of the physical environment and the way that this has in its own right precluded a natural amalgamation of tropical territory into larger, independent nations. The primary effect has to do with island formation. The tropics are more "blue than green" and the relative dearth of land area and huge expanse of ocean means that there is simply less land to go around, naturally squeezing the size of tropical nations. Equally, the relatively large area of ocean in the tropics dictates, all else being equal, that there should be more, relatively isolated islands than in the extra-tropics where the oceans are compressed by a large mass of land. If you recall from Chapter 1, the larger land area at the higher, northern latitudes is distributed over a much smaller planetary surface, due to the increasing curvature of our planet in this region, compacting the spatial relationship between terrestrial and marine. More land occupying less surface naturally decreases the achievable distance between any island and continent. The massive peak in coastline length between 55 and 75 degrees north latitude, twice the degree latitude length found in the tropics, supports the notion that the odds of finding an island at any significant offshore distance in the northern extra-tropics is far less than in the tropics, all else being equal.

At the same time, the geological processes that give rise to island formations are also less compressed in the tropics. Among the various processes that lead to island formation, volcanism is the most important. If this route to island formation is more commonplace in the tropics due to the larger marine area, this should be reflected in the number of volcanoes formed in the ocean compared to land. If we take a look at the Smithsonian's *Volcanoes of the World* database, just over half (54 per cent) of the 406 volcanoes registered as currently active are located in the tropics, but 72 per cent of these are associated with islands or submarine features, such as seamounts or guyots. This compares to 65 per cent of extra-tropical volcanoes that are insular or submarine. If we extend the list to include all of the volcanic formations in the database thought to have been active sometime over the last 10,000 years (the Holocene period), we find that 58 per cent of the features in the tropics are associated with islands or submarine features, compared to just 40 per cent of a similar number in the extra-tropics (Venzke 2013). The disproportionate size of the tropical ocean gives rise to a larger number of volcanic islands and coral atolls, but at a lower density compared to the smaller extra-tropical oceans. This lower density has led to a greater isolation of islands or island clusters – as measured in distance from the nearest continent – in tropical oceans.

While a few of these islands have sought independence or are considering it, a large number remain firmly dependent on their former colonial parents for their administration. Many of these are uninhabited nature reserves, with a few being radioactively contaminated former nuclear testing sites (e.g. Howland Island). All of these scattered islands, mainly in the Pacific, are otherwise too small, too remote and without the freshwater resources needed to support human populations (e.g. Easter Island) for any great length of time. Still others were uninhabited when Europeans arrived for the first time (e.g. Reunion, Bermuda, St. Helena, Pitcairn Islands, Falkland Islands, South Georgia, Tristan de Cunha) and the residents have preserved strong social and cultural ties to parent countries that also continue to subsidize their economies, despite the considerable distance separating them.

Factors or events leading to a specific outcome are often too numerous for any particular one to be properly elevated as a singular explanation. But the way in which tropical empires were first formed and then dismantled, combined with the isolating effect of the region's expansive oceans and more numerous island territories, set the scene for a modern political geography consisting of a larger number of smaller nations. There are now three times as many independent island nations in the tropics and twice as many dependent territories (35 independent, 25 dependent) compared to the extra-tropics (10 and 13), while the number of countries on the continents is roughly the same between the two zones. This is despite the much larger land area of the extra-tropics. The average size of the modern tropical nation (523,000 km^2) is now nearly half of those located in the extra-tropics (984,000 km^2).[7] If we consider only the biologically productive tropical forest nations, the average size declines even further (446,000 km^2) and if we remove Brazil, a tropical forest titan that alone accounts for 17 per cent of the terrestrial tropics, this average drops to nearly a third of the extra-tropics (329,000 km^2).

The age of nations

Ethiopia is an ancient nation. Autonomous control by local rulers has lasted more than two millennia. Forming part of the ancient kingdoms of Kush, it is the land of the Queen of Sheba and the purported resting place of the Ark of the Covenant. Known as Abyssinia in English, Habesh in early Arabic, the modern nation of Ethiopia most likely traces its origins to the second century and the Kingdom of Aksum (Marcus 2002). The evolution of its rule thereafter – from kingdom to sultanate to empire and, recently, to democracy – has outlasted continuous attempts by internal factions, neighbours and European empires to wrest control of its land from its long line of rulers. It escaped the final colonial scramble for Africa during the nineteenth century and was the only African member of the League of Nations and the first African and a founding member of the United Nations.

Thailand, like Ethiopia, is an old nation. Formerly known as Siam, it has nearly 800 years of continuous monarchial rule dating to the thirteenth-century Sukhothai kingdom of northern Thailand (Baker and Phongpaichit 2009). In the mid-fourteenth century, the city-state kingdom of Ayutthaya expanded its sovereignty over much of Sukhothai and the

Khmer lands bordering it to the east across an area now recognized as southern Thailand. By the early seventeenth century, Ayutthaya had emerged as the pre-eminent powerhouse of South-East Asian trade and commerce, allowing the increasingly omnipotent Dutch VOC and other European trading interests to establish offices in the city. It traded freely with the growing European powers, pre-empting any decision to administratively control the system of vassal states that made up the kingdom through a pretext of free trade (when really they were seeking to monopolize it). A successful Burmese invasion was promptly repelled. A second invasion in 1767, however, ended in the destruction of the city of Ayutthaya proper and an unceremonious end to the kingdom's royal line as the last monarch starved to death while hiding from the searching Burmese soldiers. Ayutthaya did not recover from this fatal blow, but the Burmese invaders were evicted and the vassal states again annealed into a single kingdom, with a new administrative capital at Rattanakosin Island on the Chao Praya River, also known as Bangkok. A period of rapid modernization and increasing European education of the royal family members throughout the nineteenth century assisted the country to carry out a clever practice of leveraging diplomatic ties towards the maintenance of its own independence during the peak of global colonialism.

Both Ethiopia and Thailand suffered military occupation at various points in their history – Ethiopia by Italy and Thailand by Burma – and were forced to hand over territory as part of negotiations aimed at maintaining their political independence, but against the odds they managed to remain autonomous. I say "against the odds" due to the fact that if we take a look at the coterie of 105 tropical nations that blanket the modern political geography of the tropics, these two alone do not owe their sovereignty to a release, either violent or peaceful, from colonial control held by one of ten nations at one point or another since the sixteenth century.[8] The remaining 103 former colonies arrived at their independence as early as 1804 (in the case of the Republic of Haiti) and as late as 1984 (the Sultanate of Brunei), but with the majority of states, particularly in Africa and Asia, not achieving independence until the middle of the twentieth century. The continental Americas, largely controlled by Spain and Portugal, would loosen the ties a hundred or so years earlier.

One undeniable outcome of colonialism is the relative nascence of self-governance in most tropical nations. The average age of former tropical colonies is 75 years. Many are much younger. Belize is 35 years old and the Marshall Islands only 30. But the youthfulness of the tropics' political geography is particularly striking when it is contrasted with the maturity of extra-tropical countries. The average age of these nations, 85 in total, is 227 years. If we only consider the 29 members of the relatively well-to-do club of nations,[9] the Organisation for Economic Cooperation and Development (OECD), this average rises to a staggering 465 years, a six-fold increase over the former tropical colonies. This is despite a couple of members, the Czech Republic and Slovakia, having re-booted independence relatively recently through a mutually agreed divorce. Japan, Denmark, France, Hungary and Great Britain, even the tiny principality of San Marino, can legitimately claim to have maintained their independence as a nation for more than a millennium. Only Ethiopia might arguably lay claim to the same degree of longevity in the tropics.

Young and violent

My arrival in Colombia aboard an Avianca flight from Spain was met with the in-flight announcement that the plane would be diverted to Cali to drop off the goalkeeper returning with the national football team. They had been knocked out of the World Cup finals in Italy after having qualified for the first time in nearly 30 years, but managed to make it through the group stage, drawing with the eventual champions, West Germany, along the way. They had lost in the first knock-out stage, but it seemed like a victory to a nation that was firmly gripped by football fever. The flight across the Atlantic was filled with innumerable conga trains, duty-free bottles of brandy and chants from the large contingent of supporters that were accompanying the team on their return trip.

I had the fortunate opportunity to participate in a collaborative programme of tropical forest ecology training and research in the country and the flight from Spain was the middle leg of a non-stop 24-hour journey that would take us from the wet and chilly east of England to the provincial capital of Leticia on the southernmost border of Colombia and then a four-hour boat trip up-river to the Amacayacu National Park.

The flight was filled with relaxed, friendly Colombians celebrating their own measured victory returning to a country that was far from relaxed. It was 1990 and Colombia was experiencing the worst violence in its history. Trapped between the machinations of the world's wealthiest criminal, Pablo Escobar, a fierce war for control of the burgeoning cocaine trade between the Medellin and Cali drug cartels, and a continuation of a long-running insurgency by the leftist Fuerzas Armadas Revolucionarias de Colombia, or FARC, Colombian society was reeling. This beautiful country had recently been crowned with the incommodious title of "murder capital of the world". Numerous bombings and a heavy army presence in many parts of the capital, Bogotá, drew a smothering veil over the magnificence of the ancient city, once home to the Muisca people, seat of the Spanish Viceroyalty of New Grenada and capital of Gran Colombia. Conditions in the surrounding countryside were even more perilous to those without significant protection. Nestled in the upper reaches of the Magdalena Valley, the city is surrounded by a unique neotropical alpine habitat and (once) snow-capped peaks that is typical of most of the largest and oldest cities in Andean Latin America. But it was also the most common target of a very visible and sustained campaign of bombings and killings by the Medellín cartel. Cars would speed through the streets with a singular hand thrust out of the window holding a small sign with three letters – DAS[10] –notifying the public that they were agents of Colombia's security service and that they were en route to another security crisis. DAS agents were on edge. In 1989, the Medellín cartel had detonated a bus containing over 1,000 kilograms of explosives at the entrance of DAS' headquarters with lethal impact. Foreigners were warned against travelling to the country and it was clear that our presence, while welcomed with a brave face by our Colombian hosts, was unusual. Kidnapping for ransom payment was common, particularly by FARC. The perilous triangle of violence created by insurgents, government forces and narco-traffickers, had left Colombians to live with a new social normality laced with almost daily instances of violence. The economy, despite the nation's broad base of natural resources, significant agricultural fertility and well-educated population, was in tatters. Colombia was at war. By the beginning of the new millennium, however, the Colombian government had turned things around. The Colombian drug cartels' capacity for violence had been severely degraded, though much of this arguably has only been transferred to other nations in the region, such as Honduras and Mexico. And in 2016, it appeared that Latin America's longest-running insurgency was about to call it a day after government forces spent a decade gaining the upper-hand in the largely jungle-based conflict. Colombia's economy was growing at a healthy pace and finally emerging from the long *sombra de violencia.*

Yet, the violence that dominated Colombia's recent history is far from exceptional. Analyzing data compiled by the Center for Systemic Peace (CSP)[11] on major episodes of political violence shows that tropical forest nations have suffered some form of major conflict more frequently and for longer periods than extra-tropical nations. Figure 8.2 summarizes how pervasive conflict has been in tropical countries since the end of the Second World War compared to extra-tropical states, particularly those within the OECD. Of the three states considered in the CSP assessment to have been in a constant state of conflict since 1945, two are tropical nations – India and Myanmar[12] – and most of the countries afflicted with civil conflict for at least half of the period since 1945 are tropical.

Tropical nations have suffered this prolonged, and often intense, political and social conflict during both transitional and post-independence periods. These conflicts, waged primarily over land and resources through political, ethnic and religious divisions, have dominated the history of most nations in the tropics since their independence. Fuelled by colonial geometries that typically paid consideration to imperial rivalries rather than local allegiances, decolonization often left sworn enemies as fellow citizens. Young nations already encumbered with these pre-existing social divisions and antagonism towards neighbours over map-line boundaries were also beset by the global sparring between Cold War powers.

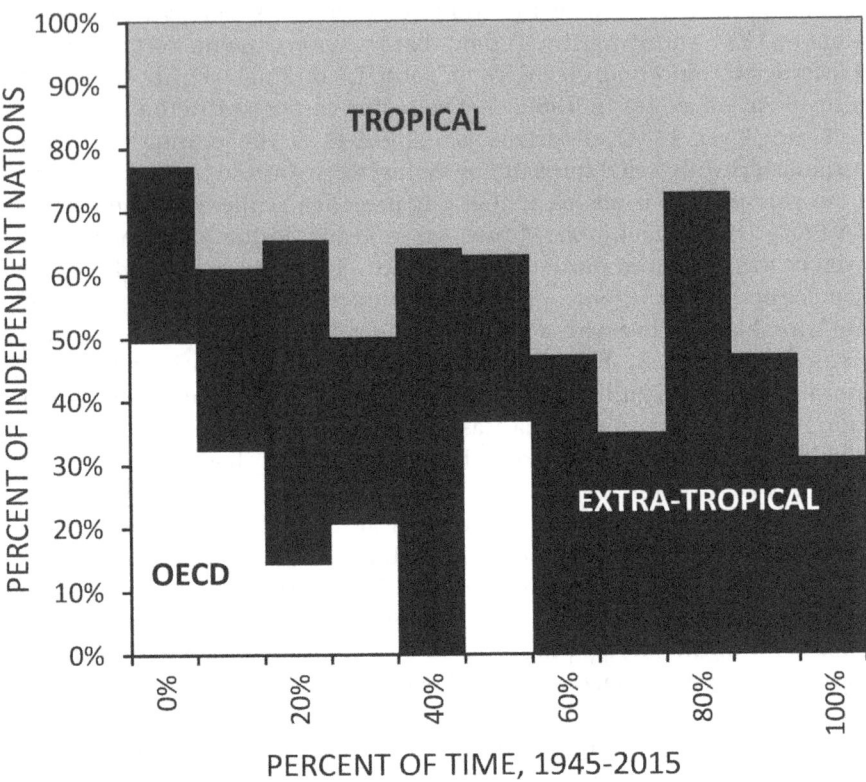

Figure 8.2 The amount of time spent in conflict since 1945 in relation to the percent of nations in tropical, extra-tropical and OECD groups.

Many tropical nations became proxy battlegrounds between the two competing, global ideologies. In Colombia, FARC were Marxist insurgents aiming to topple the Colombian government. They found their beginnings after a programme of *desarrollo economico acelerado*, instituted in the 1960s, witnessed the forced sale of large areas of land held by small, subsistence farmers to large agri-business in an attempt to make the national agricultural production more competitive and efficient. This initiative echoes the process of land clearances under the Enclosure Acts that successfully improved agricultural productivity by consolidating holdings in Great Britain and Ireland during the seventeenth and eighteenth centuries and similar insurgencies in the region plagued Nicaragua and Peru for decades. A series of conflicts in Indochina culminated in the Vietnam War and many wars that led to independence – such as the Mau-Mau uprising against the British Empire in Kenya, the Indonesian war of independence from the Netherlands, the Malayan conflict and the Cuban

Revolution – ultimately pitted Cold War adversaries against one another, although never directly.

Of course this is not to disregard similar periods of intensely violent conflict in extra-tropical nations. The French monarchy, despite its obvious ideological mis-match with the democratic intent of the belligerent American colonies, aided their rebellion primarily in an attempt to damage their greatest rival at the time, the British. The machinations of the USA and USSR in the tropics were often plied towards similar effect. The difference between conflict in tropical and extra-tropical is time and stability – the conflicts that have afflicted tropical nations are more recent since the countries of this region are often much younger. They are also often much smaller in size and conflicts in this case often consume the entire populace and economy. Conflicts are also more likely to resurface after a period of dissipation since many tropical nations have been left with few of the social and political institutions that have allowed many extra-tropical countries to stabilize and recover from periods of intense conflict. Many states, according to assessments carried out by the CSP's Montgomery Marshall and Ted Gurr, have devolved into anocracies, where political rule is incoherent, ineffective and inherently unstable (Marshall and Gurr 2005). Anocracies sit uncomfortably between autocratic (such as Syria or China) and democratic (mainly OECD nations) systems that dominate the extra-tropics, being quasi-authoritarian and prone to selective suppression of civil institutions and liberties. Frequent outbreaks of social conflict arise as a consequence, squeezing the valuable time available for economic recovery. And when it comes to the impacts of prolonged conflict, the time to recover from the loss of economic capacity, investment and infrastructure due to conflict looms large in measuring the wealth of nations.

The economic wealth of tropical nations

Tropical nations are thus, on average, smaller and younger than extra-tropical nations. In part, this can be attributed to the geographic distribution of land and sea across the surface of our planet. But most national boundaries were born from lines drawn on maps during the colonial era and frequently, often catastrophically, contested following the sudden disappearance of the heavy-handed administrative restraint applied by otherwise disengaged imperial governments. Haiti and the Dominican Republic, Colombia and Panama, Sudan and South Sudan – there are many examples. Many land borders in the tropics remain under dispute today – Guyana and Venezuela, Belize and Guatemala, Papua New Guinea and Irian Jaya. These contestations have extended to maritime rights too. Instances where these are under dispute invariably point to historic map lines as a source of disagreement. China's acceleration of claims to the South China Sea, based originally on a 1935 map then revised unilaterally in 1947 after the Japanese surrender, is one of the more recent examples of simple cartographic constructs adopted as *prima facie* evidence of territorial jurisdiction. In many ways, modern tropical nations continue to contest borders using the very same cartographic geometry employed by their former colonial rulers to substantiate competing claims without considering the widespread irregularities and frequent unilateral approaches taken in the survey and map-making processes. But when there are simply more nations, there are more borders and more opportunities for disagreement.

Averages paint a picture of central tendency, not the full range of individual outcomes and some tropical nations, such as Brazil, through happenstance of having a different colonial ruler to its neighbours, or greater desire to acquire further land area post-independence, have become sovereign behemoths able to rival the largest and most populous in the extra-tropics. But these are few (Brazil, India, Indonesia) and the lion's share of tropical nations are relatively small. It is not hyperbole to state that an inviolable truth for most modern tropical nations is that the sovereign state and everything it implies – land, resources, security and governance – have been shaped, more than in any other region, by their colonial antecedents and the effect these have had on the size of the country and the time its citizens have had to adjust to life after empire. The intense Cold War rivalries often subjected these new nations to an ideological "with us or against us" choice where neutrality and third-way politics were suffocated from the first day of self-rule, furthering instability and the path to anocratic governance.[13]

Why do these attributes of size and age matter? The size of a nation shapes a country's economic foundation – its natural resources – and the position of this inheritance in relation to those of other modern nation-states. Natural resources and the environment that has spawned them figure uniquely in a nation's economic prosperity. They represent a minimum material value that can generate income despite the size and success of the domestic economy. Unfortunately, as we will see in the next chapter, an inability to escape an over-reliance on natural capital can create an autonomous poverty machine that leaves most of the population in an inescapable trap, cursed by the resources that should improve, not retard, their prospects.

The age of a nation hints at how well government has done at defending these resources and the time it has had to build up institutions that can counter the predisposition to anocratic governance. Nations endowed with larger land areas simply have a greater probability of containing a natural asset with marketable value at some point in time, whether these are arable soils, minerals, hydrocarbon deposits or timber stands. Global market-makers have always had an uncanny way of creating economic opportunities from unexploited natural materials, either through new applications (rare-earth minerals in semi-conductor technology or botanical products in medicine) or by turning a local use into a global market (chocolate, nutmeg, bananas, rubber, palm oil). Older nations have had more time to build up internal market demand for these products, to consider the marketable value of natural assets – both their own and those of other nations – and to assess how best to put them to profitable use given the inevitable social, economic or environmental limitations that every country must overcome. The markets of older nations are also better structured, more balanced and diversified and often underwritten by the assurances of governmental protection of private assets, creating a platform for more confident trading of these assets. Consequently, as we will again see in the next chapter, they benefit from an increased availability of capital and financial liquidity, often fleeing tropical nations where less confidence is placed in the security of ownership.

Figure 8.3 relates the land area of individual nations to their gross national income (GNI) at purchasing power parity (PPP). Purchasing power parity is a measure of total economic output with an adjustment made to its dollar-denominated value based on currency exchange rates. It provides a more balanced measure of economic activity in relation to prices in a country rather than the relative strength or weakness of a nation's currency against the US dollar. These adjusted values are expressed in a synthetic currency, an international dollar.[14] The figure illustrates how the size of a nation's economy is positively related to its land area. The GNI shown is for the year 2010, but the relationship remains largely the same regardless of the year selected.[15]

Separating this pool of nations according to those that are primarily located in the tropics, those that are not and those that are not but are also members of the OECD is most revealing. It shows how the area–income relationship is more pronounced for tropical nations, largely due to the abundance of small tropical islands and how OECD nations on the whole manage to generate more GNI per unit area of land than those in the tropics. It also highlights the significant disparity in the size of economies in tropical countries of very similar size. For example, if we compare the economy of the Democratic Republic of Congo at just under 37 billion international dollars in 2010 with that of Indonesia, a nation of similar size but with an economy 50 times larger, we can see that the relationship between the size of the country and its economy is positive, but within considerable deviation from the trend. Equally, the economy of a relative minnow in the tropical world, Trinidad and Tobago, is of equal size to the Democratic Republic of Congo (DRC), although the latter stands as one of the largest nations on the planet in terms of land area. Perhaps the thesis that a larger size begets a greater natural resource endowment and larger economies is not true. The amount of variation in GNI that is explained by land area is relatively modest at 63 per cent for tropical nations and even less (44 per cent) for OECD countries. Are there other attributes that might better explain differences in GNI?

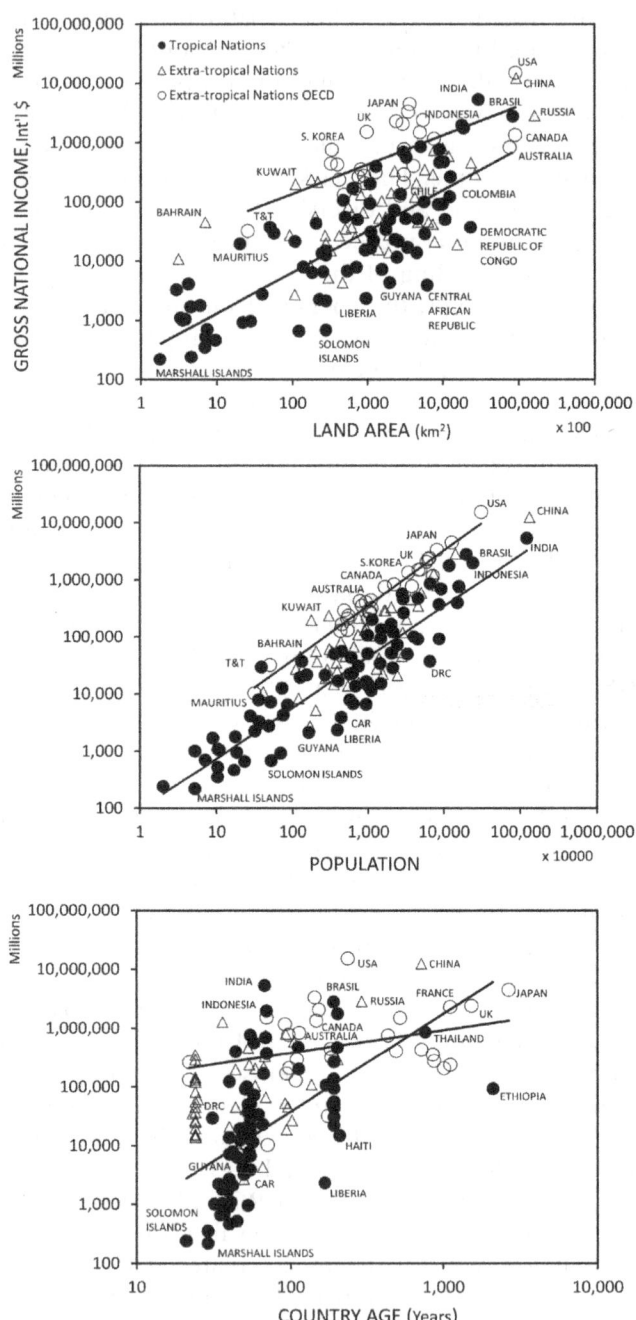

Figure 8.3 The relationships between economy, land area, population and country age for tropical and OECD nations.

An alternative approach to measuring a country's size is based on population. If we relate the GNI of each nation to its population size, a much stronger relationship between the two is apparent (Figure 8.3). An increase in the number of people living in a country also appears to provide a more consistent explanation of changes to GNI, accounting for 82 per cent of variation in the case of tropical nations and nearly 96 per cent for OECD member countries. This relationship also shows the same consistently larger GNI of OECD nations over those in the tropics as was apparent when compared against land area. The difference is staggering. An OECD nation on average had an eight-fold larger economy in 2010 than a tropical country with a similarly sized population.

We can get a much better view of how population partially negates the effect of land area on the size of nation's economy by expressing GNI on a per capita basis. The positive relationship between land area and the size of a nation's economy disappears altogether, suggesting that much of the effect of land area on the size of the economy is attributable to an increase in population. In other words, bigger countries have more people and it is population size that accounts for more, but not all, of the difference in GNI between nations within the same group. Using this metric, a country the size of Mauritius, at 2,030 square kilometres can have a GNI per capita that is similar to that of mighty Brazil. Trinidad and Tobago, a small Caribbean island with very large oil and natural gas deposits, exceeds all other tropical nations when the size of its economy is considered as a function of its population size rather than land area.

The larger effect of population on the size of a nation's economy should not come as much of a surprise. Land and its wealth of natural resources are a critical factor of production that underpins and drives economic output, but people are the economy. We take the raw resources that the environment provides and apply our intellect to develop myriad applications of these materials, expanding employment and economic activity. We add economic value to the basic resource. Importantly, we also consume the products of this effort. All of these – acquisition, application, production and consumption – contribute to a nation's GNI in what many would call a *virtuous circle*. A country with a large land area may have abundant natural resources, but without the people needed to convert these into products and consume these in exchange for the goods and services that others in turn purvey, the contribution that a nation's soil, timber, water and mineral assets can make towards economic growth remains partially latent.

The environmental wealth of nations

I say partially since the natural environment contributes significantly to our economic well-being via its natural state. Our natural surroundings normally deliver soil, water, and air of a quality that supports healthy human life and these services are only lost when they become polluted or degraded. Once they are polluted, we must incur costs to return them to their natural state or accept that the costs of contamination will be borne through our health and general well-being. The knock-on impact of accepting pollution is a reduction in economic activity as the effects of poor environmental quality take their toll on human productivity. These costs are transferred counter to the flow of economic benefits in the virtuous circle and this invariably hits the poorest in society the hardest in what one might rightly term as an *iniquitous circle*.

Our natural environment also provides a vast pool of organisms forged in an evolutionary crucible over hundreds of millions of years. Many of our brightest and best achievements in the agricultural, medical and materials sciences have been directed by the initial model provided by a wild plant, animal or microbe. Natural productivity continues to deliver a myriad of commercially important plants that contribute to economic activity across the planet, whether these are in the form of staple foods, such as the banana, rice or sugar or recently commercialized specialty products, such as coconut water, chia seeds and acai juice. The natural environment provides us with tangible services delivered virtually cost-free – a significant value if you consider how much we would have to spend in order to re-establish the baseline environmental functions they provide. In this way, a large part of the natural environment contributes passively to national economic output, a distinction that is critical in understanding its value to the long-term economic health of not only a nation, but all nations. Gross national income – in fact most orthodox measures of economic activity and wealth – do not capture these benefits since the services they provide are not transactional and consequently not valued. If we consider these environmental functions and their value to society, the sizes of nations and their economies change. As the basis for all life, one of the most important of these functions is primary productivity.

The productive size of nations

The map lines that delimited the territory of nations in the aftermath of empire did more than allocate land area to the new cohort of emerging nations. They also determined who would control the planet's biological productivity, and by proxy, the distribution of warm temperatures, freshwater resources and other environmental attributes that variously engender high biological growth. In this regard, the size of the average tropical forest nation is much larger than any cartographic scale conveys.

We can roughly quantify this disparity by re-examining the same high-resolution digital coverage used in Chapter 4 to assess global net primary productivity (NPP), but allocating these values based on the modern political geography of our planet. The outcome of this exercise reveals which countries currently control the bulk of annual planetary productivity and which have a high intensity of NPP. The map of total estimated NPP for each nation depicted in Table 8.1 reveals the massive trough in productivity share that runs along a spine of African nations, through Europe and the Middle East and into central Asia. This dearth is not strictly due to variation in the natural conditions of temperature and moisture governing the distribution of NPP, but the interaction between these environmental effects and the regional preponderance of small nations. Mega-nations in the extra-tropics, such as Russia, China, Canada, the USA and Australia capture a very large proportion of total annual NPP because they have large land areas spanning many degrees of latitude. But in every instance except the USA, the large extra-tropical nations lose position when moving from a rank-order based on land area to one based on total annual NPP (Table 8.1). Six of the 20 largest nations on our planet in terms of land area are tropical forest nations, but if we switch the metric to the amount of carbon estimated to be fixed annually by this land area, 13 of the top 20 positions are occupied by nations with significant tropical forest cover. Fifth in size but first in productivity, Brazil, home to the largest share of the Amazon Basin, fixed nearly six billion metric tonnes of carbon on average each year between 2000 and 2006. That is more than China and the USA combined, but across only 44 per cent of the land area. Globally, Brazil controls an estimated 12–13 per cent of annual productivity.[16] There is no doubt that Brazil is exceptional, due to its size and happenstance of historic geography. But when we speak of natural productivity, it is not an isolated case. The DRC occupies most of the Congo Basin – the largest area of contiguous tropical forest in Africa. It ranks fifth among nations in terms of productivity, accounting for 5 per cent of global NPP each year in a territory that amounts to just under 2 per cent of the total sovereign land area (excluding Antarctica). Indonesia, Colombia, Venezuela and Tanzania join the list of nations with contributions to global NPP double or triple the proportion of land area they control (Table 8.1).

Land area plays a very large role in determining which nations are now the main stewards of the world's biological productivity, but many countries without the vast spaces of Russia or Brazil are productivity powerhouses in their own right. These much smaller nations register amounts of carbon production for each square kilometre of land area three to four times the levels achieved on average across many of the largest nations. Russia, the USA, Canada and Australia do not make the top 50 places. Of the 20 most efficient nations in terms of NPP, all contain tropical forests and many have relatively small land areas. Guatemala, Guyana, Laos, Jamaica and Gabon are among the many small tropical forest countries that loom very large in terms of their average NPP density (Table 8.1). Because of their size, they can fit within pockets of ideal temperature, sunlight and moisture critical to primary production, reducing spatial variation in productivity. As the land area of a nation increases, so too does the variation in environmental conditions that it is likely to capture.

Table 8.1 Top 20 nations ranked by descending land area and the change in rank when classified by total annual bioproductivity (National NPP) and bioproductivity on an equal-area basis (production intensity)

Rank	Nation	Land area (km²)	Nation	National NPP (Petagrams C/yr)	Δ Rank	Nation	Production intensity (tonnes C/km²/yr)	Δ Rank
1	Russian Federation	16,400,000	**Brazil**	**5.98**	↑4	**Guatemala**	**1,184**	↑99
2	China	9,327,430	Russian Federation	4.24	↓1	Guyana	1,144	↑82
3	United States	9,158,960	United States	3.37	↔	Laos	1,138	↑76
4	Canada	9,093,510	China	2.62	↓2	Congo, Republic	1,115	↑77
5	**Brazil**	8,459,420	**DRC**	**2.34**	↓7	Honduras	1,062	↑90
6	Australia	7,682,300	Canada	2.10	↓2	Colombia	1,059	↑20
7	**India**	2,973,190	Australia	1.93	↓1	Jamaica	1,038	↑144
8	Argentina	2,736,690	**Indonesia**	**1.62**	↑7	Gabon	1,037	↑67
9	Kazakhstan	2,699,700	**India**	**1.26**	↓2	**DRC**	**1,034**	↑3
10	Algeria	2,381,740	Colombia	1.10	↑16	Nicaragua	996	↑84
11	Sudan	2,376,000	Mexico	1.10	↑3	Rwanda	965	↑129
12	**DRC**	2,267,050	Argentina	1.04	↓4	Venezuela	962	↑20
13	Saudi Arabia	2,149,690	**Peru**	**0.97**	↑6	Burundi	927	↑126
14	**Mexico**	1,908,690	Angola	0.94	↑8	French Guiana	926	↑94
15	**Indonesia**	1,811,570	Venezuela	0.85	↑18	El Salvador	919	↑128
16	Libya	1,759,540	Tanzania	0.66	↑15	Uganda	905	↑66
17	Iran	1,636,200	Bolivia	0.58	↑8	**Indonesia**	893	↓2
18	Mongolia	1,566,500	Zambia	0.58	↑20	Vietnam	888	↑46
19	**Peru**	1,280,000	Ethiopia	0.56	↑9	Malaysia	880	↑44
20	Niger	1,266,700	**Myanmar**	**0.53**	↑19	Papua New Guinea	875	↑33

Notes

Nations with tropical forests are highlighted (**bold**), as are "mega-productives", based on their presence in all three rankings (underlined).

This is the case with Brazil. Despite containing the largest share of the Amazon basin within its borders, the southern half of the country receives considerably less rainfall and experiences lower temperatures than its northernmost states and this effect reduces the average density of NPP. Producing on average in the order of 700 tonnes of carbon per square kilometre each year, Brazil stands forty-fifth among nations with the highest NPP densities.

Only two nations rank highly amongst all three of these different metrics of size – land area, total productivity and production density – the Democratic Republic of Congo (DRC) and Indonesia. Straddling the equator, they contain relatively large land areas with above average forest cover that is sequestering carbon on average at greater intensities than in most other nations. The confluence of large land area and high NPP density means that these two nations more than any other, apart from the Brazilian Amazon due solely to its size, act as lynchpins in the global distribution of NPP. As a consequence, they are particularly important in the sequestration of carbon from the atmosphere and the wholesale and permanent replacement of these forests with less productive systems will inevitably precipitate environmental changes within and beyond their borders.

A paradox of wealth

It would not be unreasonable for us to conclude that nations endowed with such high levels of natural productivity would be better off economically than those dominated by colder or drier ecosystems. After all, was it not this elevated bioproductivity that set the European nations on the path to expanding economic fortune and enrichment centuries ago? Yet, when I explore the statistical veracity of this premonition, I find no relationship between the amount of gross national income produced by a country for each square kilometre of territory and its production density. They appear entirely independent. The reason is clear – natural productivity like the solar radiation and rainfall that give rise to biological activity are not considered economic factors of production unless they are used to produce something else. In financial analysis, they are considered risks – as when rainfall fails or unseasonably cold temperatures arise – or as costs – as when forests need to be cleared or wetlands drained as part of the process of development, leaving the denuded space as the input valued economically.

Perhaps it is the amazing efficiency of photosynthesis and its ability to readily convert solar to chemical energy that has rendered natural productivity so cheap to the point of being valueless in the eyes of classical economists. It lacks scarcity value. But from an environmental viewpoint, this productivity forms a critical part of the complex interchange of carbon, oxygen, nitrogen and other elements occurring between land, sea and air that modulates changes to the global climate. Humans have entered this interchange in a very large way, not simply by our presence, but through our growing numbers and the rapid technological advancement that allow us to alter vast swathes of natural habitat using tools and techniques capable of doing the work of hundreds in a mere fraction of the time. The palaeontological record suggests the planet has never seen anything like us. Ultimately our success is altering planetary processes, particularly through changes to the mélange of various gases making up our atmosphere. If you recall from Chapter 2, these gases absorb energy at different wavelengths, regulating both incoming short-wave solar and out-going long-wave heat radiation. We are altering the balance of these energy absorbing gases and the amount of energy retained in the atmosphere is increasing. In Chapter 3, we saw how tropical forests dominate global primary productivity. Consequently, they play a pre-eminent role in absorbing gases when they are growing (via photosynthesis) and emitting gases when they are cleared (via combustion and decomposition). The permanent loss of forests also reduces the capacity of the tropics to absorb and store these gases in biomass since alternative land uses that lead to deforestation, such as farming, mining or cattle-ranching, inevitably reduce the storage function of the vegetation cover while often emitting larger volumes of greenhouse gases, such as methane.

Attempts to redress global warming due to accelerated greenhouse gas emissions have led some countries and companies that emit large amounts of carbon dioxide to begin structuring environmental options contracts with heavily forested tropical nations. These contracts aim to reduce the sizeable greenhouse gas emissions that accompany deforestation in the tropics by paying stewarding nations to forego land use options that would lead to this outcome. Efforts to develop a global market in these option contracts have largely failed to achieve meaningful trade volumes or prices,[17] but direct, bi-lateral agreements have met with greater success in part because they simpler. These options in effect bind the supplier (of forest) to maintain the supply over a specified term. The supply is quantified at the start of the contract and an allowable annual loss limit is negotiated. In return, the option holder agrees to pay annual fees to retain the contract. Monitoring of deforestation, often using remotely sensed satellite or airborne imagery, is conducted to validate that the supplier has not violated the agreed level of allowable forest losses specified in the contract. Upon inspection, if the terms (or tolerances) are met during the specified period, the fees are paid and the contract can be renewed.

Many of these recent agreements have valued the sequestration value of intact forests at two to five US dollars per tonne of CO_2-equivalent of avoided emissions. This equates to about US\$0.54 to US\$1.36 per tonne of carbon fixed through NPP. In theory, this amount should be close to the amount of economic damage attributable to each tonne of emissions and the quantity needed to generate an incremental rise in temperature with the putative environmental impacts that follow. In practice, no one really knows the true economic damage that is wholly attributable to the global rise in temperature because many other components of the global climate system are also at work and untangling the relative effects with complete confidence is not possible. Only model estimates of damage can be generated. Payments to avoid emissions in tropical forest nations are more likely to reflect how a custodial nation estimates foregone net revenue from alternative, emission-generating land uses and what the options contract buyer estimates would be their costs in meeting regulatory requirements to reduce or offset the emissions they generate. Price discovery should occur within the range where both parties are financially advantaged from the contract.

If we take the prices paid through recent contracts and apply these to the national NPP of countries, it gives us some idea of the value that resides with tropical forest nations. Globally, the average annual value of NPP between 2000 and 2006 amounted to a sliver over US$26 billion in real terms (2006 dollars) when valued at the more conservative price of US$0.54 per tonne of carbon. Tropical forest nations account for US$13.8 billion, or about 53 per cent of this amount. Of the remainder, the amount allocable to the OECD nations stands at a mere US$5.3 billion or 20 per cent of the global service value of NPP in sequestering carbon.

These figures are a probably an over-estimate, however. The reason for this over-estimation rests in the fact that the main value of NPP to carbon sequestration is attached to long-term storage. If carbon is sequestered through plant growth only to see this growth subsequently lost, as is the case with harvested annual crops or grazed pasture, the service value largely disappears. Forests store more carbon over longer periods of time than almost any other natural habitat and the main value rests with these ecosystems. The total value of national NPP calculated above is derived from satellite-borne sensors that do not discriminate between different vegetation types. A global forest-based estimate, removing the contribution of urban, agriculture and grassland cover types, gives a more accurate measure of the potential value of forest-based NPP as a carbon sink. Restricting calculations to the forest coverage described in Chapter 3, the amount of service value for the planet is closer to 16 billion US dollars per year, of which about 9.5 billion is attributable to tropical nations – a still sizeable amount. Around half of this value is shared by nations in the Americas, a third by countries in the Asian tropics and a fifth by those in Africa. The Democratic Republic of Congo accounted for more than half of the estimated two billion dollar annual value of NPP attributable to tropical nations in Africa.

In 2010, global GNI was worth a mind-bending 85 trillion international dollars. The relatively small pool of OECD nations – 27 in total – generated nearly 41.5 trillion dollars in economic activity. While 15 per cent of the world's countries accounted for nearly half of global economic activity in 2010, 88 tropical forest nations made up only 21 per cent of this figure – 18.5 trillion dollars – in stark contrast to their roughly 50 per cent contribution to carbon sequestration service value (59 per cent if forests alone are considered).

Although population size clearly trumps land area when it comes to assessing the value of annual economic activity, all of the depictions of the relationship between land area, population and economy are consistent in that they illustrate a singular point – tropical nations vary in their economic success but are chronically stunted relative to the OECD member nations and a number of other extra-tropical nations. In every instance, the smallest economies are those found in countries with tropical forests, particularly in the South Pacific and tropical Africa. Although the size and age of a nation plays an important role in determining economic activity, precisely why the economies in these regions and the tropics in general continue to lag behind their extra-tropical neighbours remains paradoxical – the most biologically productive and diverse nations on our planet are also economically some of the poorest.

This paradox of wealth formed during the earliest days of global trade and the great global exchange that followed in its wake. It reflects the processes that shape our world and the path we are following in a desire to continuously pursue an increasingly interconnected environment and society. This interconnectedness goes beyond the century and a half of sweeping changes that have taken place to the way we communicate and travel since Paxton's Crystal Palace and the Great Exhibition of 1851 ushered in the new industrial way of doing things. To understand how the tropics fit – and have always fitted since the revelations of de Gama and Columbus – in a global system of accelerating exchange, we need to examine the main working parts that make up the engine driving this exchange.

With vast financial resources ploughed into development in the tropics and the enormous markets developed around the natural resources of tropical nations, serious questions need to be asked regarding why so many nations in this region, despite their critical role in providing important inputs to the global economy and stewarding some of the world's most bioproductive environments and oldest cultures, remain so stubbornly underdeveloped? We have examined how several European nations were able to leverage tropical bioproductivity towards larger, more diverse, stable and creditworthy economic systems. But in the aftermath of independence, why hasn't bioproductivity had the same impact on tropical nations? Is the hyper-productivity of the tropics somehow retarding economic development? Could it be that there is a fundamental trade-off between the amount of natural and financial capital a nation can contain in a global economic system that only now is beginning to value the services that nature provides and has yet to properly account for these in financial transactions? Or have tropical nations been cleverly managed from the sidelines of global economic growth by countries that depend on their energy, geology, natural productivity and biological diversity, but, on paper, do not control their decision-making? I describe in the next chapter the three fundamental flows that I believe stand centre stage in the modern tragedy and triumph that is the tropics and how they reveal a great deal about why many tropical nations may never reach the myriad of milestones that have been laid out before them along a path that is presumed to lead to sustainable prosperity. Unless major changes are made to the way these flows are managed.

Notes

1 Belize became an independent country in 1981 with the territorial dispute fully unresolved and the Maya Mountains have taken on a central role in the disagreements concerning sovereignty. The border was established by treaty in 1859 between the unrecognized colony of British Baymen, Great Britain and Guatemala. The Baymen would be accessioned to the UK as a British crown colony three years later.

2 Australian oaks belong to the genus *Eucalyptus* (Myrtaceae) and *Grevillea* (Proteaceae), two genera that are currently only known from Australasia and the south-west Pacific regions. Australian robins are member of the Petroicidae, a family restricted to Australasia. "Oaks" and "robins" are distantly related to those found in Great Britain.

3 James I, by all accounts, did not like Ralegh, perhaps due to his role in popularizing tobacco in England. His dislike of smoking was made clear in his 1604 treatise *A Counterblaste to Tobacco*.

4 The political geography of our planet as it existed in 2017. Countries are classified by the dominant presence (the wet tropics) and absence (the dry tropics) of natural forests and whether they are found predominantly within or outside the tropics of Cancer and Capricorn and on continents or islands and then territorial status and membership in the OECD to arrive at the three main groups of nations – tropical, non-OECD ($n=105$) extra-tropical, non-OECD ($n=58$), extra-tropical, OECD ($n=28$) – used for comparisons in this chapter and Chapter 9.

5 In 1961, a UN special committee known as the Committee of 24 was established with the objective to support and monitor decolonization processes. The committee is comprised mainly of former tropical colonies, but also – oddly – the Russian Federation and China. The focus is on self-governance in 17 territories currently administered by the UK, France, USA and New Zealand. Votes on territorial status carried out in Palau, the Marshall Islands, Tokelau, Puerto Rico and the Falkland Islands, however, suggest there is not currently a strong appetite for independence among many small, island territories.

6 The sharp peak wedged between 40 and 50 degrees is formed by the nations of Europe and ex-Soviet territories.

7 The size of tropical nations is bi-modal. One peak (or mode) is formed from the cluster of nations occupying the continents and another is composed largely of tropical islands.

8 These include Portugal, Spain, Great Britain, France, the Netherlands, Belgium, Germany, Denmark, Italy and the USA. Independence "by proxy" of one colony via another (e.g. Papua New Guinea via Australia) is attributed to the original colonizing country (e.g. Great Britain).

 9 Mexico, the only tropical nation that is also a member of the OECD, is grouped with the latter for comparative purposes.

10 Departmento de Administrativo Seguridad – it was disbanded in 2011 and replaced by the Agencia Nacional de Intelligencia Colombiana, or ANIC.

11 Integrated Network for Societal Conflict Research www.systemicpeace.org/inscrdata.html. Details of methods used to collect and describe information on conflicts are described therein.

12 India's conflict with Pakistan is over the non-tropical Kashmir/Jammu provinces in its northern border territory.

13 The Non-Aligned Movement was established in 1961 and would ultimately see membership from nearly every tropical country.

14 World Development Indicators Database. http://data.worldbank.org.

15 The year 2010 was chosen since it is the most complete dataset available for recent years.

16 Based on the analysis of the merged land/ocean NPP cover from the OSU Ocean Productivity Unit for the years 2000 to 2006. The total global NPP derived from this analysis was 48.25 petagrams C per year, a figure largely consistent with other studies (e.g. Field *et al.* 1998; Del Grosso *et al.* 2008). Consult Zhao *et al.* 2005 for further information on the data used here.

17 Carbon credits for offsetting emissions are referred to as Certified Emissions Reductions (CERs) and are currently exchanged through various markets. Launching of the European Climate Exchange (ECX), among others, has created a platform for the trade of CERs in tandem with allowances issued within the EU. EU Allowances (EUAs) are formed through calculation of emission reduction targets, parcelled into units and distributed throughout member state entities. Entities emitting less than their allocation are then free to sell these to other entities emitting more than their allocation. This leads to price discovery of emission allowances. Similarly, the market exchange of CERs has led to a parallel market price for carbon credits. Trading volumes for CO_2 emissions exceeded five billion tonnes (CO_2 equivalent) in 2009 and 2010. Prices for CERs at that time ranged between €10 and €12 per metric ton. EUA prices ranged between €12 and €14 per metric ton. However, carbon credits generated by some types of projects traditionally accredited under the Clean Development Mechanism (CDM) component of the Kyoto Protocol, such as hydroelectric projects, cannot be traded on the exchange and constant changes to regulations regarding the use of CER purchases as a means of offsetting emissions have driven down prices to €0.25–€0.45 per metric ton by 2015–2017. Additionally, there has been some recent concern generated in regard to the illicit trading of EUAs on the ECX through recycling of permits already sold. This sort of activity combined with continued regulatory uncertainty and oversupply has debased carbon credit value and lead to significant price decline as confidence in EUAs/CERs collapses.

9 A braided flow of resources, money and people

As we have seen, colonialism was about acquiring and controlling means of natural production. I have attempted in the previous chapters to show how Europe's pre-sixteenth-century bioproductivity problem was characterized as one of both capacity and diversity and how colonizing the tropics worked to great effect towards eliminating a problem that was, above all else, a function of geographic isolation. Through colonial administrations and government-licensed companies, European nations of limited natural means were able to expand and rapidly diversify their economies. We have seen through the lens of pre-revolutionary America in Chapter 7 that imperial governments were determined to direct the colonial economies towards primary production of those natural resources that best suited the economic purposes at home, whether this enhanced food security, reduced reliance on belligerent neighbouring nations, created trade monopolies or generated public tax revenue. The expansion of credit – both public and private – was arguably the most important factor supporting globalization because it facilitated exchange and expenditure while underpinning the military expansion needed to protect ever-expanding commercial networks.

Globalization is often described through a series of "revolutions" linked to the development or mastery of certain skills or technologies. But beneath these changes rests a fundamental geographic imbalance in available energy driven by a chronic surplus in the tropics and the chronic seasonal deficit built up across the extra-tropics. I have tried to describe how this imbalance naturally advantaged the tropics and how geographic surpluses and deficits compelled Europeans (and later Americans, Japanese and, now, Chinese) towards exchange with the region in order to grow. I have also briefly illustrated how this energy leads to increased temperatures, but that nutrients and moisture – equally needed to maximize bioproductivity – are not always present. While tropical oceans are warm and account for the bulk of the global marinescape, most waters lack the nutrients that are needed to sustain phytoplankton growth and the cooler oceans at temperate latitudes are actually much more productive. Similarly, the outer tropics are on the receiving end of atmospheric flows originating near the super-humid equator. Moving poleward at a westward slant due to the Earth's spin, the rapid rise depletes it of moisture, leaving only a dry air mass to descend upon the regions straddling the Tropics of Cancer and Capricorn. The areas in this belt are warm and may offer the nutrients needed, but lack water. This is particularly true of the west coast of continents, altering the latitudinal distribution of bioproductivity in relation to the eastern continental margins where north–south changes are more gradual. If we consider how global society operated before transportation was revolutionized by the arrival of the steam engine, it is clear how geographic differences in bioproductive capacity would have shaped what and how much people ate, what they had to trade and, to a large extent, how much commerce could feasibly be conducted. Prior to the twentieth century advent of oil and natural gas, globalization and the world economy were driven by biological growth.

This immense measure of tropical production and its transfer has rarely been brought into play when discussing the growing fortunes of imperial Europe. For example, Great Britain imported approximately 31 million metric tonnes of raw sugar from the tropics for domestic use between 1730 and 1880, based on the values presented in Chapter 7. This amounts to around 122 trillion kilocalories imported over the period.[1] To put this in context, the amount of energy being delivered to the British consumer via cane sugar imports grew from 10 per cent of the total amount supplied from domestically produced wheat flour in 1700 to 40 per cent in 1830 and a staggering 80 per cent of this amount by 1880. An additional area larger than the county of Suffolk, or around 7–9 per cent of England's arable land at the time would need to have been sown with wheat to supply the same amount of calories from flour as were being delivered each year, on average, through imported sugar.[2] This specific case of cane sugar tells us that the amount of arable land capable of delivering the calories needed to sustainably support a growing British population at the time was simply not available – a condition that was repeatedly raised by economists throughout the period, most notably Malthus. As we have seen, a sustainable growth in food supply is intrinsic to increasing a population and its work productivity. When the population increases, so too does the size of the economy and, *inter alia*, public revenue, assuming an efficient system of tax collection. When linked efficiently, these increases – food supply, population, revenue (credit) – created a foundation for economic and social growth that led to the institutions and devices we associate with a modern nation. In Europe, tropical bioproductivity and the indigenous people that had learned to harness its potential played a significant role in facilitating this growth through a lop-sided exchange that diversified European markets while mono-culturing production systems in many of their colonies.

The close of the Second World War signposted the end of the imperial phase of globalization. In the preceding chapter, I discussed the immediate impact this gradual cord-cutting process has had on the age and size of tropical nations and the high propensity for conflict following release from colonial administration. I also attempted to demonstrate that the economic size of these countries remains anomalously small relative to OECD nations even after taking into consideration the effects of national land area and population size. This lag in economic expansion has occurred despite the immense bioproductive capacity that many of these new countries have in relation to their land area. If tropical bioproductivity was pivotal in the march towards a globalized economy, why have their economies not taken off since independence? Is it simply a case of age, conflict-induced capital destruction, or is an over-reliance on this capital – the resource curse – surreptitiously putting in place a ceiling on economic development? In this chapter, I make an effort to explain how the ghost of a colonial past and post-independent future have conspired to create poverty traps across much of the tropics in recent times and how the geographic expansion of these traps has created a new, braided system of counter-flowing resources, people and money.

The flow of resources

Statistics from the Food and Agricultural Organization (FAO) suggest most former colonies did not suffer the same agricultural collapse that occurred in the immediate years following the Haitian Revolution in 1805. Sugar production in both Guyana and Mauritius continued to increase after independence, only being impacted by a global downturn in commodity prices during the 1980s. Vanilla production in Madagascar more than doubled after independence in 1960, and *Hevea*-rubber production in Malaysia similarly more than doubled in the 20 years following independence in 1957. In fact, comparisons of annual agricultural production records in most tropical countries before and after both slave emancipation and political independence show no negative impact of either on this trend. In many cases, the effect was positive and accelerated production growth. But the reality of these successes is that they also perpetuated the same undiversified approach to economic growth that had been established centuries ago under colonial mandate. Many former colonies were saddled with infrastructure, institutions and networks that had been developed solely with a view to producing a limited number of agricultural and mineral commodities at very large scales. Production of coffee, tea, cocoa, rubber, palm oil, cotton and sugar had been consolidated in colonies where land suitability, labour and transport worked to best effect and government approval was secured. The great crop swap led to the re-distribution of cash-crop production across colonial holdings in order to take advantage of differences in work force availability, to avoid natural pests and diseases, and make use of suitable land and climate.

This swap created winners and losers, but the lines between these two are muddled. Let me explain. Tropical nations emerged from colonies that were demarcated on maps and negotiated between European empires. Early colonial development was left in the hands of licensed entrepreneurs and chartered companies who used whatever means necessary to expand crop production or extract mineral wealth. During the later stages of colonial rule, the process of developing natural resources became better organized. Assessing the economic potential of each colony became more scientific and systematic by the late nineteenth century, driven by series of land capability and suitability assessments, natural resource surveys and mineral exploratory studies. Colonial administrations, once dominated by the military and customs, sprouted departments responsible for agriculture, forestry and geological resources. These investments were built around a singular purpose – to supply raw materials to the administrating power for home consumption, and re-export, often in the form of more refined products. Records suggest that colonial expenditure was largely in line with the value extracted through cash-crop exports (Mitchell 1995, 2003) and that much of this was allocated towards further developing, defending and taxing this production (Subrahmanyam 2006). For example, the British government allocated a mere 0.3 to 0.5 per cent of its total annual gross expenditure each year to colonial and foreign offices throughout the nineteenth century (Mitchell and Deane 1962) while collecting between 25 per cent and 30 per cent of its annual revenue from customs and excise duties related to the trade and consumption of tropical products, mainly cane sugar. Simplifying production from its colonies in the tropics increased the efficiency of this process from the perspective of home-nation trade, consumption and tax collection but had a stultifying effect on post-colonial development since the capital invested during the colonial era was largely undiversified and pre-conditioned towards the imperial home nation. Furthermore, this lack of diversification was commodity-bound, leaving most former colonies heavily exposed to globalized market prices and speculation, sometimes to devastating effect as employment and investment collapsed. In Cuba, the early twentieth century witnessed a massive boom in sugar exports during the First World War as the beet sugar supply from Germany diminished, only to suffer a catastrophic impact on national employment and public revenue after the stock market collapse of 1929 and ensuing Great Depression. Sugar output declined from around 4 million tonnes per year in 1918–1929 to about 2.5 million tonnes from 1931 to 1940. Government revenue declined from an average of 82.5 million pesos to less than 52 million pesos over the same period (Mitchell 2003). While certain individuals and companies benefited tremendously from the intensification of agricultural production in former colonies, most of the newly independent nations instead saw a decline in public revenues and increased rates of unemployment since there was little alternative economic capacity outside the commodity-focused industries. Newly independent economies lacked diversification due to the legacy of colonial structuring. This model of primary production supplying former colonial markets was a simple extension of the economic system that had been put in place as part of the *raison d'être* for colonial development.

Value chains and comparative advantage

Much of the value of tropical commodity production continues to be lost through export of primary materials to nations with well-developed economies where refinement and processing takes place to create a diverse range of higher-value end-products. Take the case of cocoa. It is a tropical crop that, unlike tobacco, cotton, avocados or maize, cannot be grown under natural conditions in the extra-tropics. According to the FAO, 60 nations – all ex-colonies apart from Thailand – produced nearly 110 million metric tonnes of dried cocoa beans over the period from 1985 to 2015. Nearly two-thirds of the fermented and dried beans produced over this 30-year period were exported. The balance of this production was exported as refined constituents: cocoa butter, solids (powder) and paste (liquor), produced through the winnowing, roasting, pressing and heating of the seed endosperm and integument to separate cocoa butter and cocoa solids in varying proportions. These are then used directly in the manufacture of chocolate, as well as a constituent in various other food, medical and beauty products.

The global trade in confectionary products is estimated at $90 billion and the chocolate, cocoa and sugar industry in the European Union alone employs over 190,000 people and produces an annual turnover of €42.3 billion and nearly €11 billion in added value.[3] It is not surprising then to discover that tropical countries that account for 100 per cent of cocoa bean production only account for 40–50 per cent of the global exports of cocoa paste, butter, powder and cake and a mere 9 per cent of chocolate products (FAO 2016b). OECD nations exported nearly 75.5 million tons of chocolate between 1985 and 2015 and, amazingly, nearly 6.5 million tons of unprocessed cocoa beans through re-exportation. The overwhelming majority of these refined products are exported by industries based in Europe – particularly the Netherlands, Germany, Belgium, France, the United Kingdom and the USA – that have established long histories of production from former tropical colonies (ICO 2014). The dominance of extra-tropical nations in the global trade in cocoa products is in part understandable since they are also the world's largest consumers of chocolate and cocoa, but this also allows extra-tropical nations without the natural means to produce cocoa to also carefully recycle much of this value internally, growing their own economies while limiting the impact the cocoa industry could have on economic growth in tropical countries. In effect, these national industries continue to appropriate tropical bioproductivity by carefully controlling the supply chain network linking tropical producer with global consumer. Global increases in cocoa bean prices, expanding areas under production or improving yields are the only means of increasing total revenue to growers that are also subject to significant exogenous impacts on production through disease outbreaks, weather, competing land uses and conflict. Prices may respond to these impacts, but primarily due to a decrease in production, eliminating any revenue boost for those growers affected by these events while chocolate manufacturers can simply add on any added costs of their production to the price of their retail product. Only competition works against these increases. The growth in agricultural futures markets has mitigated some of these risks to the grower, but this also works for the downstream industries relying on this production. Importantly, failed production cannot be stored, sold, discounted or optioned, only insured, and these unique risks are borne only by growers. Production and export volumes and values suggest that this process of capturing value from primary production in the tropics is widespread. Many of the largest markets in tropical commodities, including those built around coffee, bananas, pineapples and sugar, show a pattern of declining participation by tropical nations as a raw material is refined or marketed towards various final products in a globalized marketplace (Table 9.1). Including refinements further along the value chain – such as those attached to the emollient properties of its butter and its use in lotions – reduces again the portion of total value attached to primary production.

Colonial trade networks were built on a model of importing raw resources and exporting "manufactures" to "vents" by return. Manufacturing in colonies was actively discouraged, if not banned altogether. The principle driver of this system of globalized commerce hinged on a singular fact – Europe lacked the bioproductive capacity to sustain population growth and develop a large-scale, global trade in home-grown agricultural products at the same time. Extra-regional trade in these products would eventually be left to the United States and to a lesser extent, an independent Australia, Canada, Chile, Argentina and New Zealand. Like agriculture, evidence within the logging industries show a similar pattern of trade that began with the export of raw, marginally processed logs towards extra-tropical processing into high-value sawn wood and veneers in Europe, Asia and North America (Dawkins and Philip 1998). Invariably, the economies of the poorest tropical nations became dominated by the export of raw materials. There are exceptions, such as tea, where the raw product requires only basic processing to make a product suitable for direct sales to retailers. Although this does not necessarily add any significant value to the harvested material, it does allow at least part of the retail price premium to be captured by producers. This vertical integration is often lacking, or remains inverted to the extent that the retailer has leveraged control over the producer. Today, many of the largest suppliers

Table 9.1 The change in participation moving from production to refinement of several major tropical cash-crops over the period 1993–2017

Product/origin	Production (%)	Exports (%)		
	Raw	Raw	Intermediate	Final
Cocoa	Beans	Beans	Cake, paste, powder	Chocolate
Tropical	100	91	45	9
Extratropical	0	1	2	10
OECD	0	8	53	81
Coffee	Beans, green	Beans, green	Extracts	Beans, roasted
Tropical	100	92	49	5
Extratropical – non OECD	0	1	8	4
OECD	0	7	43	91
Bananas	Fruit	–	na	Fruit
Tropical	89	na	na	87
Extratropical – non OECD	10	na	na	1
OECD	1	na	na	12
Pineapple	Fruit	Fruit	Fruit, canned	Juice
Tropical	89	80	87	68
Extratropical – non OECD	8	1	7	5
OECD	3	19	6	27
Sugar	Raw, centrifugal	Raw, centrifugal	na	Refined
Tropical	53	80	na	42
Extratropical – non OECD	25	6	na	17
OECD	22	14	na	41
Tea	Leaves, green	Leaves, dry	na	Packaged
Tropical	57	68	na	38
Extratropical – non OECD	36	25	na	14
OECD	7	7	na	48

Note
"na" indicates product not exported in this state.

of tea grown in India and supplied to extra-tropical markets are Indian-owned and trade statistics for tea reflect the growing share of revenue captured by tropical nations in this instance (Table 9.1).[4]

Preferences – but for whom?

As cash-cropping spread globally, producers within the tropics have also come into conflict with one another, principally due to policies from importing nations that are seen as trade distorting. Many former European colonies have been in receipt of preferential tariffs, subject to guaranteed quotas and minimum prices through the Lomé Convention, first signed in 1975. This group of 71 tropical countries, referred to as the ACP (African, Caribbean and Pacific) group, was promised aid and investment through the agreement, as well as a system of compensatory finance to mitigate price fluctuations in global commodity markets. While these policies provided some certainty to growers in former British, French, Dutch and Belgian colonies, they created a strong barrier to entry to the former Iberian ones that had grown major export industries in many of the most traded commodities, such as sugar, bananas and coffee. The case was eventually taken to the World Trade Organization (WTO) and the preferential treatment of ACP produce was discontinued in 2000.

Despite the intervention of the WTO and the mis-directed policy enabling unprofitable "zombie" producers to persist, the European Union replaced the Lomé agreement in 2001 with a new generalized system of preferences for ACP nations under its "Everything But Arms" (EBA) regulation. This granted duty- and quota-free access to ACP nations (plus an additional nine non-ACP tropical countries) but significantly imposed a complex of limits and quotas on some of the most significant tropical products, most notably sugar. European sugar prices are several hundred per cent greater than global market prices due to significant protectionist measures deployed under the Common Agricultural Policy (CAP). In Chapter 7 we saw how beet sugar broke the monopoly that tropical producers had held on the sugar trade up to the nineteenth century, but also how it could never achieve the greater yields achieved from sugar cane. Only the provision of subsidies and imposition of tariffs have ever allowed beet sugar to compete on the global market. Another programme, the Special Preferential Sugar (SPS) agreement, aimed at supporting sugar producers in former sugar colonies, such as Guyana and Mauritius, was in part cleverly consumed to feed new preferences under EBA (Gibb 2006). Both EBA and SPS programmes have been touted as providing support for development in poverty-stricken nations. But sugar production in most of these countries remains controlled by a very few companies or families that do not engender significant improvements in the standard of living of their larger work forces. Instead, preferences maintain an over-reliance on uncompetitive production by hyper-dominant industries with little boost to incomes or narrowing of income inequality in tropical nations. They suppress economic dynamism and the process of "creative destruction" that has proven to be an integral part of achieving sustainability in OECD nations.

The ACP, EBA and SPS programmes of special treatment highlight the post-colonial mis-direction of agricultural development in the tropics. Creating an artificial advantage in many former colonies forestalled the need to review where their economies could most "naturally" prosper after independence, given their individual circumstances. In reality, many in receipt of support lacked any comparative advantage to the many larger-scale producers in Latin America. Subsidies under Lomé supported chronic inefficiencies and a lack of capital investment in the Caribbean and Africa since the agreement defined a ceiling on production and basement on prices – scarcely an incentive to long-term foreign capital inflows, innovation or expansion of enterprise values. As we have seen in Chapter 4, industrial-scale banana producers in Latin America were certainly at times incurring hefty environmental and social costs but producers in the Caribbean and Africa simply lacked sufficient investment to compete with major producers such as Ecuador or Honduras. A similar scenario exists in relation to the mammoth Brazilian and Cuban cane sugar industries. Maintaining uncompetitive industries designed to supply raw agricultural products to developed nations for further value-added refinement has not allowed tropical nations to gain maximum economic purchase from their greatest comparative advantage – bioproductive capacity. It did allow European importers to benefit from a "captured" supply that had become entirely dependent on programme incentives offered under Lomé and its descendants rather than having to compete in the market for a share of global supply consolidated around the most efficient producers. In effect, it allowed cash-crop operations – largely still owned and controlled by European interests – to adhere to the colonial model of controlled supply but in a new, liberalized, post-colonial market.

Perversely, the high level of protection afforded agriculture in extra-tropical nations, such as the European Union's CAP, combined with arable land resources disproportionately utilized for cash-crop production, has often led to tropical nations becoming net importers of many staple food items or seen the substitution of domestic products by cheaper imports from nations with strong agricultural subsidies and support structures. This extends even to the cash-crop products that are exclusively produced in the tropics. For example, in 2015 the European Union, Canada and the United States together exported 7,100 tonnes of chocolate, 1,500 tonnes of roasted coffee and – perhaps most amazingly – 2,500 tonnes of bananas and over 13,000 tonnes of refined sugar to the Caribbean (FAO 2016b).

The liberalization of trade within a globalized marketplace has benefited some key actors in the cash-crop trade in a way not dissimilar to the former colonial system of issuing charters, warrants and licences. Sustainable development in tropical countries, however, has not improved overall as a consequence. Opaque policy structures continue to work to protect national agricultural sectors, even when these are uncompetitive, peripheral to concerns over food security and self-sufficiency or even at a notable cost to the consumer (e.g. Guyomard *et al.* 1999). Agriculture has always been at the heart of governance because basic caloric intake, like water, is not substitutable and a dearth in either leads to social unrest and instability. Yet, most tropical nations remain deeply dependent on cash-crop agriculture for jobs, income, public revenue and economic growth. Unlike extra-tropical nations, manufacturing and services account for a significantly smaller portion of GDP and the economies are dominated by sectors linked to their bioproductivity – crops, livestock, fish, fibre and hardwood (Figure 9.1). In OECD nations, the structure of

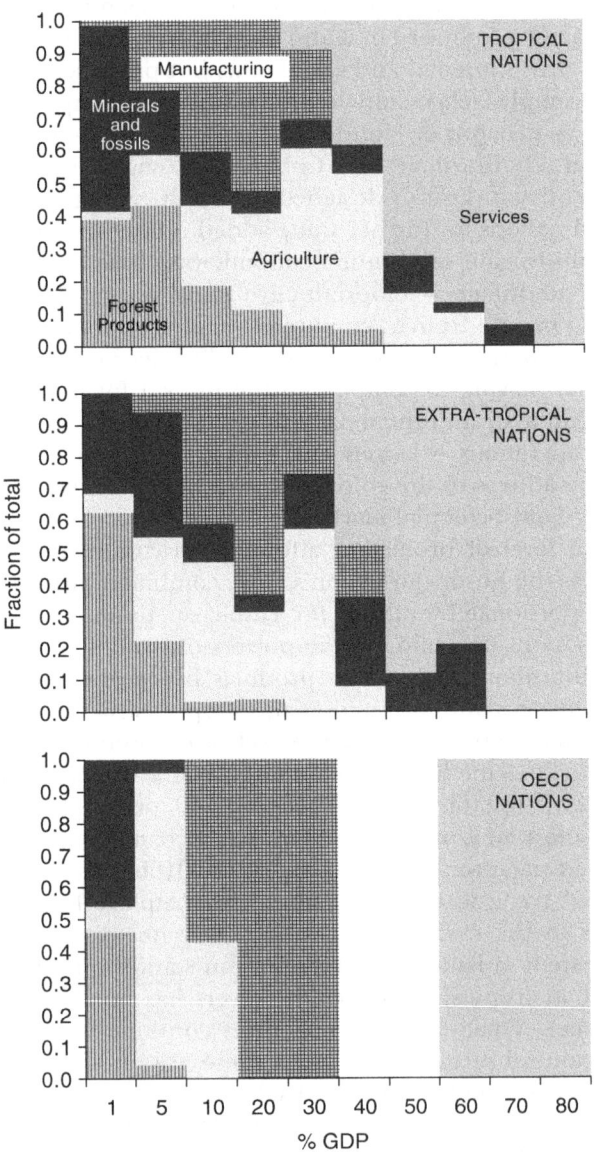

Figure 9.1 The structure of economies by contribution of different sectors to GDP.

economic activity has evolved towards a dis-associative state of service sector dominance. What I mean by this is that the service sector is uniformly dominant and not necessarily a link in the value chain that connects agriculture, fisheries and forestry with manufacturing and retail service sectors of the economy. These cross-cutting services, such as finance and insurance have proliferated, but not in tropical countries. As a part of total sector exports, these types of services are clearly the laggard when compared to other components, such as transport and communication (FAO 2016b). The economic contribution of natural resource sectors, including agriculture and forests, now accounts uniformly for less than 15 per cent of GDP while the service sector in every OECD country rarely falls below 60 per cent. Minerals and fossil fuels also can account for a very high fraction of GDP in some extra-tropical and tropical countries. The difference between the two hinges on the much greater role that agriculture and forestry typically play in the economies of the latter (Figure 9.1). A handful of tropical nations, such as Antigua, Barbados, Dominica, Grenada and St. Kitts and Nevis could be considered "OECD-like" based on the structure of their economies, in that they have few or no mineral resources, a small agriculture sector and a hyper-dominant service sector. But these very small island nations lack any significant manufacturing capability, relying heavily on tourism for GDP growth. Other tropical nations, such as Costa Rica, the Dominican Republic and El Salvador also have a small mineral and fossil fuel sector, relying more on agriculture and manufacturing, but lack a dominant services sector. Apart from manufacturing, Mexico also does not fit within the economic profile of a typical OECD country although with 62 per cent of GDP delivered through services, it is very close to the typical 69 per cent associated with most nations in the group. Figure 9.1 does not reflect the value of these various sectors. In fact, the mineral, fossils and agricultural sectors in OECD and other extra-tropical nations often far outstrip those in the tropics that have smaller economies, even after accounting for population or land area.

What do these disparate contributions to GDP tell us about the flow of resources? First, they confirm how the economies of tropical nations remain heavily dependent on plant growth and, more broadly, natural resources. They also illustrate how very few tropical nations are close to achieving a dis-associative state of service sector influence over their economies. Those that are approaching this condition tend to be very small island nations where agriculture and mineral resources are often constrained by land and population, but which offer considerable amenity value attractive to overseas tourists or secrecy for those seeking to shelter financial assets – both sources of foreign currency. Many medium-sized tropical nations may struggle to achieve a structure akin to those in the OECD because they are too large to avoid the pitfalls of an over-reliance on natural resources, but too small to scale-up these parts of the economy towards global leadership such as Brazil has achieved in sugar, soybeans, *robusta* coffee and oranges. Where tropical nations have achieved a positive trade balance, fossil fuels and minerals typically loom large among these resource-dependencies.

A fossil effect

When the petroleum industry rapidly became the new resource driver of globalization in the late 1950s it had an enormous impact on many tropical countries. Discoveries of oil deposits were meant to herald a fast-track to economic development in countries that had previously been heavily dependent on agriculture for exports. Instead of modernizing their agricultural economies, many oil-rich nations simply abandoned them altogether. For example, in Trinidad and Tobago, the collapse of "soft" commodity prices during the Great Depression incentivized a re-focus on petroleum development. Since the 1960s, production of cocoa and coffee has collapsed by more than 95 per cent and cane sugar by nearly 70 per cent. Meanwhile, oil and natural gas production grew to nearly a third of national GDP by 1980, while agriculture and forestry declined to 2.6 per cent. A similar collapse in cash-crop production occurred in Angola after oil and diamond production expanded rapidly in the 1980s – coffee declined by 90 per cent, cotton by 75 per cent and sisal by more than 99 per cent between 1960 and 2015 with the bulk of the decline occurring during the 1980s.[5] Unlike many OECD nations where agriculture and forestry continued to increase in value while declining in their contribution to GDP, in these countries they declined in both value and contribution.

There were three major problems with the rapid expansion of the mineral and fossil fuel sectors in tropical nations. First, these discoveries rapidly re-directed their economies away from the intrinsic comparative advantage that their geographic location provided them in terms of bioproductivity, towards industries where they had very little advantage. Mineralogical and fossil carbon deposits are not linked to latitude like bioproductivity. Geological resources are related to the age and morphology of landforms and the palaeoclimates and geological events that combined to create high concentrations of economically valuable minerals and fossil carbon. Tropical nations are not any more heavily endowed with economic deposits than extra-tropical ones and these do not offer a unique advantage when considering globalized markets. The use of geological resources in almost every instance was first initiated in extra-tropical regions across Eurasia. The ages we associate with the harnessing of various metallurgical techniques – the Copper, Bronze and Iron Ages – began much later in the tropics than the extra-tropics. Coal and oil arrived much later to Europe, but were first extracted and industrially processed in temperate Eurasia and North America. As a result, the major extractive industries and technologies were almost entirely developed in extra-tropical nations. Unlike tropical crop production and processing that were imbued with centuries of ethnobotanical and plant-breeding know-how, virtually every tropical nation lacked the capacity to develop these deposits and understandably had to forego a large portion of the proceeds to international companies in order to mobilize the necessary resources. But deals struck, particularly during their early development, left many oil-rich nations with far less income than one would anticipate based on global market prices.

Second, the rapid rush of public income and over-gearing towards geo-industries created conditions that ultimately would trap, rather than liberate, the economies of those nations endowed with mineral deposits. Known as the "resource curse" or "Dutch disease" (Sachs and Warner 2001), the rapid influx of petro-dollars dramatically overwhelmed the highly simplified agrarian economies, creating inflation and damaging investment in other economic sectors as local currencies strengthened on the back of oil exports, making the export of agricultural and manufactured products less competitive. In part, this explains why agricultural sectors in Trinidad and Angola began to shrink as the minerals and fossils sector expanded, but rents accrued from exports also became "easy money" for governments and well-connected local businesses while providing fewer incentives to develop alternative sectors. This funnelling of large amounts of financial liquidity through the hands of a small number of government officials invariably increases the likelihood of corruption in many oil and mineral-exporting nations, in part because of the cash flows from a few, large commercial partners and in part due to the majority of the population being agrarian, rural and poorly educated – hardly a constituency eager to intervene or capable of intervening peacefully to curtail kick-back payments, illicit transfers or embezzlement of public proceeds from oil and mineral extraction.

This third problem, corruption, is arguably the greatest among the three since it works against finding solutions to resolve the first and second problems. The weak civil society syndrome that characterizes many tropical nations eliminates an important brake on corruption that has played an indispensable role in putting a lid on illicit practices in more balanced economies. Fossil and mineral industries are undiversified compared to service or manufacturing sectors that often contained many more small and medium-sized businesses. In theory, payments from a small number of large operators are much easier to collect. This should lend greater financial security and wherewithal to public budgets and allow the funds accrued from natural capital to be more readily transformed into much-needed produced capital. To succeed, this transformation should follow various axioms of sustainable development. Among these, one in particular is crucial to success, and this is often referred to as Hartwick's Rule. Proposed by the eminent economist John Hartwick, it tells us that a sustainable economy can only be achieved when the rate at which we consume non-renewable resources, such as fossil fuels or minerals, is offset by the same rate of produced capital formation (Hartwick 1977). Produced capital is perhaps closest to that most broadly recognized by society. This consists of buildings, roads, bridges, power plants, universities and schools, hospitals, etc. – in other words, infrastructure and institutions, but also skills and social support that improve labour productivity. When the rate of produced capital exceeds the rate of resource exhaustion, savings are produced. These "genuine savings" allow nations to begin earning a yield on their investment through the support it provides to financial capital and intangible asset formation. Many critics would argue that the notion that ecosystems converted into buildings somehow improve our way of life is an oversimplification. Yet, I know of very few people anywhere that would agree to forego the physical infrastructure and life-improving services that typify most developed countries, if these are delivered from the proceeds and the benefits of this conversion are equitably shared. If these conditions are met, then it is not a question of whether to utilize natural assets, but one of whether utilizing them will yield a significant improvement in physical capital and whether they are being extracted in a manner that leaves the smallest residual impact and internalizes environmental costs (Costanza *et al.* 1997). If these impacts and costs outstrip the savings accrued through investment of the proceeds, then the resources should remain *in situ*.

Corruption is a problematic leakage in this process of transforming natural capital into produced capital since it impacts the savings rate thereby reducing the efficiency of capital transformation. It also often purposefully leaves environmental costs, particularly water and air pollution, externalized. The consequences are far-reaching as corruption neither improves agriculture and forestry nor improves the health and education of the workforce – the latter a precursor to growth in manufacturing and services that typifies high-income OECD economies (World Bank 2006). Often these leakages stimulate illicit flows of money that drain liquidity from nations or encourage "destructive" investments, but more about this later. Fossil fuel rents are not the only driver of this scenario (e.g. forestry, Repetto and Gillis 1988), but they often lead to the most spectacular incidences of mis-allocation and mis-direction of natural capital resources.

Equatorial Guinea, one of the smallest African nations, offers a shocking illustration of these mis-allocations and mis-directions and a failure to follow Hartwick's Rule. It began exporting oil in 1995 and the proceeds from this production rapidly grew to 70 per cent of the small African nation's economy by 2008. By 2004, it was ranked as Africa's third largest petroleum exporter and by 2010 its adjusted gross national income per capita was close to 17,000 international dollars, the highest on the continent (World Bank 2017). The US Energy Information Administration (EIA) estimates that 1.8 billion barrels of oil have been produced since 2000, an amount the World Bank estimates to be worth nearly

US$50 billion in current value. Roughly, these proceeds amount to nearly US$3,000 per person per year since the turn of the millennium. The nation's well-being, governed by a de facto authoritarian regime headed by a President that has remained in power since 1979, remains difficult to surmise since little information has been collected on health, education and nutrition. Yet, what is known suggests that the proceeds of the nation's oil wealth have not been wisely re-invested nor has the bulk of its population experienced an increase in standard of living consistent with the supposed boost in national income. A demographic and health survey published in 2012 indicates that access to basic facilities, such as electricity, water, sewerage and cooking fuel remains at levels typical of much poorer neighbouring countries and shows similar patterns when considering educational enrolment and health care facility attendance (World Bank 2017). The incidence of diseases such as HIV and tuberculosis has increased significantly over the same period, although overall life-expectancy too has increased steadily since 1995.

Equatorial Guinea's government established a sovereign wealth fund (SWF) in 2002. This approach to saving can often prove an ideal structure for nations endowed with significant oil and mineral wealth. It allows them to organize and manage the conversion of the proceeds towards produced capital by diversifying investment and generating higher net returns. SWFs can also effectively buffer price shock impacts that often hit natural resource-driven economies, a process called stabilization. Unfortunately, Equatorial Guinea's fund has a mere US$80 million in assets – the lowest of any of the 20 SWFs established by tropical countries since 1956, despite a nearly 600-fold larger revenue base generated from oil and natural gas. Although roads, communications and other basic infrastructural needs have expanded in the country, the bulk of the population remains rural, agrarian, uneducated and poor. The principle reason is a spectacular propensity towards appropriation of the natural resource rents by government for personal use and distribution. These leakages prompted Transparency International to rank Equatorial Guinea as the fifth most corrupt of the 77 tropical countries they examined, only ahead of Sudan, Yemen, South Sudan and Somalia – all characterized as anocracies or failed states by the Center for Systemic Peace (Marshall and Elzinga-Marshall 2017). In the meantime, OECD nations continued to deliver over US$300 per person in official development assistance between 2000 and 2016 and nearly US$30 million via various UN agencies (World Bank 2017). This movement of money – development aid flowing inwards from OECD countries while self-appropriated, natural resource proceeds flow outwards to tax-havens – typifies the financial flows involving many tropical countries where transparency and democracy remain exogenous concepts. The likelihood of successfully developing sustainable societies in the tropics that can meaningfully participate in a globalized economy appears remote as long as this template remains in place.

The flow of money

The flow of raw resources from tropical nations is deeply interwoven with international financial flows. The celebrated Colombian author Gabriel Garcia Marquez noted that everyone carries three lives: a public life, a private life and a secret life. The movement of money between nations also runs through three lives. *Public funds* are born from overseas development assistance grants, loans and debt relief. *Private funds* flow as foreign direct investment in businesses, real estate and natural resource concessions and earnings made through the export of goods and services, as well as shorter-term portfolio investments. Remittances made by nationals working in foreign countries and then repatriating a portion of their earnings to family and friends also contribute to the flow of private funds. Finally, there is the secret life of money flowing between nations as *secret funds* illegally earned, transferred or utilized, primarily by misinvoicing trade goods, but also through leakages in the balance of payments, invoice faking, misinvoicing services and intangibles and simple unrecorded cash transfers (GFI 2017). Considered together, these flows can tell us a great deal about the financial condition of tropical nations since the early 1990s in relation to their extra-tropical trading partners. I have summarized the total net value of these flows over the past quarter-century (1993–2017) for tropical and extra-tropical nations in Table 9.2.[6] I have chosen to examine these flows over the past 25 years to provide a more complete timeframe of activity for post-independent tropical nations. Annual figures over the period

Table 9.2 The cumulative amount of financial flows and net visible and true positions for the three country groups over the past quarter century, adjusted to 2017 US\$. Illicit funds are an under-estimate and only reflect the period 2004–2010

2017 US\$, billions	Tropical	Extra-tropical, non-OECD	OECD
Commerce (1)			
Foreign direct investment	(3,700)	(4,000)	5,200
Net trade in goods	15	12,400	(11,800)
Net trade in services	(1,500)	(3,000)	8,100
Total commerce	(5,185)	5,400	1,500
Remittance (2)	2,800	2,800	(7,500)
Assistance (3)			
Development aid	1,226	768	52.6
Debt relief	221	69	0.02
Total assistance	1,447	837	53
Illicit, net inflows (4)	1,850	6,200	600
Net visible position (1 + 2 + 3)	(938)	9,037	(5,947)
Net true position (1 + 2 + 3 + 4)	912	15,237	(5,347)

were then inflated to present values and then summed to present a total amount, but as if this was transferred in the year 2017. There are many other factors, such as exchange rate volatility, that shape these estimates, but I think they provide a reasonable picture of the size and direction of financial flows between tropical and extra-tropical nations. Several striking relationships are apparent.

Private funds

Tropical nations as a whole have a chronic commercial fund deficit, driven principally by large-scale foreign capital flight, but also a trade imbalance in both goods and services. The very small surplus in goods illustrates the long-term difficulty that tropical nations as a whole experience in attempting to balance the value of their agricultural, forestry and mineral exports with the added-value the economies of extra-tropical nations export to the tropics through manufactured goods. It also reflects the highly under-developed service economies of tropical nations, as illustrated in Figure 9.1. The massive 11.8 trillion US dollar deficit in goods accrued across OECD nations, however, has not been accumulated through exports from tropical countries, but from non-OECD, extra-tropical nations. This group of course includes China, a nation that has accounted for nearly half of the net trade surplus of goods from extra-tropical countries over the period, mainly due to a hyper-rise in net exports to OECD nations since 2005. This is balanced by the United States accruing a nearly 20 trillion dollar deficit trade in goods over the same period. There are a handful of exporters in the tropics that register consistent annual surpluses, although considerably smaller than that of China. But apart from Brazil, Indonesia, Malaysia, Thailand and Vietnam, most export-driven economies in the tropics – like that of Equatorial Guinea – are inadequately diversified due to a protracted failure to re-invest proceeds from their natural resource rents, the latent effects of natural resource-dependency that makes them susceptible to Dutch Disease and other factors, such as skill migration that impact economic capacity. We can also examine how nations fare in relation to the relative balance of net trade in goods and services as this relates to the total net value of all trade, both import deficits and export surpluses – something I refer to here as trade proportionality – in Figure 9.2. It again shows the disproportionate fraction of tropical nations that register accumulated deficits (–0.81 to –1 value) in both goods and services and a much smaller fraction, such as Thailand and Malaysia, that have achieved net surpluses for both (+0.8 to +1).

The OECD and other extra-tropical nations principally benefit from the flow of private funds through trade and foreign direct investment (FDI), but privately funded inflows to tropical nations occur largely through remittances. Remittances arise through the unreturned transfer of funds, principally earned by foreign migrants and sent to their originating

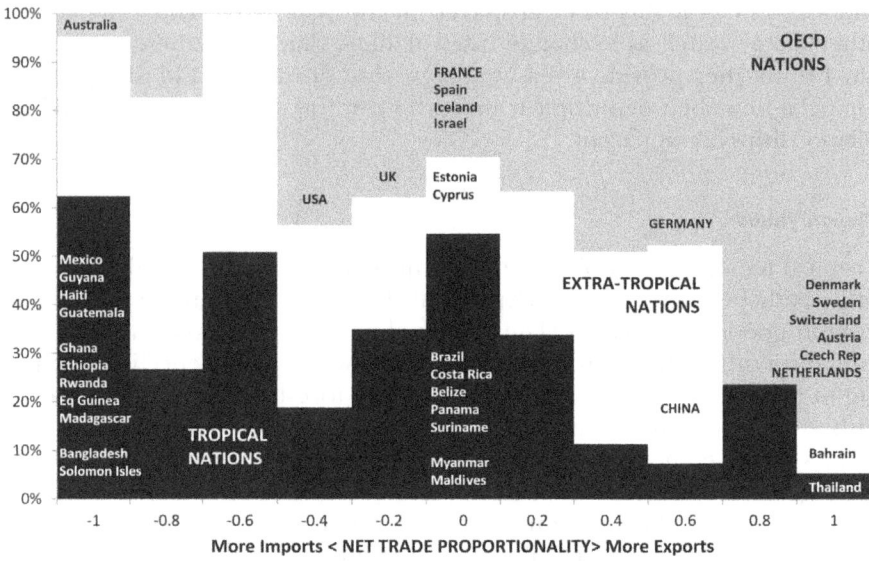

Figure 9.2 The proportionality of net trade. (−1) nations are net importers in both goods and services, (+1) nations being net exporters.

country. These are now recognized as a major source of financial capital to developing countries (Brown 2006). This would appear to render them a positive force working for improved livelihoods and economic development in tropical countries. But analyses of the impact these transfers have on the recipient country show that, while primarily positive, they do appear to be double-edged. Most studies of remittances indicate that the majority of funds can stimulate household investment in small and medium-sized enterprises, support education, act as a form of "self-insurance" against the impact of catastrophic weather events and improve prospects of exiting poverty (Yang 2008; Yang and Martinez 2005; Clarke and Wallsten 2004). Yet, other analyses suggest that this steady flow of income to households can act as drag on economic growth, by reducing incentives to seek and retain employment, thereby restricting labour supply (Rajan 2007). Remitted income can also surreptitiously lead to exchange-rate over-valuation due to the influx of foreign currency (Fajnzylber and Humberto López 2007) – similar to the effect exerted by a singular over-dependency on oil exports that can lead to Dutch Disease.

Since 1993, the total remittances of funds to tropical countries amount to at least US$2.8 trillion. The overwhelming bulk of these funds has flowed from migrants living and working in OECD countries, representing net outflows from developed nations that are three times the net surplus they built over this period through FDI and service exports (Table 9.2).

There are a few other extra-tropical countries – mainly oil-rich states with relatively small populations, such as Saudi Arabia, Kuwait, Bahrain and UAE – that originate large remittance values, but more than 95 per cent of annual remittances were transferred from OECD nations, principally from the European Union (US$4.5 trillion) and the United States (2.7 trillion), while the largest recipient nations over the period were India, Mexico, Iran and China. Of the 107 tropical countries surveyed, only 11 registered net outflows of remittances, primarily oil exporters such as Venezuela, Brunei, Gabon, Congo and Equatorial Guinea. The vast transfer of wealth from developed to developing nations can represent a significant buffer against the relatively poor economic performance in many parts of tropical Africa and Latin America – a form of quasi-development finance that is more closely linked to trade than aid (Rajan 2007). For example, the remittance of income from Salvadoreans living in the United States has accounted for half of export "earnings" and nearly a fifth of GDP (Gammage 2006). At roughly US$72 billion remitted, El Salvador ranks as the seventeenth largest recipient of these funds since 1993, despite being ranked ninety-ninth in population size and hundred and forty-eighth in land area. There are additional costs in supporting remittance flows to tropical nations in terms of social and energy costs to the originating nations, and complex, interwoven reasons why people in the tropics seek to migrate to the extra-tropics, but more about this deeper relationship later.

Public funds

As with remittances, tropical nations as a group have received significant sums of official development assistance (ODA), amounting to nearly one and a quarter trillion US dollars flowing to the region since 1993, almost entirely from OECD nations. This is nearly double the amount disbursed to extra-tropical nations and 25 times that distributed within the OECD group itself, primarily to Turkey and Eastern Europe. Among tropical nations, Nigeria (US$60 billion), Democratic Republic of Congo ($60 billion), Vietnam (60), Ethiopia (58), India (56), Tanzania (52), and Bangladesh (45) have received the largest total disbursements over the past 25 years.

Apart from ODA, developed nations have also forgiven significant portions of loans owed to them and the multi-lateral financial institutions, such as the World Bank, IMF and other lenders that they underwrite. Since 1993, nearly US$270 billion in loans to 78 different tropical nations have been restructured or permanently retired before full re-payment, a ten-fold greater level of debt relief than has been granted to OECD member Turkey (see Table 9.2). The largest recipients in the tropics of this relief include Mexico, Nigeria, Ivory Coast, Ethiopia, Nicaragua, DRC and Vietnam. Relief to extra-tropical nations is considerably smaller, although two countries are among the largest global recipients – Argentina (32 per cent) and the Russian Federation (17 per cent) – account for half of the total debt forgiven within this group.

The ultimate objective of ODA should be to catalyze progress towards sustainability, particularly economic self-sufficiency. Significant ODA investment in countries such as China (US$43 billion), Vietnam (60), India (56), Ghana (31) and even Poland (17) would suggest that public funding can lead to significant economic success. However, it is difficult to dis-entangle the flow of public funds from those arriving privately through FDI, trade and remittances. All of the above-mentioned countries and many others have benefited from net positive financial flows from at least one source of private funding. Asian economies that have experienced the most rapid economic growth and modernization over the past quarter-century, such as Malaysia, Vietnam, Thailand, China and even Singapore, have also been net recipients of private and public funds across all channels – FDI, trade, remittances, ODA and debt forgiveness, with a strong preference towards private funding when compared to some of the poorest nations in Africa and Latin America. At odds with the public money purposed towards supporting sustainable development are the massive flows of illicit finance circulating through the same recipient countries.

Secret funds

According to Global Financial Integrity (GFI), a non-governmental think-tank focusing on international financial transparency and trade probity, the flow of illicit finance is estimated to have accounted for 14 to 24 per cent of all trade conducted by developing countries between 2004 and 2015 (GFI 2017). Most of these illicit flows originate through trade misinvoicing. Misinvoicing arises primarily when the value or quantity of an exported or imported good is recorded in a way that does not reflect the true market value of the consignment. Typically, misinvoicing would be arranged between exporter and importer, sometimes through shared ownership – a process referred to as transfer pricing, between criminal enterprises or through collusion with corrupt government officials. The primary purpose of transacting these secret funds is to reduce or avoid tax payments through customs duties and resource royalties or to launder money acquired through criminal enterprise. Naturally, the principal impact of secreting value through misinvoicing trade between countries is a loss in national tax receipts and government revenues. This is particularly pronounced in tropical nations that remain heavily reliant on *ad valorem* revenues such as customs and excise duties and resource rent payments, rather than property and income taxes, to fund their public expenditures.

The estimated total net inflow of illicit money in tropical nations between 2004 and 2015 amounts to nearly US$2 trillion, three-fold greater than the estimated value into OECD nations but nearly four-fold less than believed to be distributed net of outflows to other extra-tropical countries.[7] The bulk of these net inflows can be attributed to fewer than a dozen tropical countries, including Panama (US\$488 billion), India (462), Indonesia (205), Thailand (192), Brazil (136), Philippines (130) and Mexico (103). The staggering sum estimated to flow into Panama is particularly anomalous given the size of its population and economy relative to the other countries. If estimates are reasonably accurate, nearly US\$14,000 per person in secret funds has flowed on average into Panama annually between 2005 and 2014. In India – the next highest net inflow recipient – the amount averages to a mere US\$37 per person per year. In Indonesia it is \$85 and Brazil receives on average around \$68. Why is Panama misinvoicing such a large amount? A report published by the Financial Action Task Force of Latin America, or GAFILAT, indicates that illicit funds flowing into Panama arrive largely in an attempt to flee taxes in other countries or are linked to drug trafficking, smuggling, corruption and other criminal offences (GAFILAT 2018). One conduit to entry for these funds would be the misinvoicing of legitimate imported goods. Seeking a haven from taxes is not necessarily criminal and there are many legitimate reasons that money flows into Panama and other tropical nations with offshore banking facilities, particularly where these are used to domicile holding companies that comply with national tax laws. Secret money is not necessarily illicit, but it is attempting to avoid something and this process by its very nature lacks transparency. Akin to the twentieth-century Swiss banking system, secrecy invariably invites illicit flows.

The growth in offshore banking has been particularly strong in small tropical countries where land area, natural resources and population are insufficient to generate meaningful economic growth. These nations lack the comparative advantage in bioproductivity of their regional neighbours due to their relatively small land area and populations. They have successfully jumped to an economy based principally on services and consisting mainly of intangible assets – attributes more commonly applied to most OECD nations – by providing tax-advantaged domicile to funds in return for fees. Offshore banking is an efficient means of spurring economic development in these small nations, assuming that the proceeds, like those from natural resources, are re-invested into productive and human capital that benefits the larger population. Banking does not require excessive use of freshwater or marine resources, nor create the usual dilemma of environmental impact that accompanies most overly successful tourism economies in small island states. But many OECD nations have taken issue with the offshore banking community since its facilities are often viewed as tax-avoidance schemes that deprive governments of revenue. Normally, misinvoicing deprives governments of an important source of funding, but this practice must be separated from offshore banking facilities that generate significant public income to host nations. Both, however, are intended to avoid public inspection for the purpose of taxation and so banking secrecy invariably attracts illicit funds. The critical question for small tropical nations is how they can find a sustainable path to development if this window of opportunity is shut altogether rather than reformed towards an accountable level of transparency. Shuttering of offshore banking by OECD nations does re-invite an image of colonial-era control on manufacturing and cash-crop mono-culturing aimed solely at the development of the controlling nation that stands to lose tax revenue. At the same time, publicly funded ODA and forgiven debt flow into many tropical countries unable or unwilling to expand public spending in support of improved standards of living. The consequence of this countervailing policy of "deny but support" development leaves most tropical nations with few clear avenues toward organic growth in their economies. They have become trapped in a new set of post-colonial "economic boxes".

Economic boxes

Economic boxes are both externally and internally reinforced barriers that prevent step-wise diversification of economies. As we have seen, economic diversification is key to long-term development because it allows for greater synergies to develop between a larger number of enterprises and attenuates the impact of economic shocks on overall investment, employment and public revenue. In developing tropical nations, these economic boxes include the following: (1) cash-crop dominance with little value-addition, (2) fossil fuel or mineral export dominance, often suffering from Dutch Disease and/or the resource curse, (3) runaway government corruption that curtails public re-investment and takes on unsustainable debt, (4) an inability to balance visible trade due to an over-reliance on foreign manufacturing and services, (5) a lack of foreign direct investment and heightened capital flight due to instability or concerns over the rule of law and/or (6) a chronic over-dependency on development assistance and debt relief. All of these alone or in combination also increase the allure of illicit funds. In most cases, deep income inequality arises and prospects for improvement to the lives of the poorest remain distant. Frontline public employees are paid poorly or deployed for private purpose, fostering corruption. Confined within these boxes, most populations have three choices: (1) tolerate the existing conditions, (2) attempt to break out through social, political and economic action, or (3) leave. Part of the reason tropical nations have a collective history of conflict since independence has been attempts to escape externally reinforced boxes through path #2 which, in the process, simply construct new, internal ones. For example, within a decade of Haitians gaining their independence from France in 1805, one of the revolutionary leaders had installed himself as emperor, built new castles and re-instituted a system of forced labour aimed at generating cash-crop income for his new regime. His "reign" did not last long, but it established a precedent that would be replicated across much of the post-colonial tropics at one stage or another and remains so today in some countries. For many, escaping the economic and social confines of their homeland only arrives by leaving it.

The flow of people

Migration has become somewhat of a sensitive topic, understandably so. Yet, modern humans have been migrating across our planet since our earliest ancestors first left Africa several hundred thousand years ago. European expansion globalized this process, accelerating exchange of people and cultures across all of the continents. Moving to seek a change in conditions and resources better capable of sustaining individuals and populations is of primacy in the natural world. Most animals that aggregate into herds, flocks, schools or pods will naturally seek out new sources when bioproductivity diminishes due to seasonal declines in GDD, sea temperatures, nutrients or moisture, much as Europeans did in response to their own regional constraints. An increase in population size only acts to increase migratory pressure since resources will be exhausted faster, assuming a relatively constant rate of consumption across all members. People migrating today look towards improved employment, nutrition, education, security, property tenure, healthcare and other types of intangible capital that some nations offer in much greater amounts than those they are leaving. After the onset of the Great Famine in 1845, Ireland's population had declined by nearly half from an all-time peak of 8.2 million by the time the Republic declared independence from Great Britain in 1921. The famine was caused by widespread infection by the Mexican fungus *Phytophthora infestans* which led to chronic failure in a genetically invariant variety of potato that had become a staple food for the rural poor. Nearly two million of Ireland's population migrated as a consequence, although other factors that led to widespread reliance on potato cultivation in the first place, such as land tenure, social mobility and wages, also urged people to move to the Americas, northern England and Australia.

The reasons that people migrate today are not that different than in the past – escaping famine and conflict, seeking new resources, education and security. In other words, moving in an attempt to reduce risk and increase resource availability. What has changed is the global population. Prehistoric migration occurred in the context of a global population of a few million. At the time of European colonization, the population was nearing 500 million and Ireland's great migration occurred when there were an estimated 1.25 billion humans. Today, migration is occurring across a planet with a population of 7.4 billion. Modern transportation is delivering migrants faster and across greater distances than it ever has and it has never been as easy for most migrants to explore the prospective environment of the recipient country, to maintain connections with people in their homelands through the internet once they have arrived and to remit money effortlessly through digital money transfers. These developments and others have accelerated the rate of global migration over the past 25 years. Between 1990 and 1995, the global population of migrants was growing by 5.3 per cent. By 2000, it was growing at 7.4 per cent, then in 2005, 10.4 per cent. By 2015, the number of people migrating was growing by 12.5 per cent.[8] In 1990, the United Nations estimate that there were just over 152.5 million migrants in the world. By 2017, this estimate stood at nearly 258 million.

The impact of tropical nations upon this accelerated global flow of people has been more significant than any other exchange. Taken collectively, the tropics have acted as the primary source of migrants over the past 25 years, supplying the overwhelming majority of arrivals in extra-tropical nations, particularly OECD member states (Figure 9.3). Migration within the region – mainly between adjoining countries – is comparable with patterns of movement between nations in the extra-tropics, but the flow between the two regions is strikingly disparate. By 2017, people moving to the tropics from the extra-tropics accounted for only 8 per cent of the total flows, with the bulk of migrants leaving tropical countries for

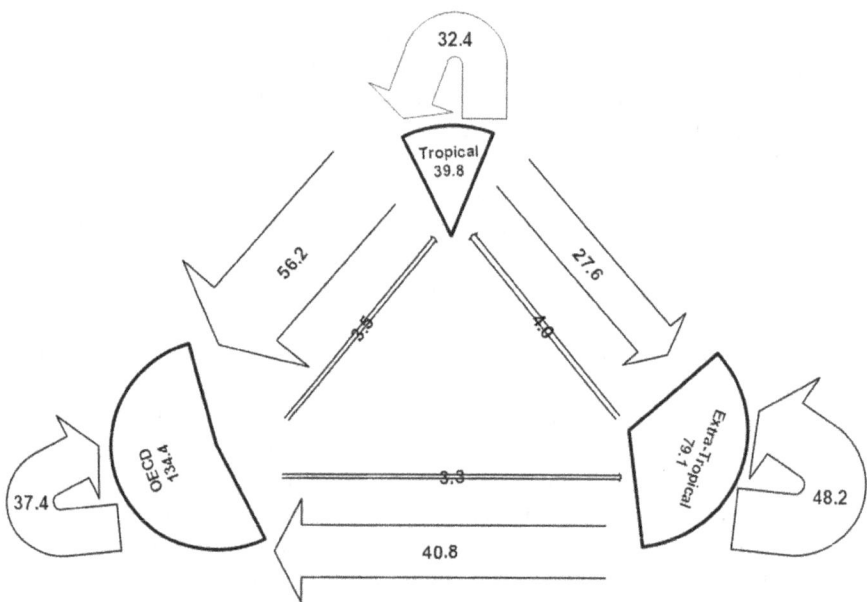

Figure 9.3 Stock and flow of "lifetime" international migrants between countries in the tropics (TCs) and those in the extra-tropics (ETCs, OECD).

Note
Numbers in millions.

those in the OECD, primarily North America and Europe (UNDESA 2017). This pattern of net migration out of the tropics has remained consistent over the last quarter-century. More than half of the world's migrants, about 134 million, now reside in OECD nations with about 42 per cent of these arriving from tropical countries. This pattern of poleward movement of people occurs across our planet, but not uniformly. A much larger fraction of the population in the neotropics has migrated to the extra-tropics than from Africa, South and South-East Asia or Polynesia. This is most pronounced in the emigrant flows from the small nations in the Caribbean and Central America where nearly 20 per cent and 10 per cent of the resident populations are living in another country, almost entirely in the United States, Canada and Western Europe. In South America, emigrants equate to only around 3 per cent of the resident population and are more evenly distributed between other tropical countries, the OECD and the temperate southern hemisphere. Guyana and Suriname, however, register some of the highest emigration ratios with over three Guyanese living overseas for every five residents and just over half as many Surinamese residents live elsewhere as live in the home nation. Belizeans, Grenadians, Jamaicans, Puerto Ricans, Trinidadians and El Salvadoreans likewise register high proportions, with at least one in five nationals living in an OECD country in 2017. Dominica's emigrant population is as large as that residing on the island itself and Montserrat's is 3.5 times the number, although the latter is largely attributable to the devastating 1995 eruption of the Soufrière Hills volcano and near total destruction of the capital city of Plymouth. Analysis of the UN data on migrant stocks identifies the northern hemispheric Americas – between 0 and 50 degrees north – as the most integrated trans-national region on the planet based on emigrant proportions. This is almost entirely a uni-directional movement out of the tropics.

In terms of its total migrant stock of nearly 36 million, emigration from the neotropics is twice the size of that from sub-Saharan Africa, but amounts to less than three-quarters of the number emigrating from South and South-East Asia. Despite a major increase in the rate of emigration of Africans across the Mediterranean since 2010, less than 1 per cent of the sub-Saharan African population, around four million, had emigrated to OECD nations by 2017. Departures from tropical Asia and Melanesia were equally low as a fraction of the resident population, accounting for a mere 2 per cent with more than half of the 16.3 million settling in the OECD arriving from India, Malaysia, Philippines and Vietnam. Emigration from Asia and Africa is less concentrated on OECD nations than in the Americas. In sub-Saharan Africa, movement between tropical nations occurs more than twice as often as migration to the extra-tropics, while departures from tropical Asia are 3.5 times greater to extra-tropical destinations, but with less than half of these being OECD members. Nearly 20 million departed for other extra-tropical countries, particularly in the Middle East where there is a high demand for construction and service sector workers.

The influx of migrants from the tropics has transformed the demographics of some extra-tropical nations while barely altering that of others. As a fraction of the total population, four OECD nations stand out in particular. The migrant community in the United States is the largest in the world and accounts for a higher fraction – nearly 15 per cent – of the total population than anywhere else. Canada, Australia and New Zealand also register double-digit representation of people that were foreign born inside the tropics. Seven European nations register migrant populations at or about 5 per cent of their broader population, including the UK, Switzerland, Sweden, Spain, Portugal, Norway and the Netherlands. But beyond these 11 extra-tropical nations, there are few others that reach these proportions. Most notable is the extremely low representation of tropical migrant communities in north Asian nations of China, Korea and Japan. There, migrants from tropical nations – mainly Indonesia, Bangladesh, Malaysia and Thailand and Brazil, account for less than 0.1 per cent of their total populations, despite registering similar population densities to those found in many Western European nations.

A braided flow

There are both positive and negative consequences of immigration to both originating and recipient nations. The most salient of these revolve around work, security, finance and energy.

The motive to move from the nation of our birth can take many forms, but the prospect of employment and often a nearly instantaneous rise in standard of living appears to loom large in the decision of most prospective migrants. But there are many other reasons too, stemming from family and marriage, health and retirement, and property and investment decisions to a dire need to escape intense and life-threatening violence, deep and institutionalized prejudice or systemic hunger and disease. Not all immigration is economic and not all migrants are refugees. But beyond the underlying motive to move are the practical implications of migration to the economic and social conditions in source and sink countries.

So what are the consequences to the source and recipient nations? In many OECD countries, the host nation has become the de facto agent responsible for developing a source nation's intangible capital value with the view that a fraction of this value will be monetized and repatriated through remittances. The costs of education and training, health, security and legal rights protection, among other factors of intangible capital production are in effect "outsourced" to the host nation in return for an increased supply of labour. Nations such as the Philippines, El Salvador, Mexico and Bangladesh have institutionalized the emigration of workers with the view to increasing remittance income and using its financial power to leverage development and spur economic growth. Remittances add to the cost of "outsourcing" by adding the opportunity cost of lost savings and consumption to the intangible capital cost of supporting immigrants in the recipient country. Moreover, not all remittances are necessarily derived from earned income, particularly where migrants are able to borrow funds in the recipient country and transfer the principal to their home nation in support of family, to pay debts or to invest (Gammage and Schmitt 2004). But the positive impact that migration has on alleviating poverty pressure through overseas employment and remittances also can have a detrimental impact on sustainable development. The remittance of funds back home also can stimulate further demand for emigration from the migrant's home country – a process referred to as chain migration. Part of the accelerated rate of emigration from the tropics over the past 25 years is attributable to this effect, as the growing pool of migrants attracts even more migrants through social connections and the heralding of economic success stories.

Increased migration followed by increasing remittances and increasing ODA does not foster greater national economic independence or diversification. Allowing another nation to take responsibility for the development of its citizens mitigates the need to follow Hartwick's Rule and re-invest proceeds of natural resource rents and commodity exports towards long-lasting increases in their nation's intangible capital account. This dynamic increases the politically tolerable levels of allowable corruption. Assistance and debt forgiveness should aim to provide "breathing space" for national governments to reform. Normally, the risk of financial illiquidity is enough for reforms to be undertaken. But if deep, underlying flows of remittances and illicit finance compensate for risk to a nation's visible balance of payments, then there is little incentive for corrupt governments to exercise greater financial integrity.

National governments have heralded the flow of public funds to under-developed countries as a necessary measure in combating poverty, hunger and economic underperformance for decades. Since 1970, the United Nations has suggested that OECD nations deliver 0.7 per cent of their GNP as ODA. Only Sweden, Denmark, Norway, the Netherlands and the UK claim to have achieved this aid target. Most modern developed nations regularly incorporate ODA into their annual budgets and many allocate significant administrative and political resources to its disbursement, but decisions made to commit funding to various countries and programmes remain broadly uncoordinated, decision-making opaque and the purpose open to significant mis-direction based on special interests. More importantly, much of the allocated finance is too large, outstripping the availability of meaningful in-country support projects, particularly when these are structured around typical timeframes for completion of less than three years. Most of the poorest or smallest tropical nations lack the institutional capacity and trained counterparts to effectively implement development assistance projects towards meaningful conclusions. In some instances, development assistance has taken the form of simple cash payments to eligible families, echoing the flow of money via remittances with the similar risk of distortion to local economies.

Para-colonies and quasi-colonies – is this the future of globalization?

Much of the narrative surrounding globalization swirls around a central pillar of greater distribution of wealth and improved governance and environmental stewardship. Yet, as I compare what we know through the lens of history with the modern movement of resources, finance and people, I cannot help but conclude that much of the tropics has not noticeably changed its relationship with the extra-tropics. Many of the most economically expedient and morally unjustifiable practices have been officially extinguished but, despite vast sums having been pumped into tropical countries via various public and private sources of funding, most of the tropics continues to house the economically poorest part of our global population. While incomes and standards of living have improved in many countries, the gap between the tropics and extra-tropics has not noticeably narrowed. Agriculture remains dominated by cash-crop practices not far removed from those once utilized in colonial plantation systems, employing workers at the most basic level, but not expanding employment through value-adding industries that could make more of the tropics primary comparative advantage – bioproductivity. This lack of expansion may be a function of the young age of most former colonies and given time, they will begin to show the promise that is implicit in their bioproductive potential. But the flow of money and people tells us that something more insidious looms over many nations in the region, focused on capital flight, illicit funds and chronic net emigration. Unlike pre-twentieth-century Europe, most of these flows are not due to bouts of famine, but to a perpetual lack of re-investment by governments that often control the overwhelming bulk of national natural resources. Underlying the challenges sits an immoveable object that must be re-shaped if tropical nations are ever to reach their full potential – the social contract.

A social contract unfulfilled – hysteresis or wild-west?

We have seen how Europeans were isolated geographically from the high bioproductivity of the tropics and suffered recurrent setbacks in developing an agricultural foundation capable of sustaining population growth up to the mid-eighteenth century. I presented evidence in the previous chapters describing the important role public debt has played in the development of European economies and the symbiotic relationship that national credit formed with global economic expansion and control of tropical bioproductivity. We have also seen how this history of colonial control and the happenstance of geology have given birth to a large number of relatively small tropical nations with a short lifetime of self-rule and a predisposition towards armed conflict. Many elements of the culture and social undertakings of the former inhabitants were relegated to subsidiary components of a new nationality based on European institutions and geographies. The overall effect in most instances has been a dampening of economic size relative to OECD nations of similar land area and population. Just as a black hole disrupts the fabric of space-time, hysteresis continues to exert an invisible influence over economic development in tropical nations. The well-known French sociologist Pierre Bourdieu viewed this hysteresis effect as the collective impact of a past way of doing things on the ability to manage current conditions – a lingering of past ways of behaving or functioning once a new set of conditions has arisen (Bourdieu 1977). More broadly, hysteresis is a structural lag in the cause–effect relationship. It often exerts its greatest influence through arcane institutional arrangements and legal frameworks that remain in place, but are not fit-for-purpose towards sustainable development. But as time passes, so too should have relative effects of hysteresis. Modern institutions have also failed in tropical nations due not to some strange attractor tugging relentlessly towards the past, but to a lack of pro-public decision-making by the individuals tasked with their stewardship. In both cause and consequence there is a failure of the social contract.

A war of all against all

Thomas Hobbes was one of the earliest political thinkers to take on the question of governance and its role in avoiding catastrophic conflict or what he referred to as the *summum malum* – the greatest evil – of humanity. While Hobbes' *Leviathan* in many ways would not resonate with modern notions of democratic governance – for example, he resolutely defended the absolute rule of monarchs over both state and Church and decried any rights of subjects to change the form of government – the underlying treatise was that governance, or Commonwealth as he referred to it, was in all of its forms a measured balance between the rights and responsibilities of both sovereign and subjects. In effect, a social contract has been agreed upon and both parties need to stick to the terms of the agreement. Violence and its worse manifestation, "a war of all against all", he argued would occur when one or both parties did not adhere to the terms agreed since this was the natural state of humanity. For the time – *Leviathan* was published in 1651 – this was a progressive work and one of the earliest attempts at formalizing social contract theory. Other social philosophers – Locke, Rousseau, Kant – would later build upon and refine many of Hobbes' earlier notions. In particular, they would thoroughly reject his promotion of monarchy as the simplest – and therefore superior

– form of government. They would argue that subjects have natural rights ("inalienable rights" in the US Declaration of Independence; "human rights" in the UN Convention) and that democratic forms of governments were the best means of ensuring these rights were respected. Locke's and Rousseau's theories were particularly provocative for the time, given the form of government that prevailed in Great Britain and France during the eighteenth century. They would ultimately act as a back-drop to the upheaval that would shake the political establishment to its core in both North America and Europe.

Yet, none considered the plight of the tropics at the time nor ascribed concepts of social contract, natural rights or democratic governance to the people of the region. There were practical reasons for this, despite the realities of the cruel trade in labour and commodities that dominated the social and economic conditions of tropical lands. Their revolutionary ideas concerning labour, property and legal rights were, like the signing of the Magna Carta by King John 400 years earlier, a controlled dismantling of the political tree commencing with the leaf and twig but never progressing much further than the branch. This remains so, despite the fact that a lack of universal adherence to the social contract could be considered as the true root cause of many of the chronic problems plaguing nations today. The rights of those that can afford to defend them have improved, but for those subjects absent the wherewithal to do so, not a great deal has changed.

The adoption of much of Locke's and Rousseau's political thinking was arguably facilitated by the frenzied expansion of competing European interests into the tropics and the rapacious trade in commodities that ensued. In order for their political concepts to gain traction, nations needed the economic boost that would allow the lower economic classes of these nations to find their way into the higher ones without disrupting the existing allocation of capital that was largely in the grasp of a small number of well-connected families. The vast territories in the tropics presented the perfect means to expand the economic capital necessary to afford both upper and lower classes the opportunity to further their economic fortunes without seeking a re-balancing of the inequitable distribution of wealth within their own countries. It is said that high tides lift all ships, but it is also true that for some tides to rise, other must fall and as the economic fortunes of Europeans rose as a consequence of the tropical trade, the circumstances of indigenous societies, no matter how brutal or inequitable they were traditionally, could hardly be worse. It is not surprising that in the aftermath of the American and French revolutions little changed in their dealings with their own tropical possessions and the business they conducted with those of other European nations. The costs of meeting the social contract in enlightened societies of Western Europe and North America were, in part, simply transferred to indigenous peoples with whom no such implicit agreements needed to be agreed.

The social contract

The list of tropical forest nations that have suffered through years, sometimes decades, of conflict over natural resources is very long. Often it is the monopoly of financial benefits by the ruling class that is the spark. At a given pressure point, the all-important social contract breaks, violence erupts and civil loss ensues. This social contract – fundamental to the continued sovereignty of most nations – is an implicit agreement of its citizens to abide by rules and obligations set out by the government on the understanding that these are designed and implemented towards their collective prosperity and long-term security. It is couched in the social and political philosophy of men such as Bacon, Hobbes, Locke and Rousseau. They unlocked the secret of political stability – seeking not to violate the "natural rights" of a nation's subjects while accepting that the mission of governance must be to further their long-term well-being. The seventeenth-century political philosopher Thomas Hobbes set out the early framework on which the modern social contract is assembled through his notion of Commonwealth:

> The only way to erect such a Common Power, as may be able to defend them from the invasion of Forraigners, and the injuries of one another, and thereby to secure them in such sort, as that by their owne industrie, and by the fruites of the Earth, they may nourish themselves and live contentedly; is, to conferre all their power and strength upon one Man, or upon one Assembly of men, that may reduce all their Wills, by plurality of voices, unto one Will: which is as much as to say, to appoint one man, or Assembly of men, to beare their Person; and every one to owne, and acknowledge himselfe to be Author of whatsoever he that so beareth their Person, shall Act, or cause to be Acted, in those things which concerne the Common Peace and Safetie; and therein to submit their Wills, every one to his Will, and their Judgements, to his Judgment. This is more than Consent, or Concord; it is a reall Unitie of them all, in one and the same Person, made by Covenant of every man with every man ... This done, the Multitude so united in one Person, is called a COMMON-WEALTH.
>
> (*Part II, Chapter XVII, Leviathan*)

John Locke and Jean-Jacques Rousseau, principal among others, built upon and refined the central ideas of Hobbes' *Leviathan*, fundamentally shaping modern Western political and legal systems.[9] Failure to make good on the most basic promises to "defend them from invasion ... and injuries ... and to secure them ... [through] their owne industrie, and fruites of the Earth, [the ability to] nourish themselves and live contentedly" inevitably provokes violent dissent and this worryingly happens more often in tropical nations than anywhere else. For a number of reasons, the way that natural resources, or as Hobbes put it "fruits of the Earth", are managed and who stands to benefit from their use in these countries now rests much closer to the centre of the social contract than in extra-tropical countries.

Sadly, most tropical nations rank very poorly in terms of corruption and, by proxy, the effort made by their government to fulfil obligations towards the social contract. According to the Corruption Perception Index (CPI) scores assigned to countries in 2017 by Transparency International, tropical nations as a group are perceived to be twice as corrupt, on average, as members of the OECD and 20 per cent more corrupt than other extra-tropical countries.[10] The average CPI score for tropical nations does mask the relatively high scores of several members. Many of these are islands with very small GDP contributions from minerals, fossils and agriculture and relatively large service sectors, such as Barbados, Dominica, St Lucia, Grenada and the Seychelles but they also include countries such as Botswana and Costa Rica. Many of these nations rank higher than several of the least well-perceived OECD nations, such as Turkey, Hungary and Poland. Oddly, when comparing the levels of ODA distributed against the CPI scores of recipient nations in the tropics, there appears to be an increase in the funding level with a decline in CPI score, or an increase in the level of perceived corruption.[11]

The social contract is interminably woven with the stewardship of natural resources since these account for the bulk of public revenue, economic activity and household livelihoods in the majority of tropical nations. Whether they are expressed as the fraction of GDP (Figure 9.1), the proportion of employment, amount of public revenues or the relative size of the rural population, it is clear that a failure to steward the use of agricultural land, forests, water and minerals towards meaningful improvements of social conditions leads to conflict and large-scale emigration. Neither is conducive to the accrual of the physical and intangible capital needed to facilitate broader participation in a globalized marketplace. Conflict destroys capital without transforming it into something else and excessive emigration – despite the significant remittance inflows – deprives most tropical nations of many of their most highly skilled, trained and intellectually gifted citizens that are often the best drivers of innovation and business development. Remittances can also lead to false economies at home, propped up by the strength of the source economy and simply leading to price inflation in the recipient economy.

The capital outflows from OECD nations in the form of ODA, forgiven debt and remittances have been substantial over the past quarter-century. In part, the intent of these flows is to buffer the lack of FDI and negative export balances in services that most tropical nations have accrued since independence. But it is in the interest of developed and developing nations alike to find a path towards eliminating subsidies and curtailing strong, uni-directional flows of resources, money and people. Dynamic, bi-directional flows that typify exchange between many advanced economies enhance economic performance by filling capital deficiencies as they arise in relation to demand. Advancing sustainable agro-economies through better technical and business management practices should be a priority since the ability to grow plants and identify new biological products remains the singular greatest advantage that the tropics retain over extra-tropical nations, where land is more abundant, physical and intangible capital is accrued more easily and marine productivity is higher. Derivative industries, capable of manufacturing products from agricultural and forest-based materials, need to proactively make markets, stimulate demand for their produce and scale-up to a level capable of supplying regional and global consumers. This requires capital investment, and ODA should be directed towards enhancing the capabilities of this sector and the ancillary sectors needed to achieve sustainability. Good governance and institutional integrity are unavoidable preconditions to any progress and governments capable of demonstrating an ability to maintain the social contract with their citizens should be given the highest priority in the allocation of public-private investment. Sadly, this does not currently appear to always be the case and the prospects of sustainable development in these nations remain beleaguered by counter-productive flows, latent effects of colonialism and high levels of corruption and poor governance, often facilitated by a lack of consensus and coordinated effort by the international community towards the establishment of a social contract process. I acknowledge that this is easier said than done, but neutrality by multilateral and bi-lateral institutions in the face of public mal-stewardship of resources and mis-allocation of financial proceeds must change before the natural patrimony of tropical nations is terminally squandered with little social, human or physical capital to show for it.

Notes

1 1 kcal = 4.184 kilojoules = 1 "calorie".
2 Annual sugar consumption statistics as described in Chapter 7. Wheat yields, acreages and arable land in Great Britain are based on multiple sources covering the same period. Where possible, averages are taken of these estimates. See Overton and Campbell (1996), Glennie (1988), Allen (2005).
3 Based on EUROSTAT statistics for 2009–2015 and World Cocoa Foundation reports.
4 Estimates of packaged tea value are based on sales of black tea from the 20 largest companies globally, including subsidiaries owned by Unilever (Lipton, Lyons, PG Tips, Scottish Blend, Red Rose Canada), Tata Group (Tetley, Tata Tea), Twinings, Typhoo, Wissotzky, Akbar, Nestle, Bettys & Taylors (Taylors, Yorkshire Tea), Çaykur and Madhu Jayanti International. Group assignment is based on national location of headquarters.
5 Based on data from FAOSTAT (FAO 2016b) and the World Bank (2017).
6 Portfolio funds are not included here since these can be highly volatile or short-lived. Illicit or hidden financial flow values are for the period 2004–2014 only and represent the mid-point between GFI's low and high inflow and

outflow estimates to arrive at a net median inflow. Values for all other sources of funds are from the WBDI database and represent annual data between 1993 and 2017. They have been re-valued into 2017 US dollars using the WBDI GDP deflator, a measure of inflation. Note that annual data for all countries are not available for all years, but in most variables the missing values are for small economies making up less than 3 to 5 per cent of global value.

7 Net inflows here represent the median of differences between low and high inflows and outflows. As such, they are broad estimates and the magnitude of difference between groups is more informative rather than the specific values.

8 Data are from Table 1 of the UN Population Division's compilation of international migrant stocks for 2017 (UNDESA 2017). A migrant is someone living for a year or longer in a country other than their place of birth. Consequently, migrant numbers used here include foreign workers and students, but not tourists, foreign-aid workers, temporary or seasonal workers or military personnel.

9 Hobbes' *Leviathan* (1651), Locke's *Essay Concerning Human Understanding* (1689) and *Two Treatises of Government* (1690) and Rousseau's *Discourse on the Origin and Basis of Inequality Among Men* (1754) and *The Social Contract, or Principles of Political Right* (1762) are compulsory reading for anyone that wishes to understand our modern systems of government and how these are linked, *inter alia*, to a nation's natural resources.

10 Methods underlying the calculation CPI scores can be accessed at www. transparency.org.

11 Using the net development assistance for 2017 from the World Bank Development Indicators database.

References

Acemoglu, D. and J.A. Robinson. 2012. *Why nations fail: the origins of power, prosperity and poverty*. Profile Books Ltd, London.

Achard, F.C. 1809. *Die europäische Zuckerfabrikation aus Runkelrüben, in Verbindung mit der Bereitung des Brandweins, des Rums, des Essigs und eines Coffee-Surrogats aus ihren Abfällen*. Verlag Bartens, Berlin.

Adams, F.U. 1847. *The seven books of Paulus Aegineta translated from the Greek: with a commentary embracing a complete view of the knowledge possessed by the Greeks, Romans, and Arabians on all subjects connected with medicine and surgery*. The Sydenham Society, London.

Adams, F.U. 1914. *Conquest of the tropics. The story of the creative enterprises conducted by the United Fruit Company*. Doubleday, Page & Co., New York.

Alfani, G. and C. Ó Gráda (eds). 2017. *Famine in European history*. Cambridge University Press, Cambridge.

Alfani, G. and C. Ó Gráda. 2018. The timing and causes of famines in Europe. *Nature Sustainability* 1: 283–288.

Allen, M.F. 1991. *The ecology of mycorrhizae*. Cambridge University Press, Cambridge.

Allen, R.C. 2005. *English and Welsh agriculture, 1300–1850: output, inputs, and income*. Unpubl. Paper, Nuffield College.

Anderson, A. 1764. *An historical and chronological deduction of the origin of commerce: from the earliest accounts to the present time. Containing, an history of the great commercial interests of the British Empire. Vol. I–IV*. Millar, Tonson, Rivington, Baldwin, Johnston, Hawes, Clarke, Collins, Longman, Dodsley and Horsfield, London.

Anon. 1713. *The assiento; or contract for allowing to the subjects of Great Britain the liberty of importing negroes into the Spanish America. Sign'd by the Catholick King at Madrid, the twenty fixth day of March, 1713. By Her Majesties special command*. John Baskett, London.

Anon. 1732. *Some considerations humbly offer'd upon the bill now depending in the House of Lords, relating to the trade between the Northern Colonies and the Sugar-Islands. In a letter to a noble peer*. N.p., London.

Anon. 1811. *The Parliamentary history of England from the earliest period to the year 1803. Vol. VII. A.D. 1714–1722*. T.C. Hansard, London.

Anon. 1844. *Accounts and papers of the House of Commons. Vol. 19. Slave trade; instructions to naval officers. Session 1 February–5 September 1844*. N.p., London.

Anon. 1876. Sugar statistics. *The Sugar Cane* 8: 47.

Ashworth, W.J. 2003. *Customs and excise: trade, production and consumption in England 1640–1845*. Oxford University Press, Oxford.

Baker, C. and P. Phongpaichit. 2009. *A history of Thailand.* 2nd ed. Cambridge University Press, Cambridge.

Beer, C., M. Reichstein, E. Tomelleri, P. Ciais, M. Jung, N. Carvalhais, C. Rödenbeck, M.A. Arain, D. Baldocchi, G.B. Bonan, A. Bondeau, A. Cescatti, G. Lasslop, A. Lindroth, M. Lomas, S. Luyssaert, H. Margolis, K.W. Oleson, O. Roupsard, E. Veenendaal, N. Viovy, C. Williams, F.I. Woodward and D. Papale. 2010. Terrestrial gross carbon dioxide uptake: global distribution and covariation with climate. *Science* 329: 834–840.

Behrenfeld, M.J. and P.G. Falkowski. 1997a. Photosynthetic rates derived from satellite-based chlorophyll concentration. *Limnology and Oceanography* 42: 1–20.

Behrenfeld, M.J. and P.G. Falkowski. 1997b. A consumer's guide to phytoplankton primary productivity models. *Limnology and Oceanography* 42: 1479–1491.

Beveridge, W.I.B. 1991. The chronicle of influenza epidemics. *History and Philosophy of the Life Sciences* 13: 223–234.

Black, F.L. 1992. Why did they die? *Science* 258: 1739–1740.

Board of Trade. 1843. *Commercial tariffs and regulations of the several states of Europe and America.* Part VI. *Holland.* C. Whiting, London.

Board of Trade. 1869. *Annual statement of the trade and navigation of the United Kingdom with foreign countries and British possessions in the year 1868.* G.E. Eyre & W. Spottiswoode, London.

Bolton, R. 1979. On coca chewing and high-altitude stress. *Current Anthropology* 20: 418–420.

Bourdieu, P. 1977. *Outline of a theory of practice.* Cambridge University Press, Cambridge.

Boxer, C.R. 1957. *The Dutch in Brazil 1624–1654.* Clarendon Press, Oxford.

Braudel, F. 1982. *Civilization and capitalism 15th–18th century. Vol. II. The wheels of commerce.* Collins, London.

Braudel, F. 1984. *Civilization and capitalism 15th–18th century. Vol. III. The perspective of the world.* Collins, London.

Bray, W., C. Dollery, G. Barnett, R. Bolton, F. Deltgen, D. Dufour, J.M. Hanna, A. Henman, T.C. Lewellen, M.A. Little, E. Picón-Reátegui, A. Fuchs, L.P. Spear, V. Valiente and T.G. Vitti. 1983. Coca chewing and high-altitude stress: a spurious correlation [and comments and reply]. *Current Anthropology* 24: 269–282.

Brennan, E.B. and N.S. Boyd. 2012. Winter cover crop seeding rate and variety affects during eight years of organic vegetables: i. cover crop biomass production. *Agronomy Journal* 104: 684–698.

Bridgewater, S.G.M., N.C. Garwood, H. Deplooy, H.P. Morgan and N. Wicks. 2007. Belize's *Chamaedorea* conundrum. *Palms* 51: 187–196.

Bridgewater, S.G.M., P. Pickles, N.C. Garwood, M. Penn, R.M. Bateman, H.P. Morgan, N. Wicks and N. Bol. 2006. *Chamaedorea* (Xaté) in the Greater Maya Mountains and the Chiquibul Forest Reserve, Belize: an economic assessment of a non-timber forest product. *Economic Botany* 60: 265–283.

Briffa, K.R., P.D. Jones and M. Hulme. 1994. Summer moisture variability across Europe based on the Palmer drought severity index. *International Journal of Climatology* 14: 475–506.

Briggs, C. 2008. The availability of credit in the English countryside, 1400–1480. *Agricultural History Review* 56: 1–24.

Broadberry, S., B. Campbell and B. van Leeuwen. 2008. *English agricultural output 1250–1450: some preliminary estimates.* Unpublished MS.

Broadberry, S., B. Campbell, A. Klein, M. Overton and B. van Leeuwen. 2011. *British economic growth, 1270–1870: an output-based approach.* Unpublished MS.

Brochado, J.P. 1977. *Alimentaçao na floresta tropical.* Caderno no. 2. Universidade Federal do Rio Grande do Sul, Instituto de Filosofia e Ciências Humanas, Pôrto Alegre, Brazil.

Brown, S.S. 2006. Can remittances spur development? A critical survey. *International Studies Review* 8: 55–75.

Brundrett, M.C. 2009. Mycorrhizal associations and other means of nutrition of vascular plants: understanding the global diversity of host plants by resolving conflicting information and developing reliable means of diagnosis. *Plant and Soil* 320: 37–77.

Burnett, D.G. 2000. *Master of all they surveyed. Exploration, geography and a British El Dorado.* University of Chicago Press, Chicago.

Burns, E.B. 1970. *A history of Brazil.* Columbia University Press, New York.

Butel, P. 2002. *The Atlantic.* Routledge, New York.

Cabo Alonso, A. 1998. Formación histórica de las dehesas. Pages 15–42 in C.G. Hernández Díaz-Ambrona (ed.), *La dehesa. Aprovechamiento sostenible de los recursos naturals.* Editorial Agrícola, Madrid.

Campbell, B.M.S. 1988. Towards an agricultural geography of medieval England. *Agricultural History Review* 36: 87–98.

Campbell, B.M.S. 2006. *English seigniorial agriculture, 1250–1450.* Cambridge University Press, Cambridge.

Carmichael, A.G. and A.M. Silverstein. 1987. Smallpox in Europe before the seventeenth century: virulent killer or benign disease? *Journal of the History of Medicine and Allied Sciences* 17: 147–168.

Carswell, J. 1960. *The south sea bubble.* Stanford University Press, Stanford.

Carter, P. 1987. *The road to Botany Bay. An essay in spatial history.* Faber & Faber, London.

Cartwright, D.E. and R.D. Ray. 1991. Energetics of global ocean tides from Geosat altimetry. *Journal of Geophysical Research* 96: 16897–16912.

Chang, M. and B.D. Tapley. 2004. Variations in the Earth's oblateness during the past 28 years. *Journal of Geophysical Research* 109: B09402. doi:10.1029/2004 JB003028.

Clarence-Smith, W.G. and S. Topik (eds). 2003. *The global coffee economy in Africa, Asia, and Latin America 1500–1989.* Cambridge University Press, Cambridge.

Clark, G. 1999. Too much revolution: agriculture in the industrial revolution. Pages 206–240 in J. Mokyr (ed.), *The British industrial revolution: an economic perspective.* Westview Press, New York.

Clark, G. 2004. The price history of English agriculture, 1209–1914. *Research in Economic History* 22: 41–124.

Clark, G. 2009. *A farewell to alms: a brief economic history of the world.* Princeton University Press, Princeton.

Clarke, G. and S. Wallsten. 2004. *Do remittances act like insurance? Evidence from a natural disaster in Jamaica.* World Bank, Washington DC.

Cleland, W. 1719. *The present state of the sugar plantations consider'd; But more especially that of the island of Barbadoes.* J. Morphew, London.

Clement, C.R. 1999. 1942 and the loss of Amazonian crop genetic resources. I. The relation between domestication and human population decline. *Economic Botany* 53: 188–202.

Coates, A.G. and J.A. Obando. 1996. The geologic evolution of the Central American isthmus. Pages 21–56 in J.B.C. Jackson, A.F. Budd and A.G. Coates (eds), *Evolution and environment in tropical America*. University of Chicago Press, Chicago.

Coe, S.D. and M.D. Coe. 1996. *The true history of chocolate*. Thames & Hudson, London.

Colledge, S. and J. Conolly (eds). 2007. *The origins and spread of domestic plants in southwest Asia and Europe*. Routledge, London.

Cortés, Hernán. 1485–1547. *Cartas y relaciones de Hernan Cortés al emperador Carlos V*. Edited by Pascual de Gayangos. 1866. A. Chaix, Paris.

Costanza, R., J. Cumberland, H. Daly, R. Goodland and R. Norgaard. 1997. *An introduction to ecological economics*. St. Lucie Press, Boca Raton.

Cózar, F.E., J.I. González-Gordillo, X. Irigoien, B. Ïbeda, S. Hernández-León, A.T. Palma, S. Navarro, J. García-de-Lomas, A. Ruiz, M.L. Fernández-de-Puelles and C.M. Duarte. 2014. Plastic debris in the open ocean. *Proceedings of the National Academy of Sciences* 111: 10239–10244.

Crafts, N. and N. Wolf. 2013. The location of the UK cotton textiles industry in 1838: a quantitative analysis. *EHES Working Papers in Economic History* 45: 1–26.

Crawford, J. 1852. History of coffee. *Journal of the Statistical Society of London* 15: 50–58.

Crosby, A.W. 1972. *The Columbian exchange: biological and cultural consequences of 1492*. Greenwood Press, Westport.

Crossland, C.J., B.G. Hatcher and S.V. Smith. 1991. Role of coral reefs in global ocean production. *Coral Reefs* 10: 55–64.

D'Alpoim Guedes, J., G. Jin and R.K. Bocinsky. 2015. The impact of climate on the spread of rice to north-eastern China: a new look at the data from Shandong province. *PLoS ONE* 10: e0130430.

Dale, R. 2004. *The first crash: lessons from the South Sea Bubble*. Princeton University Press, Princeton.

Dale, R.S., J.E.V. Johnson and L. Tang. 2005. Financial markets can go mad: evidence of irrational behavior during the South Sea Bubble. *Economic History Review* 18: 233–271.

Dargie, G.C., S.L. Lewis, I.T. Lawson, E.T.A. Mitchard, S.E. Page, Y.E. Bocko and S.A. Ifo. 2017. Age, extent and carbon storage of the central Congo Basin peatland complex. *Nature* 542: 86–90.

Davies, P.N. 1990. *Fyffes and the banana: Musa sapientum. A centenary history 1888–1988*. Atholone Press, London.

Dawkins, H.C. and M.S. Philip. 1998. *Tropical moist forest silviculture and management. A history of success and failure*. CAB International, Wallingford.

Deerr, N. 1950. *The history of sugar. Vol. I–II*. Chapman and Hall, London.

Del Grosso, S., W. Parton, T. Stohlgren, D. Zheng, D. Bachelet, S. Prince, K. Hibbard and R. Olson. 2008. Global potential net primary production predicted from vegetation class, precipitation, and temperature. *Ecology* 89: 2117–2126.

Diamond, J. 1997. *Guns, germs and steel: the fate of human societies*. W.W. Norton & Co., New York.

Diaz, R.J. and R. Rosenberg. 2008. Spreading dead zones and consequences for marine ecosystems. *Science* 321: 926–929.

Dodsley, J. 1800. *The annual register, or a view of the history, politics and literature for the year 1795*. T. Burton, London.

Dowell, S. 1888. *A history of taxation and taxes in England from the earliest times to the year 1885.* Longmans & Green, London.

Draper, F., K.H. Roucoux, I.T. Lawson, E. Mitchard, E. Honorio Coronado, O. Lahteenoja, L. Torres Montenegro, E. Valderrama Sandoval, R. Zarate and T. Baker. 2014. The distribution and amount of carbon in the largest peatland complex in Amazonia. *Environmental Research Letters* 9: 124017. doi:10.1088/1748-9326/9/12/124017.

Dupâquier, J. 1988. *Histoire de la population française, t. 1: des origines à la Renaissance.* Presses Universitaires de France, Paris.

Duyvendak, J.J.L. 1939. The true dates of the Chinese maritime expeditions in the early fifteenth century. *T'oung Pao* 34: 341–413.

Ebert, C. 2008. *Between empires: Brazilian sugar in the early Atlantic economy, 1550–1630.* Brill, Leiden.

Edwards, B. 1743. *The history civil and commercial of the British colonies in the West Indies.* L. White, Dublin.

Edwards, B. 1798. *The history civil and commercial, of the British colonies in the West Indies, to which is added an historical survey of the French colony in the island of St. Domingo.* B. Crosby, London.

Egbert, G.D. and R.D. Ray. 2000. Significant dissipation of tidal energy in the deep ocean inferred from satellite altimeter data. *Nature* 405: 775–778.

Egbert, G.D. and R.D. Ray. 2001. Estimates of M2 tidal energy dissipation from TOPEX/Poseidon altimeter data. *Journal of Geophysical Research* 106: 22475–22502.

Eltis, D. 1995. New estimates of exports from Barbados and Jamaica, 1665–1701. *The William and Mary Quarterly* 52: 631–648.

Eltis, D. 1999. Slavery and freedom in the early modern world. Pages 25–49 in S.L. Engerman (ed.), *Terms of labor: slavery, serfdom and free labor.* Stanford University Press, Stanford.

Entick, J. 1757. *A new naval history or, compleat view of the British marine.* R. Manby, London.

Erskine, P.D., D. Lamb and M. Bristow. 2006. Tree species diversity and ecosystem function: can tropical multi-species plantations generate greater productivity? *Forest Ecology and Management* 233: 205–210.

Fajnzylber, P. and J. Humberto López. 2007. *Close to home: the development impact of remittances in Latin America.* World Bank Conference, pp. 57–59.

Food and Agriculture Organization (FAO). 2016a. *AQUASTAT: global map of irrigation areas.* FAO, Rome.

Food and Agriculture Organization (FAO). 2016b. *FAOSTAT: agricultural production and yield.* FAO, Rome.

Fernández de Oviedo y Valdés, G. 1851. *La historia general y natural de Indias. Islas y Tierra-Firme del Mar Océano. 1557.* Publicala La Real Academia de la Historia, D. José Amador de Los Rios, Madrid.

Fick, S.E. and R.J. Hijmans. 2017. Worldclim 2: new 1-km spatial resolution climate surfaces for global land areas. *International Journal of Climatology* 37: 4302–4315.

Field, C.B., M.J. Behrenfeld, J.T. Randerson and P. Falkowski. 1998. Primary production of the biosphere: integrating terrestrial and oceanic components. *Science* 281: 237–240.

Finan, J.J. 1948. Maize in the great herbals. *Annals of the Missouri Botanical Garden* 35: 149–165.

Fonblanque, A.W. 1870. *Statistical abstract for the United Kingdom in each of the last fifteen years from 1855 to 1869*. G.E. Eyre & W. Spottiswoode, London.

Francis, S.A. 2006. Development of sugar beet. Pages 9–29 in A.P. Draycott (ed.), *Sugar beet*. Wiley, London.

Franklin, B. 1786. A letter from Dr. Benjamin Franklin, to Mr. Alphonsus le Roy, member of several academies, at Paris. Containing sundry maritime observations. *Transactions of the American Philosophical Society* 2: 294–329.

Fuller, D.G., L. Qin, Y. Zheng, Z. Zhao, X. Chen, L.A. Hosoya and G.-P. Sun. 2009. The domestication process and domestication rate in rice: spikelet bases from the lower Yangtze. *Science* 323:1607–1610.

Fuller, D.Q. 2011. Pathways to Asian civilization: tracing the origins and spread of rice and rice cultures. *Rice* 4: 78–92.

GAFILAT. 2018. *Mutual evaluation report of the Republic of Panama*. GAFILAT, Buenos Aires.

Galloway, J.H. 1989. *The sugar cane industry: an historical geography from its origins to 1914*. Cambridge University Press, Cambridge.

Gammage, S. 2006. Exporting people and recruiting remittances: a development strategy for El Salvador. *Latin American Perspectives* 151: 75–100.

Gammage, S. and J. Schmitt. 2004. *Los immigrantes mexicanos, salvadorenos y dominicanos en el mercado laboral estadounidense: Las brechas de genero en los anos 1990 y 2000*. Serie Estudios y Perspectivas No. 20. CEPAL, Mexico City.

Gibb, R. 2006. The European Union's "Everything But Arms" development initiative and sugar: preferential access or continued protectionism? *Applied Geography* 26: 1–17.

Giovannoni, S.J. 2012. Vitamins in the sea. *Proceedings of the National Academy of Sciences* 109: 13888–13889.

Glantz, M.H. 2000. *Currents of change: impacts of El Niño and La Niña on climate and society*. Cambridge University Press, Cambridge.

Glennie, P. 1988. Continuity and change in Hertfordshire agriculture 1550–1700: II – trends in crop yields and their determinants. *Agricultural History Review* 36: 145–161.

Global Financial Integrity (GFI). 2017. *Illicit financial flows to and from developing countries: 2005–2014*. GFI, Washington DC.

Goldsmid, E. (ed.). 1890. *The principal navigations, voyages, traffiques and discoveries of the English nation, collected by Richard Hakluyt. Vol. XV–XVI*. E&G Goldsmid, Edinburgh.

Gordon, A.L. and K.T. Bosley. 1991. Cyclonic gyre in the tropical south Atlantic. *Deep-Sea Research* 38 (Suppl.): S323–S343.

Gordon, P. and J.J. Morales. 2017. *The silver way: China, Spanish America and the birth of globalization, 1565–1815*. Penguin Books China, London.

Greene, J. 1996. *Interpreting early America: historiographical essays*. University Press of Virginia, Charlottesville.

Grotius, H. 1609. *The freedom of the seas, or the right which belongs to the Dutch to take part in the East Indian trade*. Translated by Ralph Van Deman Magoffin. 1916. Oxford University Press, New York.

Grove, A.T. and O. Rackham. 2001. *The nature of Mediterranean Europe: an ecological history*. Yale University Press, New Haven.

Guerra, F. 1993. The European–American exchange. *History & Philosophy of Life Sciences* 15: 313–327.

Guyomard, H., C. Laroche and C. Le Mouël. 1999. An economic assessment of the common market organization for bananas in the European Union. *Agricultural Economics* 20: 105–120.

Hammond, D.S. (ed.). 2005. *Tropical forests of the Guiana Shield.* CABI Publishing, Wallingford.

Hammond, D.S., P.M. Dolman and A.R. Watkinson. 1995. Modern Ticuna swidden-fallow management in the Colombian Amazon: ecologically integrating market strategies and subsistence-driven economies? *Human Ecology* 23: 335–356.

Hancock, J.F. 2004. *Plant evolution and the origin of crop species.* CABI Publishing, Wallingford.

Harris, I.C. and P.D. Jones. 2017. *CRU TS4.01: Climatic Research Unit (CRU) Time-Series (TS) version 4.01 of high-resolution gridded data of month-by-month variation in climate (Jan. 1901–Dec. 2016).* Centre for Environmental Data Analysis, Norwich.

Harrison, J. 1989. The agrarian history of Spain, 1800–1960. *Agricultural History Review* 37: 180–187.

Hartmann, D. 2015. *Global physical climatology.* Elsevier Science, London.

Hartmann, T. and H.-G. Wenzel. 1995. The HW95 tidal potential catalogue. *Geophysical Research Letters* 22: 3553–3556.

Hartwick, J. 1977. Intergenerational equity and the investment of rents from exhaustible resources. *American Economic Review* 67: 972–974.

Hattox, R.S. 1985. *Coffee and coffeehouses: the origins of a social beverage in the medieval Near East.* 3rd printing. University of Washington Press, Seattle.

Hays, J.D., J. Imbrie and N.J. Shackleton. 1976. Variations in the Earth's orbit: pacemaker of the Ice Ages. *Science* 194: 1121–1132.

Healey, J. 2011. Land, population and famine in the English uplands: a Westmorland case study, c.1370–1650. *Agricultural History Review* 59: 151–175.

Hennessy, A. 1993. The nature of the conquest and the conquistadors. *Proceedings of the British Academy* 81: 5–36.

Hernández-Guerra, A., E. Fraile-Nuez, F. López-Latzen, A. Martínez, G. Parrilla and P. Vélez-Belchi. 2005. Canary Current and North Equatorial Current from an inverse box model. *Journal of Geophysical Research* 110: C12019. doi:10.1029/2005JC003032.

Heywood, H. 1999. Conservation of the wild relatives of native European crops. Pages 146–147 in J. Janick (ed.), *Perspectives on new crops and new uses.* ASHS Press, Alexandria.

Hilton, R. 1985. *Class conflict and the crisis of feudalism: essays in medieval social history.* Bloomsbury Publishing, London.

Hoffman, P. 2015. *Why did Europe conquer the world?* Princeton University Press, Princeton.

Holdridge, L. 1947. Determination of the world plant formations from simple climatic data. *Science* 105: 367–368.

Hoppit, J. 2002. The myths of the South Sea Bubble. *Transaction of the RHS* 12: 141–165.

Humboldt, A. von. 1814. *Personal narrative of travels to the equinoctial regions of the new continent during the years 1799–1804.* Longman, London.

Humphries, M.O. 2014. Paths to infection: the First World War and the origins of the 1918 influenza pandemic. *War in History* 21: 55–81.

Inalcik, H. 1989. The Ottoman Turks and the Crusades, 1451–1522. Pages 311–353 in K.M. Setton, H.W. Hazard and N.P. Zacour (eds), *A history of the Crusades. Vol. 6. The impact of the Crusades on Europe.* University of Wisconsin Press, Madison.

International Cocoa Organization (ICO). 2014. *The cocoa market situation.* Economics Committee report EC/4/2. ICO, London.

James, L. 1997. *Raj: the making and unmaking of British India.* St. Martin's Press, New York.

Janzen, D.H. 2001. Latent extinctions: the living dead. Pages 689–699 in S.A. Levin (ed.), *Encyclopedia of biodiversity. Vol. 3.* Academic Press, New York.

Janzen, D.H. and P.S. Martin. 1982. Neotropical anachronisms: the fruits the Gomphotheres ate. *Science* 215: 19–27.

Jenkins, V.S. 2000. *Bananas: an American history.* Smithsonian Institution Press, Washington DC.

Jennings, P.R. and J.H. Cock. 1977. Centres of origin of crops and their productivity. *Economic Botany* 31: 51–54.

Jones, P.D. and R.S. Bradley. 1995. Climatic variations over the last 500 years. Pages 649–655 in R.S. Bradley and P.D. Jones (eds), *Climate since AD 1500.* Routledge, London.

Jordan, W.C. 1996. *The great famine: northern Europe in the early fourteenth century.* Princeton University Press, Princeton.

Kamen, H. 2003. *Spain's road to empire: the making of a world power, 1492–1763.* Penguin, London.

Kantha, L.H. 1998. Tides: a modern perspective. *Marine Geodesy* 21: 275–297.

Kenyon, K.W. and E. Kridler. 1969. Laysan albatrosses swallow indigestible matter. *Auk* 86: 339–343.

Klein, S.A. and D.L. Hartmann. 1993. The seasonal cycle of low stratiform clouds. *Journal of Climatology* 6: 1587–1606.

Klooster, W. 2016. *The Dutch moment: war, trade, and settlement in the seventeenth century Atlantic world.* Cornell University Press, Ithaca.

Knight, F.W. 1977. Origins of wealth and the sugar revolution in Cuba, 1750–1850. *Hispanic American Historical Review* 57: 231–253.

Knight, F.W. 2014. The struggle of the British Caribbean sugar industry, 1900–2013. *The Journal of Caribbean History* 48: 149.

Koeppel, D. 2008. *Banana. The fate of the fruit that changed the world.* Plume, London.

Kopp, G. and J.L. Lean. 2011. A new, lower value of total solar irradiance: evidence and climate significance. *Geophysical Research Letters* 38: L01766–01773.

Köppen, W. 1884. Die Wärmezonen der Erde, nach der Dauer der heissen, gemässigten und kalten Zeit und nach der Wirkung der Wärme auf die organische Welt betrachtet. *Meteorol. Z.* 1: 215–226.

Köppen, W. 1918. Klassifikation der Klimate nach Temperatur, Niederschlag und Jahresablauf. *Petermanns Geogr. Mitt.* 64: 193–203, 243–248.

Kröpelin, S., D. Verschuren, A.-M. Lézine, H. Eggermont, C. Cocquyt, P. Francus, J.-P. Cazet, M. Fagot, B. Rumes, J.M. Russell, F. Darius, D.J. Conley, M. Schuster, H. von Suchodoletz and D.R. Engstrom. 2008. Climate-driven ecosystem succession in the Sahara: the past 6000 years. *Science* 320: 765–768.

Kummerow, C., W. Barnes, T. Kozu, J. Shiue and J. Simpson. 1998. The tropical rainfall measuring mission (TRMM) sensor package. *Journal of Atmospheric and Oceanic Technology* 15: 809–817.

Landes, D.S. 1998. *Wealth and poverty of nations: why some are so rich and some so poor.* Norton & Co., New York.

Larson, G., U. Albarella, K. Dobney, P. Rowley-Conwy, J. Schibler, A. Tresset, J.-D. Vigne, J.E. Ceiridwen, A. Schlumbaum, A. Dinu, A. Bălăçsescu, G. Dolman, A. Tagliacozzo, N. Manaseryan, P. Miracle, L. Van Wijngaarden-Bakker, M. Masseti, D.G. Bradley and A. Cooper. 2007. Ancient DNA, pig domestication, and the spread of the Neolithic into Europe. *Proceedings of the National Academy of Sciences* 104: 15276–15281.

Le Provost, C. and F. Lyard. 1997. Energetics of the M2 barotropic ocean tides: an estimate of bottom friction dissipation from a hydrodynamic model. *Progress in Oceanography* 40: 37–52.

Levathes, L. 1996. *When China ruled the seas: the treasure fleet of the Dragon Throne, 1405–1433.* Oxford University Press, New York.

Levi, L. 1876. The mode of levying the sugar duties in France, and its influence on the sugar industries of Great Britain. *The Sugar Cane* 8: 1–29.

Li, L.-F., H.-Y. Wang, C. Zhang, X.-F. Wang, F.-X. Shi, W.-N. Chen and X.-J. Ge. 2013. Origins and domestication of cultivated banana inferred from chloroplast and nuclear genes. *PLoS ONE* 8: e80502. doi:10.1371/journal.pone.0080502.

Li, Y. and H. Gu. 2006. Relationship between middle stratiform clouds and large scale circulation over eastern China. *Geophysical Research Letters* 33: L09706. doi:10.1029/2005GL025615.

Lima, M. 2014. Climate change and the population collapse during the "Great Famine" in pre-industrial Europe. *Ecology & Evolution* 4: 284–291.

Lindenau, J.D., F.M. Salzano, A.M. Hurtado, K.R. Hill, M.L. Petzl-Erler, L.T. Tsuneto and M.H. Hutz. 2016. Variability of innate immune system genes in Native American populations: relationship with history and epidemiology. *American Journal of Physical Anthropology* 159: 722–728.

Liu, L., G. Lee, L. Jiang and J. Zhang. 2007. Evidence for the early beginning (*c.*9000 cal. BP) of rice domestication in China: a response. *Holocene* 17: 1059–1068.

Loreau, M., S. Naeem, P. Inchausti, J. Bengtsson, J.P. Grime, A. Hector, D.U. Hooper, M.A. Huston, D. Raffaelli, B. Schmid, D. Tilman and D.A. Wardle. 2001. Biodiversity and ecosystem functioning: current knowledge and future challenges. *Science* 294: 804–808.

Lovejoy, P.E. 1980. La kola dans l'histoire de l'Afrique occidentale. *Cahiers d'Etudes Africaines* 20: 97–134.

Macaulay Land Use Research Institute. 2010. *Land capability for agriculture in Scotland map. Sheets 1–7, 1:250,000.* MLUSRI, Aberdeen.

Malthus, T.R. 1798. *An essay on the principle of population, as it affects the future improvement of society.* J. Johnson, London.

Marcus, H.G. 2002. *A history of Ethiopia.* University of California Press, Riverside.

Marley, D. 1998. *Wars of the Americas: a chronology of armed conflict in the New World, 1492 to the present.* ABC-Clio, Santa Barbara.

Marshall, M.G. and G. Elzinga-Marshall. 2017. *Global report 2017. Conflict, governance, and state fragility.* Center for Systemic Peace, Vienna VA.

Marshall, M.G. and T.R. Gurr. 2005. *Peace and Conflict 2005. A global survey of armed conflicts, self-determination movements, and democracy.* CIDCM, University of Maryland, College Park.

Martin, R.M. 1834. *History of the British colonies. Vol. I–II.* 2nd ed. James Cochrane and Co., London.

Mason, S.J. and L. Goddard. 2001. Probabilistic precipitation anomalies associated with ENSO. *Bulletin of the American Meteorological Society* 82: 619–638.

Mattingly, D. and G. Aldrete. 2000. The feeding of imperial Rome: the mechanics of the food supply system. Pages 142–165 in J. Coulston and H. Dodge (eds), *Ancient Rome: the archaeology of the eternal city*. Oxford University Press, Oxford.

McCusker, J.J. 1971. The current value of English exports, 1697 to 1800. *The William and Mary Quarterly* 28: 607–628.

McCusker, J.J. 1989. *Rum and the American Revolution: the rum trade and the balance of payments of the thirteen continental colonies. Vol. I–II*. Garland Publishing, New York.

Mcevedy, C. and R. Jones. 1978. *Atlas of world population history*. Penguin, London.

McMaster, G.S. and W.W. Wilhelm. 1997. Growing degree-days: one equation, two interpretations. *Agricultural and Forest Meteorology* 87: 291–300.

Michel, R.L. and H.E. Suess. 1975. Bomb tritium in the Pacific Ocean. *Journal of Geophysical Research* 80: 4139–4153.

Miller, G.H., Á. Geirsdóttir, Y. Zhong, D.J. Larsen, B.L. Otto-Bliesner, M.M. Holland, D.A. Bailey, K.A. Refsnider, S.J. Lehman, J.R. Southon, C. Anderson, H. Björnsson and T. Thordarson 2012. Abrupt onset of the Little Ice Age triggered by volcanism and sustained by sea-ice/ocean feedback. *Geophysical Research Letters* 39: L02708. doi:10.1029/2011GL050168.

Mintz, S.W. 1985. *Sweetness and power: the place of sugar in modern history*. Viking Penguin, New York.

Mitchell, B.R. 1971. *Abstract of British historical statistics*. Cambridge University Press, Cambridge.

Mitchell, B.R. 1975. *European historical statistics 1750–1970*. Macmillan, London.

Mitchell, B.R. 1995. *International historical statistics, Africa, Asia and Oceania 1750–1988*. 2nd ed. Palgrave Macmillan, New York.

Mitchell, B.R. 2003. *International historical statistics: the Americas 1750–2000*. 5th ed. Palgrave Macmillan, New York.

Mitchell, B.R. and P. Deane. 1962. *Abstract of British historical statistics*. Cambridge University Press, Cambridge.

Monfreda, C., N. Ramankutty and J.A. Foley. 2008. Farming the planet: 2. Geographic distribution of crop areas, yields, physiological types, and net primary production in the year 2000. *Global Biogeochemical Cycles* 22: GB1022. doi:10.1029/2007GB002947.

Morelli, G., Y. Song, C.J. Mazzoni, M. Eppinger, P. Roumagnac, D.M. Wagner, M. Feldkamp, B. Kusecek, A.J. Vogler, Y. Li, Y. Cui, N.R. Thomson, T. Jombart, R. Leblois, P. Lichtner, L. Rahalison, J.M. Petersen, F. Balloux, P. Keim, T. Wirth, J. Ravel, R. Yang, E. Carniel and M. Achtman. 2010. *Yersinia pestis* genome sequencing identifies patterns of global phylogenetic diversity. *Nature Genetics* 42: 1140–1143.

Morris, I. 2010. *Why the West rules – for now: the patterns of history and what they reveal about the future*. Farrar, Straus and Giroux, London.

Mudge, K., J. Janick, S. Scofield and E.E. Goldschmidt. 2009. A history of grafting. *Horticultural Reviews* 35: 437–493.

Munk, W.H. 1997. Once again: once again–tidal friction. *Progress in Oceanography* 40: 7–36.

Naipaul, V.S. 1969. *The loss of El Dorado: a history*. Andre Deutsch, London.

New, M., M. Hulme and P. Jones. 1999. Representing twentieth-century space-time climate variability. Part I: development of a 1961–90 mean monthly terrestrial climatology. *Journal of Climate* 12: 829–856.

Newman, E.I. and P.D.A. Harvey. 1997. Did soil fertility decline in medieval English farms? Evidence from Cuxham, Oxfordshire, 1320–1340. *Agricultural History Review* 45: 119–136.

Olson, D.M., E. Dinerstein, E.D. Wikramanayake, N.D. Burgess, G.V.N. Powell, E.C. Underwood, J.A. D'Amico, I. Itoua, H.E. Strand, J.C. Morrison, C.J. Loucks, T.F. Allnutt, T.H. Ricketts, Y. Kura, J.F. Lamoreux, W.W. Wettengel, P. Hedao and K.R. Kassem. 2001. Terrestrial ecoregions of the world: a new map of life on Earth. *Bioscience* 51: 933–938.

Orejuela, J.E. 1992. Traditional productive systems of the Awa (Cuaiquer) Indians of southwestern Colombia and neighboring Ecuador. Pages 58–82 in K.H. Redford and C. Padoch (eds), *Conservation of Neotropical forests – working from traditional resource use*. Columbia University Press, New York.

O'Shaugnessy, A.J. 2015. *An empire divided: the American revolution and the British Caribbean*. University of Pennsylvania Press, Philadelphia.

Otis, J. 1764. *The rights of the British colonies asserted and proved*. J. Almon, Boston & London.

Overton, M. 1990. The critical century? The agrarian history of England and Wales 1750–1850. *Agricultural History Review* 38: 185–189.

Overton, M. and B.M.S. Campbell. 1996. Production et productivité dans l'agriculture anglais, 1086–1871. *Histoire et Mesure* 11: 255–297.

Packard, R.M. 2007. *The making of a tropical disease: a short history of malaria*. John Hopkins University Press, Baltimore.

Page, S.E., J.O. Rieley and C.J. Banks. 2011. Global and regional importance of the tropical peatland carbon pool. *Global Change Biology* 17: 798–818.

Palmer, C.A. 1981. *Human cargoes*. University of Illinois Press, Urbana. Page 237 in H. Thomas. 1997. *The slave trade – the history of the Atlantic slave trade 1440–1870*. Picador, London.

Parker, G. 1998. *The grand strategy of Philip II*. Yale University Press, New Haven.

Peltonen-Sainio, P. 2012. Crop production in a northern climate. Pages 183–216 in *Building resilience for adaptation to climate change in the agricultural sector*. Proceedings of a joint FAO/OECD workshop, FAO, Rome.

Perreijn, K. 2002. *Symbiotic nitrogen fixation by leguminous trees in tropical rain forest of Guyana*. Tropenbos-Guyana Series 11. Tropenbos-Guyana Programme. Georgetown, Guyana.

Perrier, X., E. De Langhe, M. Donohue, C. Lentfer, L. Vyrdagh, F. Bakry, F. Carreel, I. Hippolyte, J.-P. Horry, C. Jenny, V. Lebot, A.-M. Risterucci, K. Tomekpe, H. Doutrelepont, T. Ball, J. Manwaring, P. de Maret and T. Denham. 2011. Multidisciplinary perspectives on banana (*Musa* spp.) domestication. *Proceedings of the National Academy of Sciences* 108: 11311–11318.

Perry-Gal, L., A. Erlich, A. Gilboa and G. Bar-Oz. 2015. Earliest economic exploitation of chicken outside East Asia: evidence from the Hellenistic southern Levant. *Proceedings of the National Academy of Sciences* 112: 9849–9854.

Peterson, B.B. 1994. The Ming voyages of Cheng Ho (Zheng He), 1371–1433. *The Great Circle* 16: 43–51.

Philander, S.G. 1990. *El Niño, La Niña, and the Southern Oscillation*. International Geophysical Series, Vol. 46. Academic Press, New York.

Pitman III, W.C., S. Cande, J. LaBrecque and J. Pindell. 1993. Fragmentation of Gondwana: the separation of Africa from South America. Pages 15–36 in

P. Goldblatt (ed.), *Biological relationships between Africa and South America.* Yale University Press, New Haven.

Platzman, G.W. 1985. The role of Earth tides in the balance of tidal energy. *Journal of Geophysical Research* 90: 1789–1793.

Platzman, G.W. 1991. An observational study of energy balance in the atmospheric lunar tide. *Pure and Applied Geophysics* 137: 1–33.

Pollack, H.N., S.J. Hurter and J.R. Johnson. 1993. Heat flow from the Earth's interior: analysis of the global data set. *Reviews of Geophysics* 31: 267–280.

Powell, T. 1825. *Statistical illustrations of the territorial extent and population, rental, taxation, finances, commerce, consumption, insolvency, pauperism and crime of the British Empire.* Effingham Wilson, London.

Pownall, T. 1764. *The administration of the British colonies.* J. Walter, London.

Prance, G. and M. Nesbitt (eds). 2005. *The cultural history of plants.* Routledge, London.

Pulteney, W. 1776. *Thoughts on the present state of affairs with America and the means of conciliation.* J. Dodsley and T. Cadell, London.

Rabushka, A. 2008. *Taxation in colonial America.* Princeton University Press, Princeton.

Rainsford, M. 1805. *An historical account of the black empire of Hayti: the principal transactions in the revolution of Saint Domingo with its antient and modern state.* J. Cundee, London.

Rajan, R.S. 2007. Home-grown foreign aid: workers' remittances as a form of development finance. *Georgetown Journal of International Affairs* 8: 43–49.

Rama Murthy, V., W. van Westrenen and F. Yingwei. 2003. Experimental evidence that potassium is a substantial radioactive heat source in planetary cores. *Nature* 423: 163–165.

Ramenofsky, A., A. Wilbur and A. Stone. 2003. Native American disease history: past, present and future directions. *World Archaeology* 35: 241–257.

Rawcliffe, C. 2013. *Urban bodies: communal health in the late medieval English towns and cities.* Boydell Press, Woodbridge.

Ray, R.D., R.J. Eanes and F.G. Lemoine. 2001. Constraints on energy dissipation in the earth's body tide from satellite tracking and altimetry. *Geophysical Journal International* 144: 471–480.

Raynal, A. 1776. *A philosophical and political history of the settlements and trade of the Europeans in the East and West Indies.* Translated by J. Justamond. 2nd ed. T. Cadell, London.

Rebourg, C., M. Chastanet, B. Gouesnard, C. Welcker, P. Dubreuil and A. Charcosset. 2003. Maize introduction into Europe: the history reviewed in light of the molecular data. *Theoretical and Applied Genetics* 106: 895–903.

Reed, W. 1866. *The history of sugar and sugar yielding plants.* Longmans, Green & Co., London.

Repetto, R., and M. Gillis (eds). 1988. *Public policies and the misuse of forest resources.* Cambridge University Press, Cambridge.

Richardson, P.L. 1985. Average velocity and transport of the Gulf Stream near 55W. *Journal of Marine Research* 43: 83–111.

Ridpath, G. (attributed). 1700. *An enquiry into the causes of the miscarriage of the Scots colony at Darien, or, An answer to a libel entituled, a defence of the Scots abdicating Darien submitted to the consideration of the good people of England.* N.p., Glasgow.

Riley, J.C. 2010. Smallpox and American Indians revisited. *Journal of the History of Medicine and Allied Sciences* 65: 445–477.

Riviére, P. (ed.). 2006. *The Guiana travels of Robert Schomburgk, 1835–1844. Vol. I: explorations on behalf of the Royal Geographical Society, 1835–1839.* Hakluyt Society Series III, Vol. 16. Ashgate Publishing, London.

Robertson, R. 1733. *An enquiry into the methods that are said to be now proposed in England, to retrieve the Sugar Trade.* J. Wilford, London.

Robertson, S. 1885. The expense book of James Master AD 646 to 1676. *Archaeologia Cantiana* 15: 152–216.

Robinson, P.J. and A. Henderson-Sellers. 1999. *Contemporary climatology.* 2nd ed. Longman, London.

Rodrigues, P.T., H.O. Valdivia, T.C. de Oliveira, J.M.P. Alves, A.M.R.C. Duarte, C. Cerutti-Junior, J.C. Buery, C.F.A. Brito, J.C. de Souza, Z.M.B. Hirano, M.G. Bueno, J.L. Catão-Dias, R.S. Malafronte, S. Ladeia-Andrade, T. Mita, A.M. Santamaria, J.E. Calzada, I.S. Tantular, F. Kawamoto, L.R.J. Raijmakers, I. Mueller, M.A. Pacheco, A.A. Escalante, I. Felger and M.U. Ferreira. 2018. Human migration and the spread of malaria parasites to the New World. *Scientific Reports* 8: 1993.

Rodriguez, J.P. 2007. *Encyclopedia of slave resistance and rebellion. Vol. 1.* Greenwood Press, London.

Roosevelt, A.C. 1980. *Parmana. Prehistoric maize and manioc subsistence along the Amazon and the Orinoco.* Academic Press, New York.

Ropelewski, C.F. and M.S. Halpert. 1987. Global and regional scale precipitation patterns associated with the El Nino/Southern Oscillation. *Monthly Weather Review* 115: 1606–1626.

Rossiter, M.W. 1975. *The emergence of agricultural science, Justus Liebig and the Americans, 1840–1880.* Yale University Press, New Haven.

Rubel, F. and M. Kottek, 2010. Observed and projected climate shifts 1901–2100 depicted by world maps of the Köppen-Geiger climate classification. *Meteorol. Z.* 19: 135–141.

Rubial García, A. 1996. *La hermana pobreza: El Franciscanismo de la edad media a la evangelización Novohispana.* UNAM, Mexico City.

Saatchi, S.S., N.L. Harris, S. Brown, M. Lefsky, E.T.A. Mitchard, W. Salas, B.R. Zutta, W. Buermann, S.L. Lewis, S. Hagen, S. Petrova, L. White, M. Silman and A. Morel. 2011. Benchmark map of forest carbon stocks in tropical regions across three continents. *Proceedings of the National Academy of Sciences* 108: 9899–9904.

Sachs, J.D. and A.M. Warner. 2001. The curse of natural resources. *European Economic Review* 45: 827–838.

Sañudo-Wilhelmy, S.A., L.S. Cutter, R. Durazo, E. Smail, L. Gomez-Consarnau, E.A. Webb, M.G. Prokopenko, W.M. Berelson and D.M. Karl. 2012. Multiple B-vitamin deficiency in large areas of the coastal ocean. *Proceedings of the National Academy of Sciences* 109: 14041–14045.

Schomburgk, R.H. 1848. *Reisen in British Guiana den jahren 1840–1844.* J.J. Weber, Leipzig, Germany.

Schwartz, S. 2004. *Tropical Babylons: sugar and the making of the Atlantic world, 1450–1680.* University of North Carolina Press, Chapel Hill.

Scott, S., S.R. Duncan and C.J. Duncan. 1998. The origins, interactions and causes of the cycles in grain prices in England, 1450–1812. *Agricultural History Review* 46: 1–14.

Scrimshaw, N.S. 2003. Historical concepts of interactions, synergism and antagonism between nutrition and infection. *Journal of Nutrition* 133: 316–321.

306 *References*

Select Committee, House of Commons, Government of Great Britain. 1810. *Report, and accounts, from the select committee appointed to inquire into the cause of the high price of gold bullion and to take into consideration the state of the circulating medium, and of the exchanges between Great Britain and foreign parts.* Richard Taylor and Co., London.

Sheridan, R. 1974. *Sugar and slavery: an economic history of the British West Indies, 1623–1775.* Caribbean University Press, Aylesbury.

Simonsen, R.G. 1944. *História economica do Brasil: 1500–1820.* 2 vols. Companhia Editora Nacional, Sao Paulo.

Sinclair, J. 1804. *The history of the public revenue of the British empire. Vol. III.* 3rd ed. Strahan & Preston, London.

Smith, A. 1776. *An inquiry into the nature and causes of the wealth of nations. Vol. I–II.* W. Strahan & T. Cadell, London.

Smith, B. 1989. Origins of agriculture in eastern North America. *Science* 246: 1566–1571.

Smith, B.D. and R.A. Yarnell. 2009. Initial formation of an indigenous crop complex in eastern North America at 3800 B.P. *Proceedings of the National Academy of Sciences* 106: 6561–6566.

Smith, S.D. 1996. Accounting for the taste: British coffee consumption in historical perspective. *The Journal of Interdisciplinary History* 27: 183–214.

Smollet, T and T. Francklin. 1761. *The works of M. de Voltaire. Translated from French with notes, historical and critical.* Newbery, Baldwin *et al.*, Salisbury.

Stehli, F.G. and S.D. Webb (eds). 1985. *The great American biotic interchange.* Plenum Press, New York.

Stein, C.A. and S. Stein. 1992. A model for the global variation in oceanic depth and heat flow with lithospheric age. *Nature* 359: 123–129.

Striffler, S. and M. Moberg (eds). 2003. *Banana wars: power, production and history in the Americas.* Duke University Press, Durham NC.

Subrahmanyam, G. 2006. Ruling continuities: colonial rule, social forces and path dependence in British India and Africa. *Commonwealth & Comparative Politics* 44: 84–117.

Szabó, P. 2013. Rethinking pannage: historical interactions between oak and swine. Pages 51–61 in I.D. Rotherham (ed.), *Trees, forested landscapes and grazing animals: a European perspective on woodlands and grazed treescapes.* Routledge, London.

Temple, W. 1720. *Observation upon the United Provinces of the Netherlands.* Pages 7–72 in *The works of Sir William Temple. Vol. I.* Churchill, Goodwin *et al.*, London.

Tenaillon, M.D. and A. Charcosset. 2011. A European perspective on maize history. *Comptes Rendus Biologies* 334: 221–228.

Thacher, J.B. 1903. *Christopher Columbus: his life, his works, his remains. Vol. I–III.* G.P. Putnam, New York.

Thomas, H. 1997. *The slave trade – the history of the Atlantic slave trade 1440–1870.* Picador, London.

Thornthwaite, C.W. 1948. An approach toward a rational classification of climate. *Geographical Review* 38: 55–94.

Thornthwaite, C.W. and J.R. Mather. 1955. The water balance. Laboratory of Climatology, Centerton, N.J. *Publications in Climatology* 8: 1–104.

United Nations, Department of Economic and Social Affairs, Population Division (UNDESA). 2017. *Trends in international migrant stock. The 2017 revision* (United Nations database, POP/DB/MIG/Stock/Rev.2017).

US Bureau of the Census. 1975. *Historical statistics of the United States, colonial times to 1970, Part 2. Chapter Z – Colonial and pre-federal statistics.* US Government, Washington DC.

US Census Bureau. 2015. *American Housing Survey. Data tables.* US Department of Housing and Urban Development, Washington DC.

US Energy Information Administration (US EIA). 2015. *Drivers of U.S. household energy consumption.* US Department of Energy, Washington DC.

Van Groesen, M. 2011. Lessons learned: the second Dutch conquest of Brazil and the memory of the first. *Colonial Latin American Review* 20: 167–193.

Vavilov, N.I. 1992. *Origin and geography of cultivated plants.* Cambridge University Press, Cambridge.

Venzke, E. (ed.). 2013. *Volcanoes of the world, v. 4.5.0.* Global Volcanism Program. Smithsonian Institution, Washington DC.

Walsingham, T. 1574. *The English history of Thomas Walsingham, monk of St. Alban's, A.D. 1272–1422.* Published by the Archbishop Parker, London.

Wan, Z. 1999. *MODIS land-surface temperature algorithm theoretical basis document (LST ATBD).* Version 3.3. Contract No. NAS5-31370 for NASA. ICESS, University of California–Santa Barbara.

Wang, B. and Y.L. Qiu. 2006. Phylogenetic distribution and evolution of mycorrhizas in land plants. *Mycorrhiza* 16 (5): 299–363.

Wen, F., T. Bedford and S. Cobey. 2016. Explaining the geographical origins of seasonal influenza A (H3N2). *Proceedings of the Royal Society B* 283: 20161312.

Whitehead, N.L. 1987. *The discoverie of the large, rich and bewtiful empyre of Guiana, transcribed, annotated and introduced by Neil L. Whitehead.* Manchester University Press, Manchester.

Whitworth, C. 1731. *The political and commercial works of that celebrated writer Charles D'Avenant relating to the trade and revenue of England, the plantation trade, the East-India trade and African trade. Vol. I.* Horsfield, Becket, de Hondt & Cadell, London.

Wiersema, J.H. and B. Leon. 1999. *World economic plants: a standard reference.* CRC Press, London.

Wilbert, J. 1987. *Tobacco and shamanism in South America.* Yale University Press, New Haven.

Wiley, J. 2008. *The Banana: empires, trade wars, and globalization.* University of Nebraska Press, Lincoln.

Williams, M. 1989. *Americans and their forests. A historical geography.* Cambridge University Press, Cambridge.

Wilmott, D.E. 1960. *The Chinese of Semarang.* Cornell University Press, Ithaca.

Witney, K.P. 1990. The woodland economy of Kent, 1066–1348. *Agricultural History Review* 38: 20–39.

World Bank. 2006. *Where is the wealth of nations? Measuring capital for the 21st century.* World Bank, Washington DC.

World Bank. 2017. World Development Indicators. World Bank, Washington DC. http://data.worldbank.org/data-catalog/world-development-indicators.

Yang, D. 2008. International migration, remittances, and household investment: evidence from Philippine migrants' exchange rate shocks. *Economic Journal* 118: 591–630.

Yang, D. and C. Martinez. 2005. Remittances and poverty in migrants' home areas: evidence from the Philippines. Pages 81–123 in C. Ozden and M. Schiff (eds),

International migration, remittances, and the brain drain. World Bank, Washington DC.

Zhao, M., F.A. Heinsch, R.R. Nemani and S.W. Running. 2005. Improvements of the MODIS terrestrial gross and net primary production global data set. *Remote Sensing of Environment* 95: 164–176.

Zohary, D., M. Hopf and E. Weiss. 2012. *Domestication of plants in the Old World.* 4th ed. Oxford University Press, Oxford.

Zumbroich, T.J. 2007. The origin and diffusion of betel chewing: a synthesis of evidence from South Asia, Southeast Asia and beyond. *eJournal of Indian Medicine*: 87–140.

Index

Page numbers in **bold** denote tables, those in *italics* denote figures.